GEOSPECTROSCOPY

GEOSPECTROSCOPY

R.K. Tiwari

Capital Publishing Company
NEW DELHI KOLKATA BANGALORE

Capital Publishing Company
7/28, Mahaveer Street, Ansari Road,
Daryaganj, New Delhi 110 002

Geospectroscopy

Author: R.K. Tiwari

ISBN: 81–85589–17–8

Typeset by Excel Publishing Services, New Delhi
Printed in India at Chaman Enterprises, New Delhi

Preface

मत्तः परतरं नान्यत्किञ्चिदस्ति धनञ्जय।
मयि सर्वमिदं प्रोतं सूत्रे मणिगणा इव।

ŚRĪMADBHAGAVADGĪTĀ

The earth is recognised as a dynamic system not only due to its internal structural convection and magnetic field but also because of the continued spatio-temporal variation of its sub-systems like hydrosphere, cryosphere, biosphere, atmosphere and outer space environment. The modern plate tectonics paradigm suggests that shape, size and geographic distribution of continents and various processes of earth's interiors have evolved through times, as have the atmosphere and biota. The evolutionary processes appear to have been affected by external as well as internal modes. The available geological and geophysical observations suggest that various evolutionary processes of the earth are exposed to cyclic, chaotic and/or catastrophic forces of nature. Geoscientists are concerned with characterising the nature of these complex geoprocesses. Examination of cyclic pattern in the earth system's processes has a long tradition. Intriguing correlations with the earth's system periodicities and external forcing phenomena, such as galactic and orbital cycles and influence of solar and tidal effects have been suggested. While the definite physical link between the endogenic and exogenic processes on different time scales remains a tantalizing issue amongst earth's system scientists and astronomers, it is interesting to note that, in many cases, the existence of striking correlation between these phenomena has been widely accepted for the past few decades. This has also been vigorously supported by recently acquired more precise data and improved statistical methods, however, with sparking conjectures.

'Geospectroscopy' tries to provide an integrated picture of cyclic behaviour of earth system processes covering wide time spectrum/scales from a few years to a few million years. The main focus is to present evidence of periodicities in various earth system's processes and their interpretations in terms of external and internal forcing mechanisms. Both views dominate, but the association of periodic external forcing (movement of the solar system in the galaxy, impact cratering, periodic variations in the earth orbital parameters, solar and tidal influences) with the various earths system processes gaining favour in many earth processes appear to be highly credible. It is becoming more and more clear that the significant role of exogenic forcing on earth's system processes must be acknowledged at various time scales. The implications of coupling and linkages of exogenic and endogenic processes offer fascinating new challenges to geoscientists. Modern methods of time series analyses identify the pattern of periodicities of earth system's processes and provide ample evidence of rigorous test of statistical significance of observed periodicities. This, in fact, provides more confidence and motivation for better physical interpretation to understand the evolutionary history. Geospectroscopy is concerned with evaluating these causal relationships in an attempt to replace and narrow down speculations with quantitative knowledge and its wider ramification.

The book, *Geospectroscopy*, consists of 10 chapters. In chapter 1 a brief introduction of the origin and history of cyclicity on different time scales is presented. In chapter 2, various methods of spectral analyses are briefly discussed with minimum possible mathematical details. This chapter is written to provide better feeling to the readers to understand chapters 4–9, where these methods are frequently employed to various time series to detect periodicities. Chapter 3 provides review of the well-known cycles in the realm of astronomy on widely different time scales, very frequently observed in various geological and geophysical records. Chapters 4–9 are devoted to review of evidence of periodicities in various earth processes on different time scales. For instance, chapter 4 deals with long term cycles of earth's interior processes while chapters 5 and 6 are respectively devoted to long-term ocean and atmospheric and mass extinction processes. Chapter 7 deals with the evidence of orbital cycles, also popularly known as 'Milankovitch cycles', in various sedimentary records, in particular in climate records. Chapter 8 provides the evidences of solar and tidal cycles in short term coupled earth–ocean and atmospheric processes and description of various physical mechanisms. Chapter 9 is devoted to the short-term dynamical oceanic process popularly known as ENSO cycles. In chapter 10, conclusions are presented.

November, 2004 **R.K. Tiwari**

Acknowledgements

Geospectroscopy is based on substantial research of the earth by the astronomical science community and in particular relies heavily on those who contributed over the past few years to hunt for evidence of periodicities in the earth science system's record and related phenomena and attempted to establish physical linkages. I would mention here that in order to provide integrated picture of periodicity on widely different time scales and to make this book self-contained, repetition of some material is inevitable.

While writing this book, it has been my endeavour to provide an unbiased coverage of all necessary geophysical, geological, oceanographic, atmospheric and astronomical literatures and related research relevant to this book that have come to my knowledge and have been published till June 2004.

There are several illustrations, tables and quotes, taken from the work of various researchers. I feel honoured and delighted to acknowledge them. I wish to express my thanks to those publishers and individuals who have promptly and generously consented to use/reproduce their figures in this book. Especially, I am indebted to the American Association for Advancement of Science (AAAS), Macmillan Journal (Nature Publishing Group), The University of Chicago Press, Elsevier Publishing Company. Kluwer and Springer Verlag and also American Geophysical Union (AGU) and Geological Society of America (GSA) for granting permission to use their figures. I am also grateful to Elizabeth Sandler, AAAS; Helen Gainford, Elsevier; Coxhead Rois; Annemareke Roest, Springer; Perry Cartwright, The University of Chicago Press from the rights and permissions department and various publishing companies for their quick response and cooperation.

My special gratitudes to Professors Andre' Berger, Mark A. Cane, Allan C. Mix, Andreas Prokoph, Richard Ernst, Vincent Courtillot, Samuel G. Philander, Thomas Felis, Maya Elkibbi, M.E. Mann and Dr. Michael R. Rampino of USA, who have readily consented to use their figures in this book and provided their useful literatures. Separate acknowledgement is also given with figures captions. I am also thankful to professors R.D. Robert and Richard B. Stothers for providing me their computer codes for the clean spectral analysis and non-parametric time series analysis respectively. During late eighties, I had the privilege to visit the Department of Geophysics, University of Edinburgh UK, where I had the opportunity to discuss with Professors K.M. Creer and W.M. Napier at length on long-term cyclicity of the earth system processes, particularly geomagnetic reversals, mass extinctions and galactic motions of the solar system. I derived substantial inspirations and motivations from them and I sincerely thank them.

I record my gratitude to all my senior colleagues with whom I shared my knowledge and wisdom over the years, in particular, Shri K.N.N. Rao and Drs. J.G. Negi, M.S. Joshi and U. Raval. My sincere gratitude to Dr. H.K. Gupta, Secretary, Department of Ocean Development, Government of India for his encouragement and writing the foreword for this book and Drs. Hari Narain, ex Director, NGRI and V.P. Dimri, Director, National Geophysical Research Institute, Hyderabad, for their encouragement and motivation. My sincere thanks to my Ph.D. students Mrs. S. Srilakshmi and Saumen Maiti for their help during preparation of this book. In particular, I record my gratitude to Mrs. S. Srilakshmi for her help at all stages of writing this book. In addition to this, I acknowledge my indebtness to all those who generously supported in this endeavour.

At last, I am grateful to my wife Rukmini and daughters, Rashmi and Supriya for their enduring support during the preparation of this book.

Foreword

It is, indeed, my great pleasure to write a few lines about the book "Geospectroscopy" written by Dr. R.K. Tiwari of National Geophysical Research Institute, Hyderabad. Various processes of the earth are coupled and are dynamically evolving system on widely different time scales. This coupled system include not only the processes of earth's interior such as mantle-core convections and related phenomena but also its sub-system like ocean and atmosphere that have an important influence on the earth's climate and in turn, the biosphere. The term 'Geospectroscopy' tries to encapsulate the analyses of temporal variations, particularly the quasi-periodic/cyclic processes of the earth system by means of modern tools of signal processing. Geospectroscopy leads to deciphering the spectrum of each sub-system who's each period or frequency may manifest due to an underlying causative process or forcing.

Evidence of periodicity in the earth system processes has a long distinguished tradition and honourable antiquity. The subject has been a topic of debate for the past two decades or so. Fortunately this situation has changed dramatically due to enhancement in the capacity of high-resolution geological and geophysical observations and the advent of modern computer based time series analysis techniques. These techniques have provided essential ingredient to monitor the periodic behaviour of various earth system processes covering time scales ranging from few years to millions of years. Dr. Tiwari integrates these modern developments and finding in ten chapters and has been successful in highlighting wider ramifications of external and internal forcing mechanisms of Frontier Research. Geospectroscopy presents coherent and comprehensive picture of the spectrum of the Earth's various activities and their physical linkages to the external phenomena. It is a concise and well-referenced book that permits more advanced pursuit and insights to the readers of wide and diversified interest in the field of global earth system processes. In view of the importance of the interdisciplinary research, the readers can expect to find up-to-date bibliography and published scientific material on periodicity of both general and specific relevance.

Dr. Harsh K. Gupta
Secretary
Department of Ocean Development
Government of India, New Delhi

Contents

1

Introduction

INTRODUCTION

The evolutionary history of the earth is preserved in rocks and sediments. The oceanic sediments of the earth carry significant information about the interactions and linkages of the coupled geo-bio-ocean-atmospheric processes on widely different time scales. Hence the study of geological time series decoded from rocks and sediments provide useful information for the understanding of the basic trend of the earth's evolutionary history. 'Geospectroscopy' deals with various interactions of the earth and astronomical processes directly or indirectly covering wide range of spectrum in time and space. The complete spectrum of these astro-geo-interactions can be viewed more conveniently on the astronomical time scale covering different bands of frequency. Several recent novel findings would provide a scientific basis to astrophysicists and geophysicists to judge evidence of periodicity and to unify various new linkages between exogenic and endogenic processes. Evidence of periodicity and their possible inter-linkages in various geo-bio-ocean-atmospheric-processes starting from years to million years time scale appears to provide ample evidence, although some of these may be speculative, of unification between various exogenic and endogenic processes with enough scope to formulate physical mechanisms to resolve/reconcile debated scientific issues. The earth's system processes are complex in nature and appear to be associated with various internal processes of earth's interior, ocean and atmosphere and external such as galactic, solar system, bolides impact, orbital and solar and tidal phenomena. Figure 1.1 depicts integrated picture of all these various processes.

NATURE OF GEO-TIME SERIES

The earth is a dynamical system. The dynamical behaviour of the earth's system processes can be decoded and understood by analysing geo-time series. The time series of various earth's processes and their spectrum display very

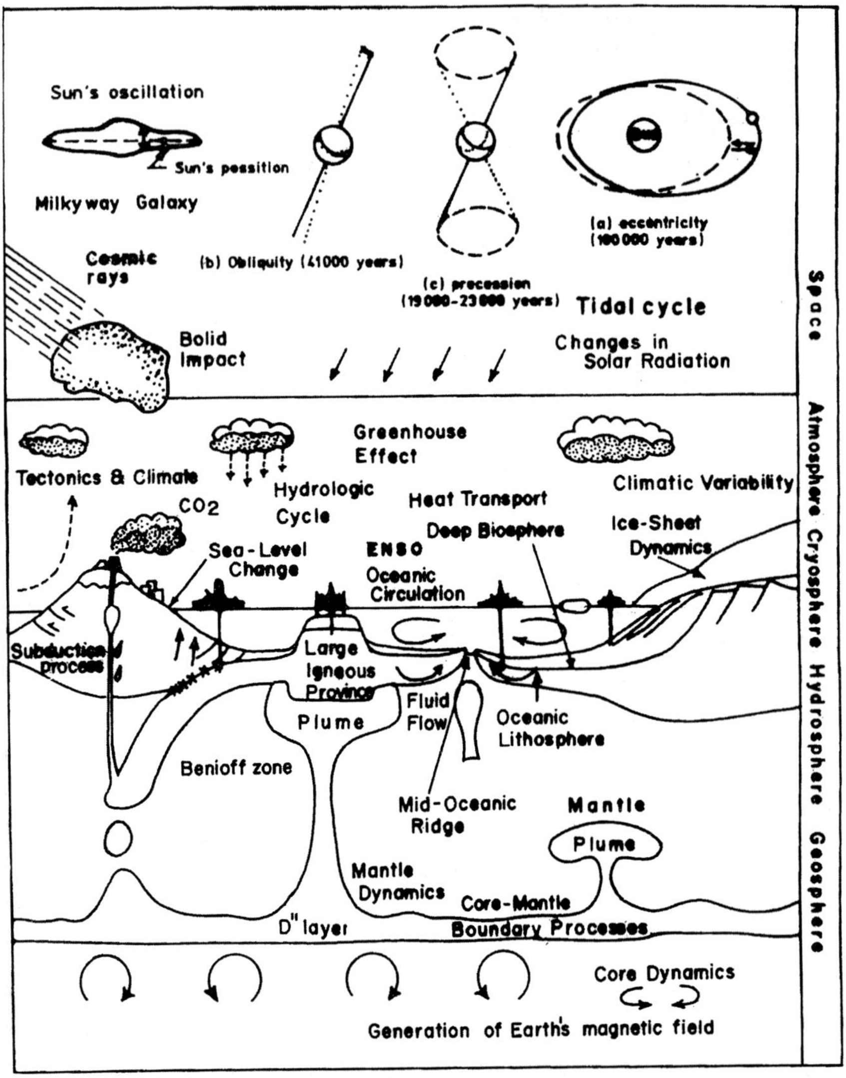

Fig. 1.1 Earth's System Components, Processes and Astronomical Phenomena

(*Source*: Modified from Initial Ocean Drilling Plan, on Earth's Ocean and Life System, Asahika Taira University, Tokyo)

complex characteristic on different time scales. Long-term geological processes are complex and generally episodic in nature. There are many sedimentary records, which are continuous but entail some gaps and hiatus. Short-term temporal records are, however, relatively better but generally found to be dominated with various kinds of noise and trends. The complex temporal variability on different spacio-time scales, thus might include random, stochastic, chaotic and periodic/cyclic behaviour. While random signals do not carry any significant information of physical relevance, stochastic signals are found to be associated with probability of large number of causative variables. A signal is said to be random if its precise value cannot be predicted in advance. Random signals are also often called stochastic processes. Recent spectral and statistical studies have shown that most of, but not all, geophysical signals are non-random. In many cases they are chaotic and in some cases quasi-periodic. There is no widely accepted definition of chaos. It might be described as a notion of disorder/irregularity and unpredictability. However, presence of chaos provides a new paradigm to all branches of science and bridges the gap between random and periodic processes. Chaotic signals contain ample evidence of determinism although in many cases they are only predictable in principle and not in practice. (Lorenz, 1993) Visual inspection of chaotic signal appears to be random but on closer examination there is considerable degree of deterministic components or underlying signal may even be quasiperiodic in nature. Evidence for chaos has now been found in many geological and geophysical processes such as mass extinctions, geomagnetic reversals, various ocean and atmospheric phenomena etc. It has also been found in the earth's orbital variations and dynamics of the solar system (Laskar 1990; 1999). The evidence of chaos in physical system, however, does not rule out the presence of periodicity. The long-term interactions between the two competing forcing processes that could be external or internal or both produce very complex dynamics leading to chaos and subsequently attain steady state equilibrium. By 'complexity' we mean here that the operating system has multiple interactions between many different systems and sub-system components.

Although, many geological episodes and environmental signals such as tectonics, volcanism, climate, sea level changes etc., display chaotic nature, they are found to be punctuated. It seems that the earth's dynamical systems oscillate almost periodically between the states of retardation and acceleration depending upon energy losses and gains around steady state. Strict periodicity is, however, a notion of order/regularity in a given physical system, which means that the underlying physical system, recur with constant time intervals. It appears that the apparent regularity (cyclicity) might be related to the inherent non-linear dynamics with interacting systems but the arbitrariness probably stems from the sensitive dependence of initial conditions. We thus may surmise that earth's system processes may combine both regular and irregular components (Tiwari et al., 1994). It is worth mentioning here that even though strict periodicity might exist in geological and geophysical time series, it is quite difficult, and rather impossible, to clearly identify or visualise it in complex and non-linear interactive dynamical earth system. Added problem comes from the poor and deplorable quality of data available for such purposes. Identifying cyclic pattern from such a complex geological time series, therefore, although of considerable importance for understanding the underlying physical processes, is quite a formidable task for the geoscientist. Strict statistical measures to test the reliability of harmonic components, therefore, become essential to judge the physical significance of the underlying processes/ phenomena.

IMPORTANCE OF GEOCYCLES

Distinguishing the nature of this complex temporal variability is useful in many ways: for instance, it would provide constraint in narrowing down the options for modellers to forecast the underlying phenomena. Evidence of identical cycles in different phenomena on similar time scales may be linked together through a common internal or external forcing mechanism. This would further provide a basic tenet to the understanding of the fundamental physical linkages among the various earth's and astronomical processes. There is another fundamental reason for resolving and investigating the orderly behaviour in earth science. While searching for long-term cyclic pattern has its significance in basic science for understanding the evolutionary history of the earth, the study of short-term global processes would provide the scientific basis for forecasting and thereby has its relevance for society by revealing the role of human impact in altering the global environment. Indeed, the complete spectral structures of these interactions are key to search for new alternative perspective to resolve/ reconcile scientific dogma. In fact the planet earth has been used as an observing platform to gain the astronomical knowledge and to study terrestrial fluctuation to provide additional knowledge of the involvement of these interactions (McCrea, 1981). It is, therefore, significant to note that the importance of cyclicity in various evolutionary records of coupled geo-bio-atmospheric system on different time scales lies in discovering evidence of hidden cyclic pattern in space-time. Searching for these astronomical cycles in various earth processes therefore appears to assume one of the major goals and noble endeavour of earth's scientists. It is interesting to note that a phrase, 'Everything on the earth operates in a cyclic manner' is cited as one of the twenty great ideas of sciences (Hazen and Trefil, 1991).

During the past two decades there has been an upsurge of interest in precise geo-data acquisition, processing and time series analyses. Attempts have also been made to confirm periodic signature in geological and geophysical records and their possible linkages with endogenic and exogenic processes. Modern time series analysis methods have played a central role to quantify the hidden information contained in the time series. During the recent years, modern spectral methods are used to identify periodicities in these records which provided high spectral resolution and accurate statistical significance. Application of multiple spectral methods, such as maximum entropy spectral analysis, multi-taper, singular and wavelets spectral methods, have considerably reduced biases and eliminated spurious result obtained from one particular method and thereby helped to evaluate stationary characteristic, spectral variance and stability of spectral components. The modern development in harmonic analyses also provided a robust statistical test, which allowed rejection of null hypothesis against commonly observed deceptive red noise in geophysical signals. These various spectral methods are reviewed in chapter 2. Various recent studies suggest that improved statistical sprectral techniques armed with statistical significance test and reasonably good detailed quality of precise geological measurement provided a clear picture of cyclic pattern in geophysical records. These geological findings indicate that various geological/geophysical processes exhibit cyclic nature with remarkable similarities with extra-terrestrial cycles on various time scales, which in turn, provide strong foundation to unify various earth system processes on a common time scale. The cyclic geo-bio-hydro-cryo-atmospheric processes in its linkages to exogenic forcing have, therfore, undergone a series of basic research and changes in scientific thoughts and attitude.

MEMORY OF THE EARTH SYSTEM

The earth is a coupled dynamical system and its various processes on a possibly all-different time scale appear to possess 'memory'. If a periodicity does exists in coupled dynamial system; common periodicy among many geo-processes on an identical time scale might indicate exchangeable response between them. Considering, in general, all time scales, by 'memory' we mean here the geological processes that affect or influence the subsequent evolution of the system periodically. There have been persistent claims of coupling and exchange of response among earth's system processes. Hence, geological history as a whole cannot be devoid of being at least a quasi-cyclic process on both large and small scale. All geological cycles have various lengths and cyclic processes, which may also be subjected to different kinds of driving forces. Indeed, one of the most intriguing concepts that have emerged from the recent analyses of earth's evolutionary record is that almost all the earth's processes (geo-bio-ocean-atmospheric system) on different time scales display quasi-periodic pattern. Based on these various information, the memory of the earth's system may be broadly divided on three time scales: (1) very long time scales (10^7–10^9 years) dealing mainly with long-term solid earth, ocean and atmospheric processes and their possible connection with plate tectonic processes (internal) and solar system's and galactic cycles, bolide impacts, cosmic flare and solar luminicity etc. (external), (2) long time scale (10^4–10^6) years processes dealing with climate and sea level changes associated with various ocean and atmospheric processes (internal) and earth's orbital cycles (external) and (3) short time scale (10^0–10^3 years) fluctuations of temperature, precipitation, drought, El Nino and storm etc. associated with various ocean-atmospheric processes (internal) and tidal and solar radiation (external). There is possibility that cycles, on all time scale, may arise due to internal and/or external forcing mechanisms, but detailed physical mechanism or mechanisms are yet to be clearly explored about their cause and effect.

ORIGIN OF GEOCYCLES

Spectrum of geo-records indicates that observed cycles appear as a possible result of interaction of internal and external forcing processes. Identical periodicities provide a testimony to the fact that either both process are driven by either a common forcing mechanism or one is the cause for the other. On a longtime scale, plate tectonic theory provides inter-relationship between earth's external and internal activities through processes widely known as 'rock cycle' (Wicander and Monroe, 2002). It is responsible for recycling rock material from surficial processes such as weathering, transportation and deposition to internal processes such as magma generation and metamorphism.

According to the plate tectonic theory, mobility of the lithospheric plates control creations of new crust and uppermost mantle at both divergent and convergent boundaries and cause recycling of old crust and mantle lithosphere at the subduction zones. For instance, heating within the earth's interior reach in convection cells that power the movement of plate and also in magma, which forms intrusive and extrusive igneous rocks. Movement along plate boundaries may result in volcanic activity, earthquake and mountain building (Wicander and Monroe, 2002). Scientists believe that long-term earth's processes such as plutonism, volcanism, metamorphism, rifting and sedimentation depicted in Fig. 1.2 recur every 30–35 m.y. time intervals. Deep mantle processes might affect the boundary conditions for outer core convection with possible influence on earth's magnetic field. There is evidence of periodic instability of the thermal boundary layer called as 'D layer' at the base of the mantle (Loper, 1984). It will lead to periodic perturbation of the mantle flow leading to changes in geomagnetic reversals frequency and several other consequences such as crustal fracturing and periodic cataclism of the activity of the mantle plumes. Some researchers are of the view that the earth's processes possess some characteristic time scale of the order of 30 m.y. and believe that cyclic patterns are maintained from within the system's energy itself. Plates descending back into earth's interior are subjected to increasing heat and pressure which may lead to metamorphism and alteration of rocks as well as generation of magma leading to another recycling of material.

These various processes are seen as single entities but are also coupled with several other bio-geo-hydro-cryo systems (Fig. 1.3). There are growing evidences to suggest that mantle dynamics and lithosphere recycling play a role in environmental change and minerals and natural resources accumulation. The interaction between the atmosphere, biosphere and hydrosphere contribute to weathering of rocks exposed on earth's surface. A longer tectonic cycle starting from the rifting of ocean basin to its closer proposed by Wilson (1966), takes about 400–500 m.y. also widely known as Wilson Cycle (Fig. 1.2).

During the past decades, however, there has been an upsurge of interest into looking at the order/periodicity/episodicity more quantitatively in these various records of tectonics and environmental activities. Several workers identified a basic quasi-periodicity of about 30 m.y. and 280 m.y. It has been

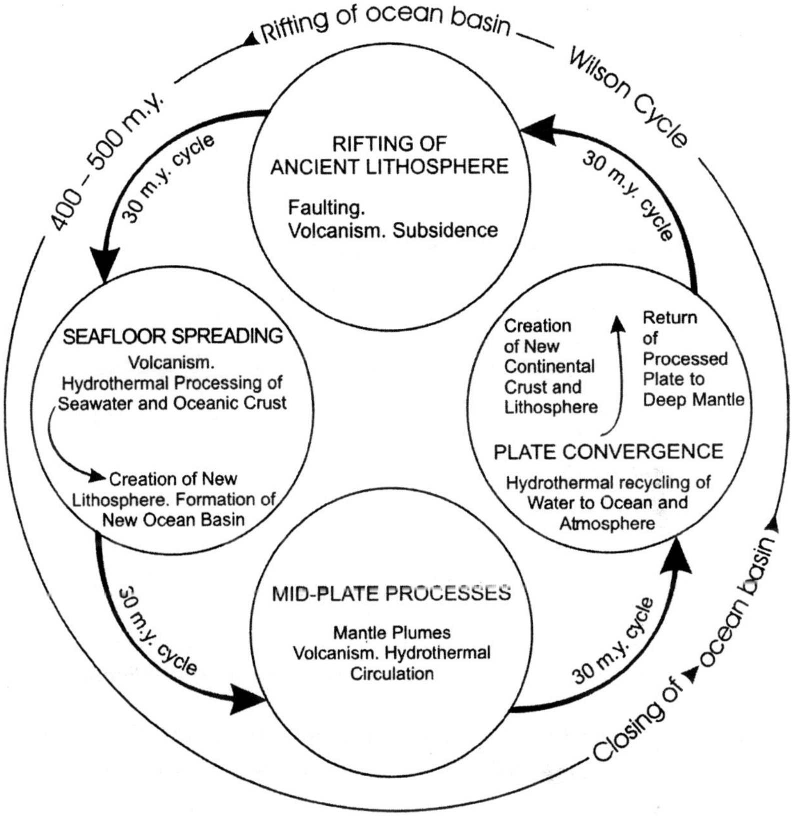

Fig. 1.2 The Solid Earth Cycle Depicting the Processes of Crustal and Mantle Evolution

(*Source*: Modified from Initial Ocean Drilling Plan on Earth, Ocean and Life System)

argued that the underlying periodicities might originate from exogenic forcing mechanism. Modern theory of bolides impact caused by galactic processes (exogenic forcing), although not proven, may provide one of the forcing mechanisms for triggering various tectonic activities, earth's interior processes, geomagnetic reversals and several other environmental consequences. Scientists believe that this kind of unusual paroxysm may increase mantle outgassing every 30 m.y. due to intense subduction and related volcanisms to raise the level of the atmospheric CO_2. It will cause 'cyclic green house' effect and other global cataclysmic episodes leading to global warming, climate deterioration and mass extinction during the geological past. The carbon cycle, including the carbonate-silicate geo-chemical cycle, has an important role in controlling the earth's climate. The complex hydrologic cycle through which water moves from the ocean to the atmosphere, to the land and back to the ocean again, is linked to cyclic temperature and CO_2 variations. As discussed

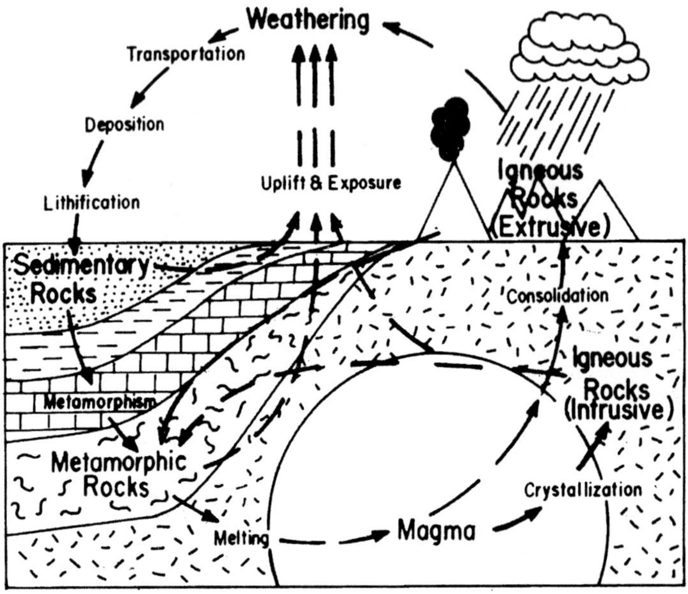

Fig. 1.3 Schematic Diagram Depicting Rock Cycle
Sources: http://www.winsocam.org/msa/k12/2kcycle/rkcycle/index.html

in chapters 4–6, a 30 m.y. cycle has been found in almost all the geo-bio-ocean and atmospheric processes along with higher order periodicities and their multiples.

Even on a relatively shorter time scale, in many cases, it has been found that evolution of sedimentary cycle and other related cyclic phenomena evolved following multiple interactions. For instance, sediment is believed to be a record of environment conditions. Hence, decoding sedimentary successions would reveal cyclic fluctuations of environmental condition. Evidences indicate that these cycles are distinctly separated by intervals of 100 k.y., 41 k.y. and 23 k.y. However, sediment analysis is a complex process in which most of the variables interact and depend upon each other. The possible role of external forcing cannot be ruled out here. Two main causes have been associated for periodic sedimentary successions (Schwarzacher, 1993): (a) Tectonic movements and thereupon fluctuations in the water table and (b) extreme changes in climate. Uplifting of some source area due to tectonic disturbances, for instance, may increase terrestrial flux and increased run-off due to increased precipitation. Increase in current activity may allow increase in storm activity leading to several other consequences. Hence, the cycle might also begin with a critical triggering of solar energy support changes or due to some exogenic triggering such as variations in the earth's orbital parameters.

On a still shorter time scale, periodic pattern evolves in earth's system processes. A model given by Rind (1999) gives more detail on how cyclic pattern evolves. Accordingly, in the beginning, the process starts with rainfall and returns to its original place with so many interactions and consequences. This complex system may be periodic, quasi-periodic and chaotic because of several interactions. In some of the cases, it might proceed to stochastic or even random oscillation. Accordingly consider how rain falling on earth contributes to growth of plants which in the process of growing transpire the moisture back to the atmosphere (Fig. 1.4). Once there, the atmospheric water vapour can form clouds, which affects the solar radiation that influences growth of plants. The atmospheric moisture also absorbs radiation from earth (as a 'greenhouse' gas), and this absorption, alongwith its cloud-generation capability, affects the temperature patterns. Temperature differences lead to pressure differences that generate wind and storms and El Nino activity; the storms provide for the rainfall, and the wind helps to produce the turbulence that enables transpiration to take place. The wind and rain also help to govern the circulation of the oceans, with their further influence on temperature. Involved in this cycle are the components of cloud physics, biological processes, solar radiation, atmospheric and ocean dynamics, all with multiple interactions. External forcing such as arising due to solar and tidal effects cannot be ruled out, which subsequently amplify the intensity of underlying processes and also act as a catalyst for initiating/triggering off already critically unstable ocean and atmospheric processes.

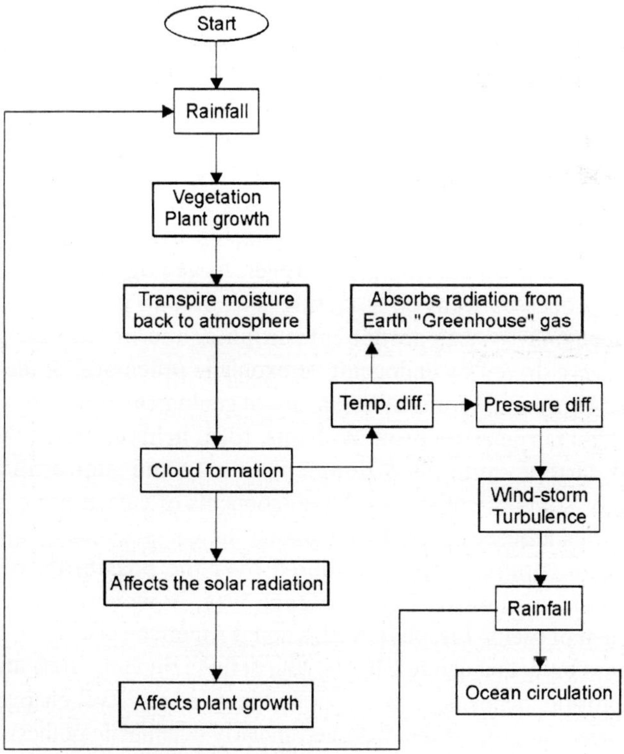

Fig. 1.4 Short-term Cycle

ASTRONOMICAL FORCING OF GEO-CYCLES: A HISTORICAL PERSPECTIVE

Mega-Geo-processes and Galactic Cycles

The search for long periodicity in the geological records has a long tradition. In the late eighteenth century, James Hutton, a Scottish scientist, in his monumental work, *Theory of the Earth* perhaps pioneered the idea that various earth processes operate in cycles. Hutton's classical theory of 'uniformitarianism' states that 'present is the key to the past'. It is based on a principle that contemporary geological processes have occurred in the same regular manner and with essentially the same intensity throughout geologic time. Another word, 'actualism' coined by some geologist also accept the basic premise of uniformitarianism. The theory is based on the premise that the processes that shaped the earth during the past are the same ones that operate today (Capedes, 1977). Some of the catastrophic events of geological processes, such as landslides, volcanic eruptions, earthquake and flooding recurring quasi-periodically are a result of mild and intense activity of forces within the closed earth system. Hence, the principle of 'uniformitarianism' means that it does not exclude sudden or catastrophic events and also does not require that the rates and intensities of geological processes be constant through time. Nevertheless the principle of 'uniformitarianism' has now been rejected on some obvious ground discussed in several books of geology.

Pioneering protagonists of long-term geo-cyclic theory include such notable scientists as the Chamberlin (1914), Joseph Barrel (1917), Joly (1923), Lull (1929), Schuchist (1932), Grabau (1940), Machado 1967. The early eighties has been another turning point after birth of the plate tectonic theory where the theory of 'catastrophism' again emerged as one of the major issues in earth science. Several researchers supported the view of 'catastrophic periodicity'. The major questions however, in geology are: (i) whether the geological processes went on smoothly with certain periodicity, or (ii) the intriguing regularity was abrupt/catastrophic, and/or (iii) the geological processes were driven by endogenic or exogenic processes. Some geologists had faith in Huttonian 'uniformitarianism' in geological processes while some others had faith in catastrophism. Williams, 1981, in his editorial, 'Megacycle' presented various opinions of long-episodicity and 'uniformitarianism'. Accordingly, some scientists developed concepts of catastrophic periodicity. These theories aroused large debate among the geo-scientists, climatologists and environmentalists. They pondered over the possibility of rhythmic catastrophism of nature. This evidence was, however, based on crude measurement of rocks. Joseph Barrell, a noted American geologist, deliberated over the possibility that nature vibrates with definite rhythms while analysing the cycles of erosion, deposition, mountain building and sea level changes. A noted German meteorologist, Alfred Wegener, in early twenties hypothesised that the continents drift about the globe on currents of the hot rock. Holmes (1929)

surmised that the floor of the ocean rose and fell periodically and coincidently matched each episode of mountain building with the same span of time. He concluded that geological changes occurred in cycles. Holmes (1929), a British geologist, proclaimed that the cosmic theory of the earth process of the recent past exhibit quasi-periodic variations of certain terrestrial phenomena on different geological time scales. Umbgrove (1947) in his classic treatise, *Pulse of the Earth* apparently found evidence of the intriguing regularity in the pattern of climate, sea level changes, tectonic and volcanic episodes of the earth's evolutionary history. He proposed a long-term geological cycle of 250 m.y. and related it to the 'cosmic year' term (the period taken by solar system to go round the galactic centre). There are other cycles and harmonics of the cosmic cycles. There has been a major debate on the kind of forcing to cause these catastrophic activities on the earth's surface. Are these forcings endogenic or exogenic, which effectively modify the response of earth-ocean-atmosphere and their evolutionary processes? Chapter 3 provides a more detailed account of astronomical cycles on different time scales.

Despite this honourable antiquity, the cyclic theory of earth processes could not win acceptance because (i) most of these changes were based on crude and fragmentary nature of geological observations, (ii) mathematical/numerical techniques could not be applied to these geological/geophysical observations and consequently crucial statistical tests failed to deduce a firm evidence of cyclicity in geo-processes, and (iii) because of weak evidence of cyclicity, the astronomical theory of earth processes could not get its due recognition. During the 1960s, however, the plate tectonic paradigm introduced a concept of new cyclic evolution of geodynamical processes starting from generation of a rift to the closure of the ocean and the creation of geo-structures in a long stipulated time period of 400–500 m.y. also widely known as a 'Wilson Cycle'. Several researchers found evidence of the remarkable regularity in the long-term pattern of climate, sea level changes, tectonic and volcanic episodes and also a clear natural division of the stratigraphic column during the earth's evolutionary history. Earth processes are influenced by the extra-terrestrial processes such as impacts of comets/asteroids on the earth and its environment. Evidence also suggests that an ice epoch may be triggered by variations of the astronomical environment encountered in the Sun's motion relative to the galaxy. Crain et al. (1969) provided evidence for periodicities in reversals record of earth's magnetic field. Williams (1981) provided a very lucid picture of long-term geocycles and their association with galactic cycles. McCrea (1981) also associated the possible cause of episodicity of various terrestrial processes with exogenic phenomena.

Negi and Tiwari (1983) provided first quantitative evidence of 285 m.y. and 32 m.y. periodicities in geomagnetic reversals record based on Walsh spectral analysis and associated these cycles to the period of solar system movement in the galaxy. Later, two groups of American scientists, one group consisting of Raup and Sepkoski (1984) from Chicago University made a remarkable discovery of a periodicity of 26 m.y. in biological mass extinctions. Another NAASA based group of geologist and astronomer, Rampino and Stothers (1984 a, b) analysed various episodic records of tectonics, sea level changes and provided scintillating evidence of periodicities of 30 m.y. and 260 m.y. and also identical periodicity in the impact cratering record of exogenic origin suggesting a possible link among theses processes (Rampino and Stothers, 1984; Alvarez and Muller, 1984). The issue of cyclicity was hotly debated (Lutz, 1985) but it was later maintained based on fuller analysis that a 30 – 32 m.y. periodicity does formally exist in reversal record (Stothers, 1986). These findings provided a crucial evidence for supporting the cyclic theory of earth's processes and turning point in the history of geological sciences. A 30 m.y. signal is important, in particular, since it is detected in almost all the processes of the earth system and also impact record and galactic processes. Napier (1989) noted that even prior to 1980, there seems to have been claim of apparent 30 m.y. periodicity, although based on qualitative evidences, in various earth and environmental phenomena (Dorman, 1968; Ager, 1978; Fischer and Arthur, 1977; Innanen et al., 1978) without any reference to its discovery in others, an independence might tend to strengthen confidence that a real signal was being detected. The major question however, at present is that whether we have now sufficient and high quality data, analysis techniques and sound physical evidences to prove or disprove the historical arguments and claim of existence or non-existence of periodicities more confidently and provide viable mechanism. As known, there is no end to such dialogues and arguments. However, the present scenario of long-term cyclicity in the earth's system processes are discussed in chapters 4–6 which provide review of recent important findings.

Environmental Changes and Orbital Cycles

In one of the major revolutions in earth's sciences that took place during the last three decades was the discovery of Milankovitch cycles in climate records. Croll (1875) was perhaps the first to suggest that the ice ages are caused/ controlled by variations in the earth's orbit. Gilbert (1895) showed connection between sedimentary cycles and orbital cycles. Bradley (1929) suggested a possible evidence of 21 k.y. cycle in a measured section that corresponds to the orbital precession of the equinoxes. Major breakthrough, however, came after the work of Milankovitch (1941) who published new calculations of the orbital variations and showed quantitatively how these determine the amount of solar radiation, which is ultimate driving force of terrestrial climate. Later, these cycles were also reported in several other records. Starting early in the nineteenth century, interest in the origin of the growth and melting of ice caps induced astronomers to make calculations of astronomical frequencies and led to suggestions about the way in which changes of the volume of ice caps could be related to astronomical influences (Imbrie and Imbrie, 1979). Several other researchers showed the relation between astronomical influences and rhythmicities in the pre-Pleistocene sedimentary record (de Boer and Wonders, 1984). Berger (1988) provided a detailed review of the astronomical theory of climate change. He has also discussed theoretical and astronomical basis, which might be useful to non-astronomers. The evidence of Milankovitch cycles in various environmental records was reviewed by de Boer and Smith. Today the idea of astronomical variables leaving traces in the sedimentary record is becoming widely accepted. With the development of analytical techniques, the effect of increases and decreases of the ice volume upon the stable oxygen isotope composition of seawater and of biogenic carbonates could be measured, and the marine sedimentary record of ^{18}O of carbonate could be related to changes in the volume of global ice caps. This allowed detailed reconstructions of the growth and melting of ice caps during the Pleistocene, the history of which is related in turn to astronomical influences (Hays et al., 1976; Imbrie et al., 1984; Fischer and Schwarzachar, 1984). In much of pre-Pleistocene geological history, polar and other ice caps have been of minor or no importance, and astronomical influences must therefore have had a more direct influence on sedimentation patterns and the environment of deposition, thus allowing their recognition (Schwarzachar, 1993). The review of these latest developments is given in chapter 7.

Ocean-atmospheric Processes and Solar and Tidal Cycles

Burroughs, 1992 has nicely brought out the historical development of short-term weather cycles and their linkages with astronomical cycles, in his excellent monograph, *Weather Cycles*. Accordingly, the noted astronomer, William Herschel (1932), was, in the early nineteenth century, the first to explain weather variation and climate change by extra-terrestrial means as early as in 1956. Several workers discussed on the short time scale the probability of astronomically driven climate change. They opined that the changes in the sun's output could influence the weather. Later, Heinrich Schwabe (1843) discovered that the number of sunspot varied in a periodic manner and provided credential support to the weather and solar cycles relationship. There are several examples of periodicities that arise in climate and weather. The search for underlying cause of these quasi-periodic fluctuations required improved basic data and modern analysis. These cyclic patterns may also reflect fundamental proposition of natural variability of global climate and tidal cycles, due to interactions of sun and moon with earth. Sunspot cycles reflect basic changes in solar behaviour and provide quite satisfactory explanation to various weather phenomena. Solar and tidal signals have now been found in many short-term environmental records such as temperature, precipitation, drought, floods, magnetic variations, and volcanic and seismic activities (Chapter 8)

The dynamics of the earth and its environmental stability can be affected by a variety of internal and external influences. Weedon (2003) in his monograph, *Time Series Analyses and Cyclostratigraphy*, provides detailed account of relations of solar and tidal cycles with short-term ocean-atmosphere and terrestrial processes. There is a strong belief that most of the earth's phenomena are governed by internal origin. However, recent investigations demonstrated that there are compelling evidences of the strong role of solar and tidal effects on the coupled earth-ocean and atmospheric processes such as in ENSO dynamics. The recent development and ongoing controversies of the role of solar and tidal cycles in short term earth's ocean and atmospheric processes is presented in chapters 8 and 9.

SUMMARY

Although there are ample evidences of periodicities on all time scales, the role of external forcing on various earth-ocean atmospheric phenomena have been one of the hotly debated issues in the earth's sciences (Bernett 1990; Garric and Hubblor, 2003). In addition to this, controversy continues regarding the statistical reliability of spectral peaks and stationary character e.g., whether the observed cyclicities cover a large portion of the observed variability or limited to a portion of this variability. There has been skeptic view over the existence of cyclicities, however, mainly because of lack of data and sophisticated techniques of time series analysis. The study of ergodic properties (stationary and non-stationary) of these records is therefore essential for understanding physical behaviour of a given record. Hence linking various terrestrial cycles with all sorts of astronomical phenomena at different time scales became quite a formidable task. However, with the late twentieth century, came high quality of geo-data based precise dating techniques and sophisticated computer based time series analyses methods and growing awareness about the controversies and skepticism of their physical interpretations. The applications of modern multiple spectral and numerical techniques to these new data provide more clearer picture of the nature of these records and the role of exogenic forcing in the earth system processes appears to become a virtual paradigm of terrestrial and exogenic connection on all time scales. Several recent findings suggest that there might be a more continuous role for exogenic forcing for terrestrial processes, although no viable mechanism was ever proposed. The astronomical link is being proposed due to the fact that the claimed terrestrial cycles were recognised as having frequencies corresponding to natural astronomical cycles. If true, linking terrestrial cycles with known astronomical cycles would provide supporting and well-established evidence and the physical basis for understanding as well as forecasting terrestrial phenomena. Some exciting examples are already there, where orbital cycles are being used for tuning geological records.

Exogenic forcing castrophism due to bolide impact cratering on the earth has been suggested as possible cause to long-term periodically surge cyclic activity on the earth. There is no accepted view about the internal/external mechanisms which can conclusively explain such recurring activities. Recently, however, it is argued that the 'geo-biological clock' of the whole earth processes may be of the extra-terrestrial origin. There is some convincing and intriguing evidence of the earth's 'cosmic cycles' which is perhaps one of the most remarkable examples of interaction of cosmic phenomena and terrestrial records.

Attempt has been made to synthesise this information and clarify about the scale of astronomical influence to geophysical fluctuations in this book. We mainly focus here on the possible existence of periodicities in various earth system processes and discuss fluctuations on scale of 10^7–10^8, 10^4–10^6 and to a lesser extent on scale of even 10–10^3 and their different geophysical mechanisms. There are persistent claims, in many cases, that many geological/geophysical environmental phenomena on all different time scale occur in cycle. Many of the earth's processes are, however, strictly periodic in nature and it could be generated by the chaotic dynamics, which contain considerable amount of deterministic component and possibility of predictability. It is also becoming more and more clear that an exogenic role must be acknowledged behind terrestrial effects. It is time now to evaluate causal relationship and to design strategies that will help us replace speculation with knowledge. Growing evidence, such as linking between various endogenic processes to exogenic forcing (solar and cosmic rays activities) on different time scales suggest that extra terrestrial phenomena are responsible for atleast some climate variability in time scale ranges from days to millennia.

With this brief introduction, we review the available information of earth's astronomical environment e.g., known astronomical cycles on different time scales in chapter 3 and examine their existence and physical relevance in claimed terrestrial processes starting from chapters 4–9. However, before doing so, it appears logical to briefly review various spectral techniques which have been used to distinguish and discover periodicities in various earth processes.

2

Statistical and Spectral Background

INTRODUCTION

This chapter is devoted to the brief description of statistical and spectral methods. Spectral methods have played an important role in geophysical and geological time series analysis. These methods have mainly been restricted to identifying periodicities and thereby provided solid basis for quantifying scientific arguments. Thus, this chapter forms the foundation for understanding the forthcoming chapters 4–9. However, we will not go much into details of mathematical rigour discussing step-by-step the various procedures of spectral analysis as it is beyond the scope of this book. We, however, briefly describe these various spectral techniques which have frequently been used for identifying periodicities from geophysical records. The purpose of this chapter is precursor to the readers to understand the subsequent chapters. We mainly focus on the procedures of identifying periodicities from geophysical records.

TIME SERIES

The important goal of geophysics is to provide quantitative information about the dimensions of depth (*Z*) and/or time (*t*) of physical system that produced it. Several geophysical signals are recorded, spatially or temporally, in the field or measured through laboratory experiments eventually as a set of sequential observations. Most of the geophysical time series of the geo-bio-ocean-atmosphere system generated on various spacio-temporal scales are analysed using theoretical and experimental tools with the purpose to understand the anomalies encoded in space/time series in terms of earth's structure and/or nature of the processes (e.g. cyclic, stochastic, chaotic and random). There are, however, several complexities which are involved in the recording system and source mechanism of these geological/geophysical records. At every time (t), a set of probabilities can be associated with the occurrence of each of the possible value $X(t)$ of the time/space dependent variable. A set of these recorded observations is called time series. In a precise statistical term, this set of random variables and associated probability distribution is called 'stochastic process'. Thus, the time series arising due to various geophysical/geological phenomena with sources in the earth's interior, ocean, astmosphere and extra-terrestrial domain could be treated as a sample of an infinite set of realisation of the underlying processes. The underlying time series could result either due to random/stochastic sources or might stem from deterministic/periodic or abrupt/chaotic processes. Only appropriate time series analyses of such complex records can tell something about the nature of such records.

TIME SERIES ANALYSIS

Methods of time series analysis provide quantitative information about the nature of geophysical and geological processes. This, in turn, provides some key properties of the system's behaviour that become basis for physical interpretation to understand the evolutionary history of the underlying processes. The purpose of time series analysis is to extract information (signals) from a background of random noise and unwanted signals using appropriate deterministic and statistical approaches.

Spectral methods, in particular, have played important role in the analysis and interpretation of geological time series. In fact, power spectral analysis of geophysical time series has been largely restricted to search for hidden periodicity. The concept of 'power' is derived from the field of electrical engineering. In time series analysis, the power spectral density function is the distribution of variance (mean square value) of time history over frequency. There are several methods now available which are used for the power spectral analysis. The spectral analysis using the Fourier transform and Blackman–Tukey methods, for instance, have commonly been applied to detect periodicity in geophysical and geological records. Later, more efficient techniques based on maximum entropy method (MEM) of spectral analysis were developed and applied invariably for resolving periodicities from geological time series. During the past few decades, many other efficient and powerful methods such as multitaper, wavelets and singular spectral analysis were developed and applied successfully to identify periodic components from a complex and noisy geophysical time series. Many geophysical signals, however, also show very sharp or transient variations. In such situations, Walsh spectral analysis is more appropriate. While the Walsh spectral analysis also works on the principle of Fourier analysis, it is indeed a computationally elegant and appropriate tool for the analysis of discontinuous geophysical processes.

PREPROCESSING CONSIDERATIONS

(i) *Sampling*: One of the important considerations prior to performing the spectral analysis is concerned with sampling of time series in a discrete data form. Generally, the sampling interval of a given record is decided on the basis of the shortest period present in the time-curve. The chosen sampling interval must be such that the original signal could be accurately reconstructed from them. The maximum rate of change of a signal is related to its maximum frequency component, called the 'bandwdth (B)'. Let us assume a time series having a highest frequency signal with one millimeter spacing between its peak and dip. According to the sampling theorem atleast one sample every $1/2B$ or two samples per cycle of the highest frequency, is necessary for the representation of a given time series (Schwartz and Shaw, 1975). Roughly speaking, a minimum of three points are needed to construct the most rapidly varying portion of the original signal. It implies that the shortest period which can be detected from the record would be twice that of the sampling interval. Then the highest resolvable frequency would be $1/2\Delta t$ (where Δt is the sampling interval), called folding or Nyquist frequency. Inappropriate choice of Δt, the sampling interval, lead to incorrect spectral estimate and hence wrong interpretation. For instance, if any signal is present with frequency higher than Nyquist frequency $1/2\Delta t$, its power will be reflected back towards low frequency resulting into the spurious power spectrum. This is called 'aliasing'. Hence, in general, the choice of smaller sampling interval would be appropriate in all spectral analysis to reduce the effect of alisaing (Bath, 1974). Alternatively, it is important to filter out frequencies higher than Nyquist frequency. A careful examination of the record is usually sufficient to make a good choice of sampling interval.

(ii) *Stationarity and Non-stationarity of Time Series*: Geophysical records are generally found to be time variant (non-stationary). However, statistical and spectral analysis are based on the assumption that a particular geotime series under study is stationary. The problem, therefore, arises as to how to comply with violation of stationarity assumptions. In practice, such data cannot be used directly for spectral or any other statistical analysis. In such cases, the stationarity of the data must be evaluated by studies of available sample time history records. The stationarity of random data can be tested by investigating a single record $f(t)$ by dividing the sample record into N equal time intervals where the data in each interval may be considered independent and stationary. Geophysical time series can be grouped into stationary or non-stationary depending on whether the statistical properties of a given time series are time invariant or not. In the case of stationary time series, the statistical properties (such as mean, variance and autocorrelation) of given time series does not change with time. However, if the time series is non-stationary, these properties will change. Most of the spectral theory is concerned with stationary time series and for this reason time series analysis requires to transform a non-stationary series into a stationary time series.

(iii) *Removal of Mean Value*: In practice, an analysis of stationary geophysical time series require that sample mean value should be subtracted from the time history involved. It is necessary that sample mean be reduced to zero before taking the power estimates. If it is not done, there is a possibility that the d.c. power may dominate at low frequencies. The removal of mean from the time series provides a better spectral representation at low frequency. However, it is not appropriate for transient data analysis (earthquake, etc.). As shown by Otnes and Enochson (1978), subtraction of mean, in case of transient data, would cause an undesirable distortion.

(iv) *Trend*: Geophysical time series are often found to be associated with seasonal effect, cyclic changes, trend (progressively increasing and decreasing) and other irregular fluctuations. Most of the geophysical/geological time series encompasses either local or long-term trend (low frequency component). Removing such a long trend, whose period is generally longer than the desired frequency, is essential for proper interpretation of the data. Otherwise, the data analysis containing such trend creates problems in later processing stages. In particular, large distortion can occur in the spectral and other statistical quantities. Presence of long-term trend in the data can completely eliminate the presence of high frequency spectral contents. Appropriate detrending of the time series is, therefore, needed to bring the data in proper form. A method of detrending is to transform the series by 'differencing' a given time series until it become stationary (Box and Jenkins, 1970), e.g., $x(t) = x_1, x_2, \ldots, x_n$ becomes $y(t) = (x_1 - x_2), (x_2 - x_3) \ldots (x_{n-1} - x_n)$. Second order differentiation of $x(t)$ is same as first order differentiation of $y(t)$ and so on. Other commonly used forms of detrending involve fitting linear, polynomial or exponential curves to the data and then computing residuals (Barton, 1983). Detrending is a powerful technique which suppresses the zero spectrum and can be a profound effect on the shape and scaling of the spectral estimates. Detrending is usually carried out by some formula which is to be determined from the data series. Least square procedure can be employed for the removal of linear as well as higher order polynomial trends. Traditionally, a simple polynomial curve (linear, quadratic and cubic) are fitted to remove the trend. The best fitted curve provides a measure of trend and residual which are the differences between the observation and corresponding value of fitted curve to provide an estimate of local fluctuations.

Alternatively, one may also use linear filter to convert the original data into another set of data. Different kinds of filters can be designed to produce desired output. In order to suppress local flucutations or high frequency variation, we generally use low-pass filters. The resulting time series is smooth data free from short term distortion. High pass filters are used to eliminate the long-term fluctuations or low frequency variations.

(v) *Prewhitening*: In many situations, geophysical spectra exhibit tendency of large power or amplitude in one part of the spectrum and very small in another. This is one of the characteristics of 'red noise spectra' which produces deceptive spectral features. The resulting spectrum shows leakage of additional low frequency energy which could be wrongly interpreted. It is, therefore, desirable to eliminate dominating spectral components before the spectrum is calculated. In order to obtain precise spectral estimates, it is necessary to eliminate side lobe effects of strong peaks in the spectrum. The processe of prewhitening is performed to eliminate the deleterious effect which is caused due to rapid spectral variations. A simple procedure commonly used, is the differencing method as discussed above in the previous section. Alternatively, each successive value is subtracted from the next after multiplying the next value by a weighting factor. (Emery and Thomson, 1997). The effect of spectral window spread the power from large peaks into adjacent frequencies. The prewhitening procedure pre-emphasises the high frequencies and suppresses the low fequencies. This procedure produces a flatter spectrum which is appropriate for physical interpretation. There are numerous literatures which suggest that after the prewhitening an inverse filter is applied which removes the effect of the prewhitening filter. (Weedon, 2003).

Windowing and Tapering

It has been generally observed that real geophysical and geological observations show discontinuous/abrupt behaviour at the end of the time series. Consequently, this transient nature of the data series introduces unwanted spectral features (leakage of power) that is often misleading.This leakage of power results due to end discontinuties. This imperfect representation of the additional power from the unwanted multiple sinusoids is called the 'Gibbs Phenomena'. The problem of leakage of power can be circumvented by reducing the oscillation gradually at both ends of the time series. This procedure is known as tapering. The procedure consists of multiplying the observations by the window weights starting at one in the centre of the observation and dropping to zero at the edges. The main objective of the application of the spectral window is to minimise the leakage caused due to finite record length and abrupt discontinuity behaviour of the signal. A number of windowing techniques are available which can be efficiently applied to remove the irregular appearance and erratic behaviour of the spectral characteristics calculated through periodogram method. The choice of spectral window depends on the requirements of resolutions, stability and smoothness of the resultant spectrum. A major disadvantage with Sine and Cosine functions are, that they do not fit well with discontinuities of time series. In order to minimise the error, it is desirable to apply spectral window in Fourier domain. However, such problem does not arise in Walsh domain, because Walsh functions are discontinuous themselves and fit better to such discontinuous functions and, therefore, computation of power in Walsh domain is devoid of such problems.

Filtering

Geophysical observations represent composite (mixture) wave-form which consists of signals of various frequencies alongwith different kinds of noise. It is sometimes desirable to extract only a certain kind of information. For instance, one might be interested only in low, high or a certain narrow band frequency component. It is necessary that signal-to-noise ratio within the pass band should be high. Filtering operations aim to retain desired signal in the spectra and filter out noise and unwanted signals. There are various kinds of filters which are designed to select desired frequency band depending upon the purpose of analyses. The most commonly used filters in time series analyses are low, high or band pass filters. All frequency-selective filters treat the band pass information as signal and rejected information in the stop-band as a noise to be removed or filtered out (Weedon, 2003; Ifeachore and Jervis, 1993). In low pass and high pass filtering, for instance, low and high frequency components are passed and the remaining components are rejected. Band pass filter is used for extracting information in a limited range of selected band of frequencies and suppress low and high frequency components (Fig. 2.1).

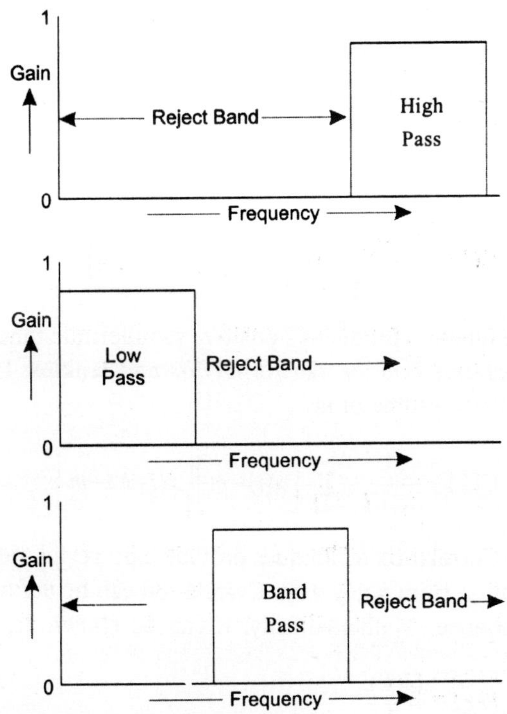

Fig 2.1 Schematic Diagram Showing High Pass, Low Pass and Band Pass Filter

Fourier and Walsh transform analyses have played a significant role in extracting signal information from such composite waveform. In frequency or sequency domain, transformed frequency components lying outside the band pass are rejected and then inversely transformed. In particular, Walsh zero-phase shift characteristic in sequency domain provides quite useful approach for the filtering of geophysical signals. These are low pass, high pass and band pass filtering also known as 'sequency limited' filtering. (Harmuth, 1970). The filtering operation in sequency domain is quite simple. It is simply accomplished by multiplying the element of a filter vector in 0 or 1, to each component of the transform. (Beauchamp, 1974). This is similar to the familiar boxcar type of filter. The output result is stepped version of the original signal. Such operations require minimum computing operations. Band pass, high pass and match filters can all be designed using same techniques.

STATISTICAL MEASURES

These statistics can be estimated easily from N data set $\{x(i), i = 1, \ldots, N\}$ using the well known relations:

Mean: For a given set of data mean can be calculated as

$$m = \frac{1}{N} \sum_{i=1}^{N} x(i)$$

Variance:

$$s^2 = \frac{1}{N} \sum_{i=1}^{N} \left[x(i) - m\right]^2$$

Covariance:

$$C(k) = \frac{1}{N} \sum_{i=1}^{N-k} \left[x(i) - m\right]\left[x(i+k) - m\right]$$

Covariance function produces a positive semidefinite function that leads to a finite non-negative Fourier transform (Box and Jenkins, 1970)
Another alternative estimator is:

$$C(k) = \frac{1}{N-k} \sum_{i=1}^{N-k} \left[x(i) - m\right]\left[x(i+k) - m\right]$$

Correlation: Correlation techniques provide a powerful means of extracting signal information from noisy data. Correlation can be defined as covariance divided by variance. Mathematically, it can be shown as,

$$r(k) = \frac{C(k)}{s^2}$$

Auto-correlation: The autocorrelation is a measure of statistical dependence

and similarity of a given function with itself at different lags. The dyadic autocorrelation defined in Walsh domain is analogous to the correlation function in Fourier analysis. However, computational procedure differs in a sense that the Walsh domain uses modulo-2 addition.
Discrete autocorrelation function in real time is given by:

$$A_F(\tau) = \frac{1}{N} \sum_{i=0}^{N-\tau} X_i X_{i+\tau}$$

Discrete autocorrelation function in dyadic or Walsh domain is given by:

$$A_W(\tau) = \frac{1}{N} \sum_{i=0}^{N-\tau} X_i X_{i \oplus \tau}$$

where $i = 0, 1, 2, \ldots$, M and $M < N$ where M is total correlation lag, $\oplus$ represents modulo 2 addition ie., binary sum without carry $0 + 0 = 0$, $0 + 1 = 1$, $1 + 0 = 1$, $1 + 1 = 0$ (Harmuth, 1970).

Similarly cross correlation function can be given as

$$C_F(\tau) = \frac{1}{N} \sum_{i=0}^{N-\tau} X_i Y_{i \pm \tau}$$

$$C_W(\tau) = \frac{1}{N} \sum_{i=0}^{N-\tau} X_i Y_{i \oplus \tau}$$

Examples: Figure 2.2 depicts the comparative example of autocorrelation in real time domain and dyadic domain using different Walsh wave forms. Comparative examples show significant differences in their computational behaviour. In real time domain this is arithmetic and in the Walsh domain this is modulo 2 additon. Auto correlation analysis of binary type signal, like reversals of magnetic field and seismic tremors in Walsh (dyadic domain) domain can be more appropriate and economical.

METHODS OF SPECTRAL ANALYSIS

Fourier Analysis

A review of statistical measure and spectral analysis of stratgraphic time series is given by Barton, 1983 and Pardo-Iguzquiza et al. (1994). The basic principle involved in the Discrete Fourier Transform (DFT) (Periodogram) analysis is to fit a sequency of sinusoids to the observed time series in a least square sense. The period becomes the integer-divisor of the length of the record. The series is thus decomposed into a set of hormonics.

In mathematical terms a discrete time series (X_t, $t = 1, 2, \ldots N^{-1}$ with an odd number of term N equally spaced at Δt and with zero mean) can be

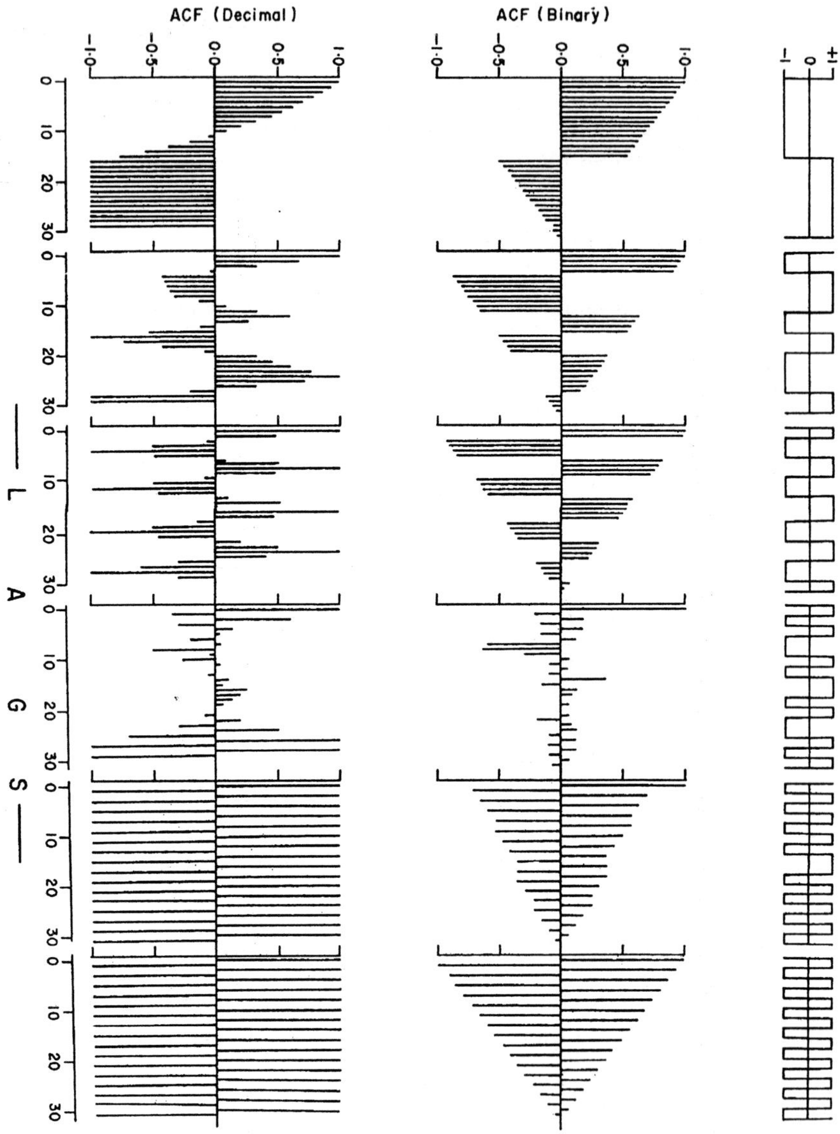

Fig. 2.2 Depicting the Comparative Example of Autocorrelation in Real Time Domain and Dyadic Domain

represented in the form of sine and cosine transform as follows:

$$x_t = \sum_j [a_j \cos(\omega_j t) + b_j \sin(\omega_j t)]$$

$$a_j = \frac{2}{N} \sum_j x_t \cos(\omega_j t)$$

$$b_j = \frac{2}{N} \sum_j x_t \sin(\omega_j t)$$

Here a_J and b_J are least square estimates of the amplitudes of the best fitting cosine and sine wave of each frequency ω_J. The upper limit for j in above equation is (N–1)/2.

Here,

$$\omega_J = 2\pi j/N\Delta t,\ j = 1, 2, 3, \ldots, (N/2) - 1$$

The case of $j = 0$ may be ignored since the mean of the series $a_0/2$ is zero.

The squared amplitude of the transform is used as spectral amplitude of the jth harmonics.

$$R_j = (a_j^2 + b_j^2)^{1/2} \quad \text{or} \quad \sqrt{a_j^2 + b_j^2}$$

and power variance accounted by each harmonics is

$$P_j = \frac{R_j^2}{2}$$

The phase of the jth harmonic is given by

$$\phi_j = \tan^{-1}\left(\frac{-b_j}{a_j}\right)$$

The above equation can be put in a computationally more convenient form (complex number form) using the Euler relation as:

$$\cos\omega t + i \sin\omega t = \exp(-i\omega t)$$

$$X_t = \sum_{j=1}^{N-1} J_j \exp(i\omega_j t)$$

$$J_j = \frac{1}{N} \sum_{N-1}^{t\,N-1} x_t \exp(-i\omega_j t)$$

where $\quad J_j = (1/2)\,(a_j - ib_j)$

The set of N complex numbers $J_0, J_1, \ldots, J_{N-1}$ is the Fourier transform of X_t. J_0 is zero mean since mean of X_t is zero. The squared modulus of the transform.

$$|J_j|^2 = J_j J_j^* = \frac{1}{4}(a_j^2 + b_j^2) = \frac{1}{4} R_j^2 = \frac{1}{2} P_j^2$$

where * denotes the complex conjugate.

The main difficulty in the spectral analysis using periodogram approach is the erratic behaviour of spectral lines and large number of computations involved in the computation of Fourier transform. However, an appropriate procedure of smoothing combined with Fast Fourier Transform (FFT) routine provides stable spectral estimates with considerable reduction in time.

Walsh Spectra

Historically, Walsh functions were first defined by an American mathematician J.L. Walsh in early twentieth century. He proposed a set of complete and orthogonal functions assuming essentially only two amplitude values, +1 and –1 and possessing many other properties similar to sine and cosine functions. Several different descriptions on the Walsh functions and their theoretical developments could be found in literature. Beauchamp (1974) and Harmuth (1970) have presented a lucid picture on this subject and have brought out anologies between the Walsh and Fourier analyses. As the Walsh spectral technique is unusual, it would be appropriate to present a systematic theoretical development prior to making any comprehensive practical application of this method to geophysical time series. In order to elucidate the theoretical discussions and to highlight the relative merits and usefulness of this method, some examples of statistical and spectral analysis on synthetic data sets are also presented. Some important considerations in the discrete spectral analysis

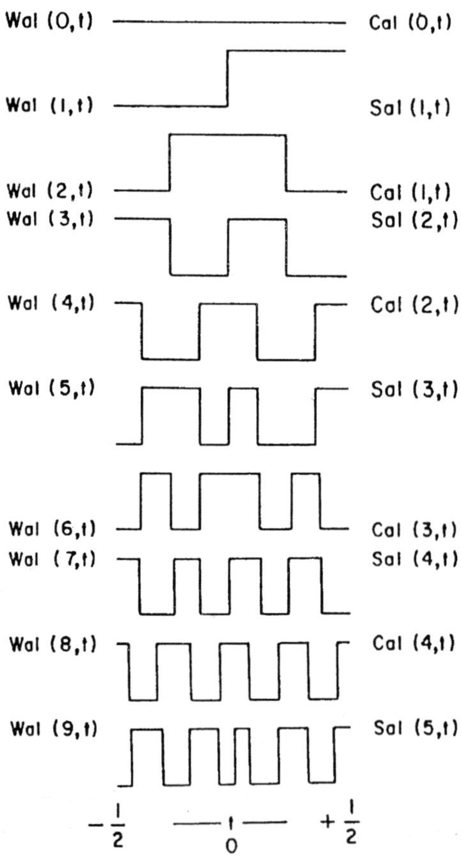

Fig. 2.3 An Example of Walsh Function

of real time series will also be discussed. Figure 2.3 shows an example of Walsh function.

The complete definition of the Walsh functions can be found in the following recurrence formula (Beauchamp, 1974).

$$Wal_P(t) = 0 \text{ for } t < 0 \text{ and } t > 1$$

$$Wal_O(t) = 1 \text{ for } 0 \leq t \leq 1$$

$$Wal_{2P}(t) = Wal_P(2t) + (-1)^P \; Wal_2 \; 2(t - 0.5)$$

$$Wal_{2P+1}(t) = Wal_P(2t) - (-1)^P \; Wal_P \; 2(t - 0.5)$$

$$(P = 0, 1, 2, \ldots)$$

where P is related to frequency and t is time period.

It implies that functions have a value +1 or –1 in the interval between $0 < t < 1$ and zero outside this interval. Each Walsh function is formed by compressing the previous $(P/2)^{th}$ function (truncated to zero) into first half of the range followed by a repetition of this sign on the second half of the range with same or opposite sign.

An alternative notation to the Walsh functions has been introduced by Harmuth (1970) in order to classify the functions in terms of even and odd waveform symmetry i.e.,

$$Wal_{2S}(t) = Cal_S(t)$$

$$Wal_{2S-1}(t) = Sal_S(t) \; (S = 1, 2, \ldots, N/2)$$

where S is 'sequency' coined by Harmuth (1970).

In the Fourier analysis, 'frequency' is a parameter which can be interpreted physically as number of cycles/unit of time. In the Walsh domain a new term 'sequency' is introduced for the generalisation of the term 'frequency' which is identical for circular functions. In a more general sense, it is defined as one-half of the average number of zero crossing/unit of time, in order to cover non-periodic functions also (Beauchamp, 1974). The term 'sequency' makes it possible to replace such important terms as 'frequency-power-spectrum by sequency power spectrum'. The concept of period of oscillation, $T = 1/f$, associated with frequency leads to the more general definition after substituting sequency 's' for frequency 'f'. Sequency domain of the average period of oscillation is given by $T = 1/S$ where T is average time separation of successive zero crossings. The sequency of Walsh domain coincides well with the frequency of Fourier domain. Some time sequency is also referred to as generalised frequency (Bath, 1974; Harmuth, 1970).

In sequency order, the components are arranged in ascending order of zero crossing. Sequency order is favoured for communication and signal processing such as spectral analysis or filtering. This ordering convention is found to be directly related to frequency where Fourier components are also arranged in increasing harmonic number (Beauchamp, 1974). The principal advantage of such an order is that the derivation of the alternate Cal and Sal function resembles that of orthogonal sine and cosine functions in Fourier analysis and hence permit suitable comparison. In the present analysis, we use sequency order convention for spectral analysis of geophysical time series.

Analogous to the Fourier power spectral analysis, Walsh periodogram requires that the sum of the squares of the Cal and Sal function coefficients of the same sequency are added for each spectral coefficient. Here only $N/2 + 1$ spectral points will be obtained with the first and last consisting of the square of the Cal and Sal functions. The equation for this can be given as follows:

$$P(0) = X_c^2\ (0,\ t)$$

$$P(K) = X^2_c\ (k,\ t) + X_s^2\ (k,\ t)$$

$$P(N/2) = X_s^2\ (2/N,\ t)$$

where $K = 1, 2, \ldots, N/2 - 1$, where $X_c\ (k, t)$ and $Xs\ (k, t)$ are the Cal and Sal coefficients of Walsh transform.

Discrete Walsh and Fourier transforms are very slow computationally and hence fast Walsh and Fourier transfom techniques are developed which require data point as a power of 2 (32, 64, 128). In many cases, however, the required numbers are not available and hence zero can be added to fulfil the requrement. It is advisable then that the effect of zero-padding should be compensated for calculating bandwidth and confidence intervals (Pardo-Iguzquiza et al., 1994).

Comparative Examples of Walsh and Fourier Spectra

Comparative study of Walsh and Fourier spectra will bring out important differences between the spectral structures of a certain wave form. It would be useful in deciding the choice of a particular spectral technique while dealing with real time series.

Figure 2.4(top), shows a comparison between Walsh and Fourier spectra of eight cycle rectangular waveforms. Walsh power spectra exhibits only one peak, as expected at 8th harmonic whereas Fourier power spectra of the same waveform exhibits in addition to the main peak additional side lobes. A more pronounced difference in Walsh and Fourier spectral structures can be seen in the Fig. 2.4 (bottom). Fourier spectra of 8 cycle triangular waveform exhibit a number of harmonics besides the main peak whereas the Walsh spectra provides a relatively simple and compact representation. Similar spectral comparison between Walsh and Fourier spectra of eight cycles sine wave is shown in Fig. 2.4 (middle). In this case the situation is reversed. Fourier power spectra exhibits a single peak at 8th harmonic whereas Walsh power spectra exhibit higher harmonics also besides main peak. It can be concluded that the Walsh spectral technique is appropriate for time series analysis which are dominated by quasi-stable/abrupt or discontinuous processes. However, if the

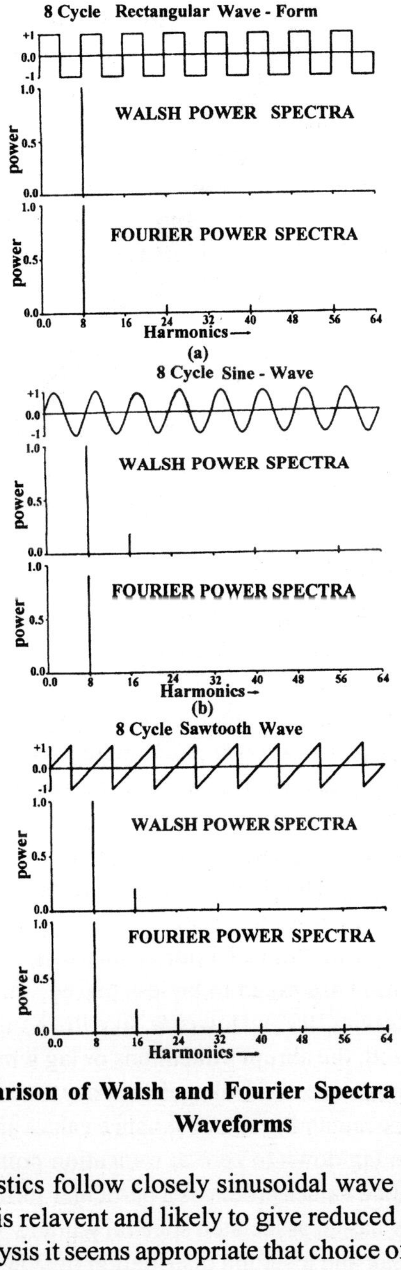

Fig. 2.4 Comparison of Walsh and Fourier Spectra of Different Types of Waveforms

signal characteristics follow closely sinusoidal wave pattern, then the use of Fourier analysis is relavent and likely to give reduced error. From the ensuing comparative analysis it seems appropriate that choice of an appropriate spectral technique should be dependent on the dominant signal characteristics of the time record. A compact and exact spectral representation could be obtained if both time functions and orthogonal functions have matching signal characteristics. It may be noted that distinct difference in the various spectral structures are clearly seen in very simple and ideal cases. It is expected to have adverse and more severe effects when dealing with more complex waveforms. It is, therefore, important to take appropriate caution while making choice of spectral techniques for the analysis of a given time series.

Walsh periodogram as discussed above is generally found to be inconsistent. It is therefore desirable to improve the stability of the periodogram. In order to circumvent the problem and achieve desired stability one can adopt the following procedure (Beauchamp, 1974 and Negi et al., 1993).

Let the data $x, \ldots, X_{N-1}$ be divided into L streches of length K each, $LK = N$. Here it is required that K is power of two. Then compute the Walsh periodogram for each section at frequency S,

$$P^{(N)}(K) = \frac{1}{L} \sum_{t=0}^{L-1} l^{M}(K, 1) \text{ or } 1 = 0, 1, 2, \ldots, L-1.$$

This equation provides more stable spectral value than the spectra calculated using the entire time series. However, resolution of spectral components will be constrained according to length of divided time record.

Blackman-Tukey Method

Blackman-Tukey (BT) method of spectral analysis is another example which has been widely applied in geophysical data analysis for identifying periodicities. The Wiener-Khinchine theorem states that the periodogram of a time series can be obtained by using the Fourier transform of the autocovariance function. Hence in BT method, the power spectra is computed by taking the Fourier transform of the autocovariance of a given time series (Barton, 1983 and references therein). The autocovariance is nothing but product of autocorrelation functions multiplied by its variance. It is, however, to be noted that highest autocovariance lag should be eliminated prior to taking transform as these values add noise in resulting spectra. The BT method has some advantages over traditional method of Fourier analysis. For instance, the spectral peaks in BT method are found to be less biased, smooth and slighly narrower (Ifeachor and Jervis, 1993). However, like the Fourier analysis, the BT method also suffers with the abrupt truncations or lag window problem in the analysis resulting in large spectral leakages.In order to reduce the spectral leakages and side lobes, the remaining autocovariance values are weighted from a maximum value at zero lag down to zero at truncation point. Each form of weighting of autocovariance values produces a particular form of lag window. Every type of lag window has an associated spectral window with a particular width of main spectral peak and a certain distribution of side lobes (Priestley, 1981; Weedon, 2003).

Cross Spectra and Coherency

It may be noted that even if two time series show an identical spectral peak, there is no reason to believe that their origin also must be the same or one is a result of the other. Hence, to come up with a clearer picture of such relationship, an alternative method of cross-spectral analysis may provide some insight of such linkage at a particular frequency, if any.

Cross spectra can be defined as (Jenkins and Watts, 1968)

$$C_{XY} = \frac{2}{T}\left[A_X(\omega) - B_Y(\omega)\right]$$

The '*X*' refers to the complex conjugate of the function C_{XY} cross spectra of the A_X *(w)* and B_Y *(w)* spectra.

Another parameter called 'coherency' provide a quantitative measure of similarity of amplitude variations in two time series determined at particular frequencies.The amplitude relationship is quantified through the coherency squared, calculated as

$$C_{XY}^2(\omega) = \frac{\left|C_{XY}(\omega)\right|^2}{A_X(\omega)B_Y(\omega)}$$

where significant coherency squared value $C_{XY}^2(\omega)$ indicates degree of dependence of one signal on another. The value of coherency ranges from 0–1.

Maximum Entropy Method

Power spectral estimation techniques discussed above are based on unrealistic assumption of periodic or null extension on any side of the observed record, hence the traditional spectral methods as discussed above have certain drawbacks, when applied to short time series. Ulrych (1972) pointed out that in addition to the spectral shift problem, there is loss of resolution wherever the data lengths are comparable to the period. Because of several such other problems a new approach for the power spectrum analysis based on maximum entropy principle has been proposed by Burg (1967). The method has become increasingly popular because of its capability in resolving spectral peaks when applied to variety of geophysical signals.

The basic advantage of Maximum Entropy Method (MEM) lies in the generation of a unique filter based upon the information contained within the available data sample. The filter is applied to the time series and serves to whiten the input data samples (Ulrych and Bishop, 1975). In general terms, the method may also be described as a data-adaptive method where it assumes neither a null nor periodic extension of the data but adapts itself to the sample of the process being analysed. The spectral techniques have been reviewed by several researchers (Barton, 1983) and most recently by Ghil et al. (2002).

The MEM involves a fundamental parametric model, which is the discrete autoregressive (AR) process, proposed by Yule in the late twenties (Yule, 1927). The method has been discussed in detail by Barton, 1983; and Ulrych and Bishop, 1975. Accordingly, the value of the variable at time *t*, *X*(*t*), is derived linearly from a purely random Guassian process according to the previous *m* values of the variable:

$$X(t) = \alpha_1\ X(t-1) + \alpha_2\ X(t-2) + \ldots + \alpha_m\ X(t-m) + b(t),$$

where m is known as the order of the autoregressive process, $\alpha_1, \alpha_2, \ldots, \alpha_m$ are the constants, b(t) is called the innovation of the process, and the mean of both the AR process and b(t) are zero. The AR process may be viewed as a prediction of $X(t)$ from the previous m values of X, and b(t) as the prediction error (Barton, 1983). Many authors (e.g. Ulrych and Bishop, 1975 and recently Ghil et al., 2002) have demonstrated that the spectrum of the mth order autoregressive process is:

$$S(\omega) = 2V_m^2 \left| 1 - \sum_{j=1}^{m} \alpha_j \exp(-2\pi i \omega j) \right|^{-2},$$

where V_m is the variance of $b(t)$. $S(\omega)$ can be evaluated at all frequencies between $\pm\omega_c$.

The computational procedure in MEM analysis involves fitting an autoregressive model to the data based on the principle that the resultant spectral estimate should be based on all the information (hence the name maximum entropy) about the series outside the observed record. The condition for this happens to be satisfied by the mth order autoregressive process. The set of coefficients $(1, -\alpha_1, -\alpha_2, -\alpha_3, \ldots, -\alpha_m)$, or more commonly $(1, \gamma_1, \gamma_2, \ldots, \gamma_m)$ where $\gamma_1 = -\alpha_1$ etc., is called the prediction error filter since the prediction error, b(t), is the convolution of this filter with the vector X(t) (Barton, 1983). The spectrum of the process is thus estimated by the equation just given above which in frequency terms is:

$$S(\omega) = \frac{P_{m+1}}{\omega_c} \left| \sum_{j=0}^{m} g_j \exp\left(-i2\pi\omega j \Delta t\right) \right|^{-2},$$

where $\gamma_0 = 1$, and P_{m+1} is a constant representing the power (variance) of the $(m + 1)$ term prediction error series b(t), i.e. the prediction error power. The derivation of P_{m+1} and then $S(\omega)$ requires estimation of the $(m + 1)$ prediction error coefficients.

The method of calculation of filter coefficients from the time function on digital computer is given by Anderson (1974). More details about this method can be found in articles by Lacoss (1971), Ulrych and Bishop (1975) and others which are particularly useful for uninitiated readers.

The MEM approach is useful when one is dealing with a segment of a sampled function which is short compared to the autocorrelation of the function. Chen and Stegen (1974) have demonstrated the effectiveness of the method when periodic components are present. The main problem in applying MEM, however, is the optimum selection of the order of the process. If the order chosen is too low, the spectrum is over-smoothened. If it is too high, frequency shifting and spontaneous splitting of spectral peaks occur. Though several attempts have been made to provide a suitable criterion for choosing the order of the processes, no general consensus exists. Best choice of criteria depends ultimately on the stability of signal characteristic and the record length. In view of the lack of general agreement in selecting appropriate order of the analysis, it is preferable to apply first Fourier or any other method to obtain first hand information about the relative power and frequency content in the data. The MEM analysis, may then be performed using a varying range of filter coefficient. MEM spectral analysis is generally performed using variable length of prediction error filter (PEF) or Akaike's final prediction error criterion (Akaike, 1969, 1974) to check the stability of spectral peaks and to avoid any spurious peak in the spectrum.

Clean Spectral Analysis

Robert et al. (1987) have presented an efficient 'clean' spectral technique especially for unequally spaced data analysis. Basically, the method revolves around computation of one dimensional deconvolution of the original time series with suitable spectral window. The detailed theoretical and computational procedure and algorithm are given in Robert et al. (1987). Clean spectral algorithm has successfully been applied to the analysis of astronomical and geophysical time series (e.g. Dreher et al., 1986; Duvall and Harvey, 1984, Negi et al., 1990, 1996). Experiments with numerical time series of synthetic data of irregularly sampled periodic signal have shown that the clean algorithm was able to recover effectively most of the lost information even for a significantly less number of data points (Vio et al., 1992).

Application of this technique has been reported by others (Drehar et al., 1986; Duvall and Harvey, 1984) for analysing astronomical data. The 'clean' algorithm removes the artefacts introduced by missing data. The method is based on a complex one-dimensional version of the clean deconvolution algorithm widely used in two-dimensional image reconstruction. Some essential features of the algorithm are presented below:

According to convolution theorem the Fourier transform of the sampled signal $D(\omega)$ is the convolution of the spectrum with the Fourier transform of the sampling function $W(\omega)$ ie.

$$D(\omega) = F(\omega)*W(\omega)$$

$$\equiv \int_{-\infty}^{\infty} d\omega' F(\omega') W(\omega - \omega')$$

Where $D(\omega)$ and $W(\omega)$ are the dirty spectrum and spectral window function, respectively.

The $D(\omega)$ and $W(\omega)$ can be calculated directly from the discrete data and are given by:

$$D(\omega) = \left(\frac{1}{N}\right) \sum_{r=1}^{N} f_r(\omega) e^{-2\pi i \omega t_r}$$

and $$W(\omega)=\left(\frac{1}{N}\right)\sum_{r=1}^{N} e^{-2\pi i\omega t_r}$$

The symmetries of the $D(\omega)$ and $W(\omega)$ are the same as that of $F(\omega)$:

$$D(-\omega) = D^*(\omega)$$

$$W(-\omega) = W^*(\omega)$$

To apply 'clean' to the above deconvolution equation, as an example, we consider a single harmonic component (cosinusoid) with a harmonic amplitude A, frequency 'ω' and phase constant 'ϕ' $f(t) = A\cos(2\pi\omega t + \phi)$. The spectrum of this cosinusoid signal is given by $F(\omega) = a\delta(\omega - \hat{\omega}) + a^*\delta(\omega + \hat{\omega})$ where $a = \left(\frac{A}{2}\right)e^{+i\phi}$

The spectrum is complex and contains a pair of spectral components at $\pm\hat{\omega}$ with equal amplitude A/2. The phases of the positive and negative spectral components are $\pm\phi$ respectively.

For spectral analysis an iterative procedure for the clean algorithm is given as follows (Robert et al., 1987):

(i) We find the amplitude and frequency of the first spectral component using the equation:

$$C^i = ga(R^{i-1};\ \omega^i)$$

where g is clean gain (0<g<2) defined by Robert et al. (1987).

and $$a(D;\omega)=\frac{D(\omega)-D^*(\omega)W(2\omega)}{1-|W(2\omega)|^2}$$

represent dirty or contaminated spectra as defined by Robert et al., 1987.

(ii) In the second step, we remove the contribution of this component, including its side lobes, by subtracting the response of the consinusoid spectrum from $D(\omega)$. This will form the first residual spectrum R'. To prevent errors from destabilising the clean procedure preferably a fraction g of the spectral component is subtracted.

(iii) In the third step, this iterative procedure is repeated to obtain successive approximations, i.e., R^1, R^2, . . . , R^i. After completion of the ith step we have a set of i spectral clean components and residual spectrum R^i. In general, the iterative prrocess for the ith residual component is given by:

$$R^i(\omega) = R^{i-1}(\omega) - [C^iW(\omega-\hat{\omega}) + C^i * W(\omega+\hat{\omega})]$$

with the notation $R^O = D(\omega)$

(iv) After determining all the spectral clean components, they are convolved with the clean beam $B(\omega)$. $B(\omega)$, which has a Gaussian shape of amplitude with $B(0) = 1$ and linear slope in phase, is chosen to fit the central peak of $W(\omega)$.

(v) Finally the clean spectrum $S(\omega)$ is given by

$$S(\omega) = \sum_{i-1}^{K}\left[C^i B(\omega-\omega^i)+C^{i*}B(\omega+\omega^i)\right]+R^K(\omega),$$

where K is number of clean components.

The clean algorithm of spectral analysis has been used extensively in geophysical as well as in the astronomical data analyses. A comparison of Fourier and Clean spectra using the synthetic data is shown in Fig. 2.5. Figure 2.5(a) shows known signal with two sinusiods embedded with noise, Fig. 2.5(b) shows

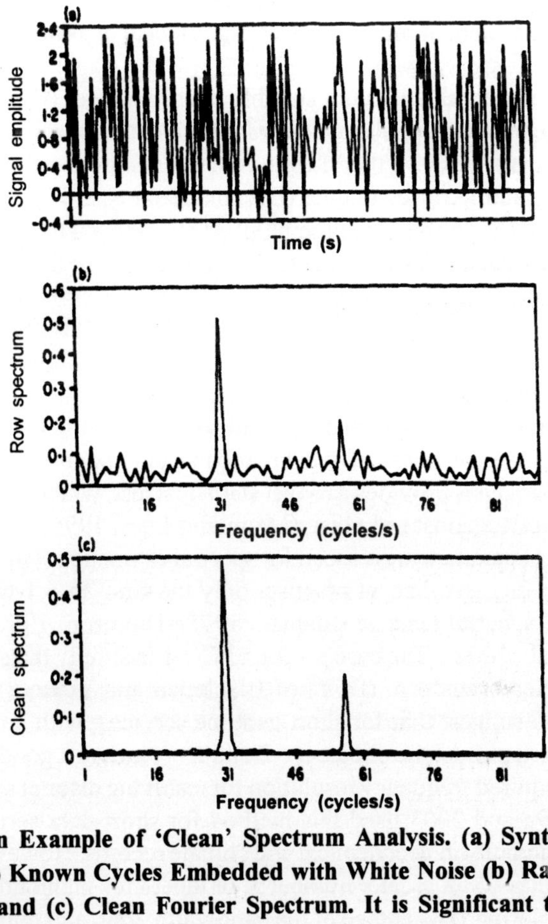

Fig. 2.5 An Example of 'Clean' Spectrum Analysis. (a) Synthetic Signal with Two Known Cycles Embedded with White Noise (b) Raw Fourier Spectrum and (c) Clean Fourier Spectrum. It is Significant to Note that Both Cycles are Clearly Recovered in Clean Spectra While the Same, Although Seen in Fourier Analysis is Surrounded with Noise.

Fourier spectrum and Fig. 2.5(c) shows clean spectra. It is clearly seen that in the noisy condition clean spectra resolve peaks better than Fourier spectra.

Multi-Taper Method

The multi-taper method (MTM) of spectral has been proposed by Thomson, 1982 and later discussed in great detail by Park, Lindberg and Vernan (1987); Percival and Walden (1993), Mann and Lees (1996); Berger et al. (1991) and Ghil et al. (2002). MTM reduces the variance of spectral estimates by using a small set of tapers (Thomson, 1982; Percival and Walden, 1993). Accordingly, the basic approach is same as that of periodogram, except that instead of using a single taper, multi-tapers calculated to the optimum degree are used to reduce leakage without increasing the estimation variance outside a prescribed bandwidth. The optimal tapers also called as 'eigentapers' belong to a family of functions known as discrete prolate spheriodal sequences (Slepian, 1978). The tapers are the discrete set of eigenfunctions that solve the variational problem of minimizing leakage outside of a frequency band with half-width equal to $p\omega_R$, where $\omega_R = 1/(N\Delta t)$ is the Rayleigh frequency, Δt is the sampling interval, and p is a suitably chosen integer (Ghil et al., 2002). Because the eigentapers are the specific solution to an appropriate variational problem (Ghil et al., 2002), the MTM has better solution to the problem than traditional nonparametric techniques (Box and Jenkins, 1970). For the spectral analysis using MTM, the data are premultiplied by orthogonal tapers constructed to minimise the spectral leakage due to the finite length of the time series, and set of independent estimates of the power spectrum is computed. Averaging over the ensemble of spectra obtained by MTM provides a better and more stable estimate. Figure 2.6 shows prolate spheroidal wave function data windows for a time bandwidth product = 4 and $k = 0, 1, ..., 4$. (Thomson, 1982).

The MTM method is particularly useful for a time series whose spectrum may contain a mixed response of both broadband and line spectral components. The harmonic analysis provides a robust statistical test, which allows rejection of null hypothesis against red noise (Mann and Lees, 1996).

The choice of parameters is crucial for spectral estimation (Ghil et al., 2002). These workers suggest that, in practise, only the first $2p - 1$ tapers provide usefully small spectral leakage (Slepian, 1978; Thomson, 1982; Park et al., 1987; Ghil et al., 2002). The case $p = 1$ and $K = 1$ is simply the single tapered Discrete Fourier Transform (DFT) of Blackman and Tukey (1958). Mann and Lees, 1996 suggest that for short geotime services, with a typical length of a few hundred points, the choice $p = 2$ and $K = 3$ offers a good compromise between the required frequency resolution for resolving distinct signals. Tiwari and Rao, 1999a and 2003 used this method for short data series to resolve significant periodicity in geochemical and climate records. However, for longer data sets one may use a greater number K of tapers for maintaining a desired frequency resolution. The choice of the bandwidth $2pw_R$ and number of tapers K thus represents the classical trade-off between spectral resolution and

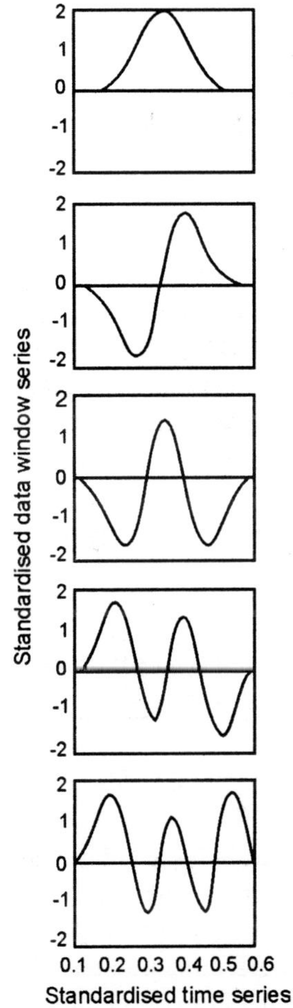

Fig. 2.6 Prolate Spheroidal Wave Function Windows for *s* Time Bandwidth Product = 4 and *k* = 0, 1, . . . , 4. (a) Function 0, *k* = 0 (b) Function 1, *k* = 1 (c) Function 2, *k* = 2 (d) Function 3, *k* = 3 (e) Function 4, *k* = 4.

(*Source*: D. J. Thomson, 1982).

stability of the spectral estimate (Thomson, 1982). Thus, the optimal choice of p and K depends, in general, on the length and other properties of the time series (Ghil et al., 2002).

Algorithms and Computational Procedures

In MTM analysis first the tapers are computed for a chosen frequency bandwidth. Then, the total power spectrum can be estimated by averaging the individual spectra given by each tapered version of the data set (Ghil et al., 2002). Accordingly, let us call $\hat{S}_k\omega \equiv |Y_k(\omega)|^2$ the kth eigen spectrum, where Y_k is the Discrete Fourier Transform (DFT) of $\{X(t)\ w_k(t)$: $t = 1, \ldots, N\}$ and $W_k(t)$ are the tapers. The high-resolution MTM is a weighted sum of the K eigenspectra,

$$S_r(\omega) = \frac{\sum_{k=1}^{K} \mu_k \left|Y_k(\omega)\right|^2}{\sum_{k=1}^{K} \mu_k}$$

Percival and Walden (1993) provide more detail for choosing the weights μ_k and other parameters. Here, the frequency resolution is $\pm p\omega_R$, which means that line components will exactly be detected as peaks or bumps of width $\pm 2p\omega_R$. This also implies that peaks can be identified with a higher resolution and greater confidence in MTM compared to the classical methods (Ghil et al., 2002). Further a more leakage-resistant spectral estimate can be computed by adjusting the relative weights of the contribution of each of the K eigenspectra. This is known as adaptively weighted multitaper spectrum and given by,

$$S_w(\omega) = \frac{\sum_{k=1}^{K} b_k^2(\omega)\mu_k \left|Y_k(\omega)\right|^2}{\sum_{k=1}^{K} b_k^2(\omega)\mu_k}$$

Here the weighting functions $b_k(\omega)$ play significant role to check the broad band leakage for a process that exhibits a fairly flat spctrum (white noise) in the frequency range of interest. According to Ghil et al., 2002, the adaptive spectrum estimate has an effective number of degrees of freedom 'ν' that generally departs only slightly from the nominal value $2K$ of the high resolution multitaper spectrum (Thomson, 1982).

Statistical Confidence Test

The objective of spectral analysis is to determine the amplitude of a line in the spectrum of a periodic time series of infinite data length.The available data are, however, of finite length which contain noise and consequently spectral analysis of these records exhibit side lobes associated with white and red noise. Spectral components emersed in such noise create problems for physical interpretation. It is essential, therefore, to examine the statistical significance or confidence level of line components appearing in the spectrum. Fisher's test (Fisher, 1929; Nowroozi, 1967; Shimsoni, 1971) has been applied invariably to calculate the statistical significance of spectral components associated with white noise processes. The spectral analyses based on autocorrelaation function uses a somewhat different method for calculation of confidence interval which is discussed elsewhere (Jenkins and Watts, 1968). Recently, Weedon (2003) has also discussed in detail various methods of statistical significance to test the statistical validity of spectral components.

The multi-taper method has been widely applied to detect periodic/harmonic components which is often used to be with *F* test for its statistical significance (Thomson, 1982). However, most of the geophysical time series are found to be dominated with red noise background, where as discussed by Ghil et al. (2002), '*F*' test fails to judiciously recognise the significance of harmonic components. Spectrum dominated with red noise background exhibits greater power at lower frequencies even in the absence of any real signals.

Mann and Lees (1996) and Ghil et al. (2002) explicitly described the method of calculation of confidence interval, against the red noise spectra, with examples. Accordingly periodic and/or quasi-periodic components in geophysical/geological time series will look like as delta function in the power spectrum of the geo-data series. In the spectrum they emerge as sharp peak, which might be confused with narrowband random contributions. Hence it is desirable to identify the presence of periodic component so they will not be misinterpreted as narrow band random components with finite power spectral densities (Mann and Lees, 1996). If periodic components in the data are intense, their presence is usually obvious. However, less periodic component in random data may not be so obvious. Hence, in practical terms, a test for periodicities usually evolves from analysis procedures which would be transformed assuming the data are random.

Due to the number of such other problems associated with statistical reliability of spectral components in geotime series analysis, Mann and Lees (1996) provided a more robust statistical significance test of spectral peaks against red noise background. In doing so they retain the pure line test of the traditional MTM procedure and test all the peaks for their statistical significance relative to the null hypothesis of a red background noise. For this purpose, they choose the first order autorgressive AR(1) process as a model for a discrete finite red noise series given by

$$r_n = \rho r_{n-1} + \beta_n$$

where $n = 1, \ldots, N$ denotes the discrete time increment in units of the sampling interval Δt, $0 \leq \rho < 1$, the lag-one autocorrelation coefficient, describes the degree of serial correlation in the noise, and β_n is Gaussian white noise sequence with variance σ^2. Here the trivial limit $\rho = 0$ yeilds a white-noise process. For the $AR(1)$ red-noise process, autocorrelation decays exponentially as a function of time $\rho_n = \exp(-n\,\Delta\, t/\tau)$ with a characteristic decay time τ. Thus if the lag-one auto correlation coefficient $\rho \equiv \rho_1$ is known, the characteristic noise decay time scale can be determined as (Mann and Lees, 1996):

$$\tau = \frac{\Delta t}{\log \rho}$$

In brief, following steps are taken to perform significance test, as discussed in detail by Mann and Lees (1996) and Ghil et al. (2002), for recognising and validating spectral components/peaks.

(i) The spectral density of this background noise is assumed to be obtained by the equation;

$$S(\omega) = S_0 \frac{1-\rho^2}{1-2\rho\cos\pi(2\omega/\omega_n)+\rho^2} \text{ for frequency } \omega$$

where the parameters S_0 is the average value of the power spectrum related to white noise process variance where, $S_0 = \frac{\sigma^2}{1-\rho^2}$.

Here $0 < S_o < 2\infty . \omega_n$ is Nyquist frequency, $\omega_n = 2/\Delta t$ and ρ is the lag one autocorrelation.

(ii) The robust estimate of the spectral background is obtained by minimising, as a function of ρ the misfit between an analytical AR(1) red noise spectrum and the adaptively weighted MT spectrum are convolved with a medium smoother.

(iii) To ensure that the globally estimated noise background is insensitive to outlier, a median smoothing operation is carried out. A median smoothing width of $\Delta\omega$ = min $\Delta\omega = \min\left(\frac{\omega_n}{4}, 2p\omega_R\right)$ is found to be generally good.

(iv) Significance level for harmonic or narrow band spectral features relative to the estimated noise background can be determined from the appropriate quantiles of the Chi-Square distribution by assuming that the spectrum is distributed with ν=2K degree of freedom (Mann and Lees, 1996).The ratio of power associated with a peak in the spectrum to the local power level of background is assumed to be distributed as $\frac{x^2}{\nu}$ and can be compared to the tablated x^2 probability distribution to determine statistical significance of spectral peaks (Mann and Lees, 1996).

Figure 2.7–top. demonstrates one application of MTM spectral analysis in a condition of signal and noise embedded time series (Mann and Lees, 1996). Figure 2.7 bottom shows multitaper spectrum of above time series (signal + noise) based as robust estimate of AR (1) background. The true 'red noise' background (heavy solid) is also shown for comparison and 90–95% and 99% confidence limits based on raw fit to spectrum.

Wavelet Spectral Analysis

Traditional method like Fourier analysis provides useful information about a signal characteristic e.g. exact 'localisation of frequency'. Fourier analysis,

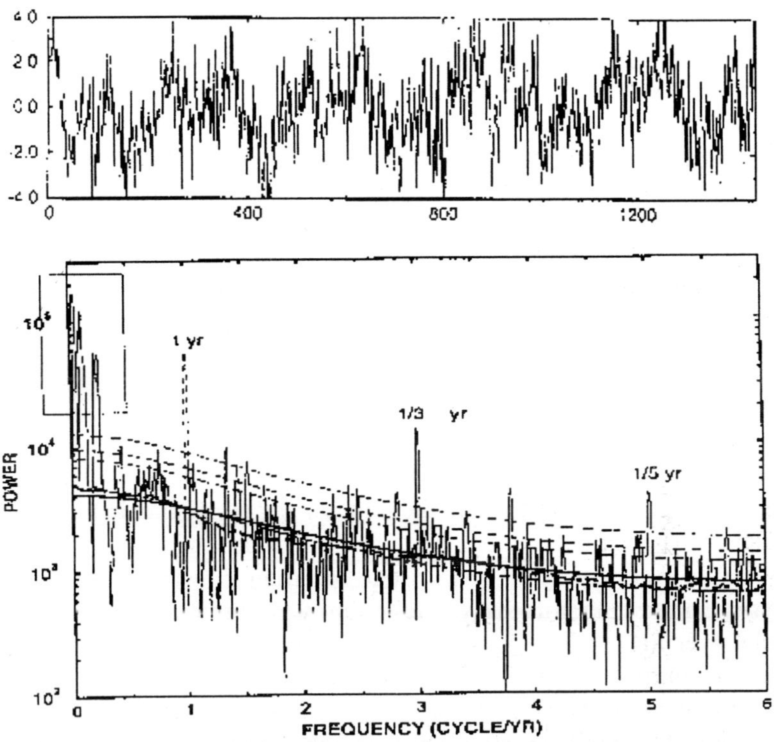

Fig. 2.7 Synthetic Time Series of Signal Representing 120 Years of Monthly Sampled Data (Top). Multi-taper Spectrum of Signal Component of Time Series Using k = 3, D = 2 with Peak Labelled by Periodicity in Years (bottom)

(*Source*: Michael E. Mann and D. J. Lees, 1996)

however, does not give any information about the stationary behaviour of observed spectral components. By this we mean to say that we do not have any information whether the identified spectral component is present throughout the evolutionary record or limited only in certain part of it, although this information is quite important for understanding physical mechanism. There are several geophysical signals, such as temperature, precipitation, El Nino and many others which exhibit time varying frequency (non-stationary) characteristics very similar to statistically self-similar and fractal signals. A wavelet based transform is a very useful tool for extracting information from such non-stationary (time-variant frequency characteristics) signal/time series which is otherwise not possible in other methods. The wavelet method works as a mathematical microscope that can focus on specific part of the signal to monitor local structures and singularies (Ghil, 2002).The purpose of wavelet transform, therefore, is to obtain the ferquency content (pattern of periodicity) of the process as a function of time. They revolve around expanding signals into waveforms whose time frequency properties are adapted to the signal's local structure (Kumar and Foufoula-Geoggion, 1997). During recent years, there have been lots of mathematical developments on wavelet analysis and the method has become a basic tool for study of not only geophysical time series but also in several other kinds of data. There is extensive literature available at present (Foufoula-Geoggion and Kumar, 1994; Holschneider, 1995; Hubbard, 1996; Kumar and Foufoula-Geoggion, 1997; Mallat 1998; Torrence and Compo, 1998; Ghil et al., 2002) on the subject.

Here, we restrict ourselves to a brief description of wavelet transform as used for time-frequency analysis. Mathematically, the continuous wavelets transform of a time series $f(t)$ can be given as:

$$W_{\psi}(a,b) = \left[\left(\frac{1}{\sqrt{a}}\right)\right] \int_{-\infty}^{\infty} f(t)\psi\left[\left(\frac{t-b}{a}\right)\right] dt$$

Here ψ is the base wavelet function (in this case, Morlet function), with a length that is much shorter than the time series $f(t)$. W stands for wavelet coefficient. The variable 'a' is a scale factor that determines the frequency (or scale) so that varying 'a' gives rise to spectrum 'b' which is shift of the analysis window in time so that varying b represents the sliding method of the wavelet over $f(t)$.

For the purpose of spectral analysis, large number of wavelet bases have been introduced. However, Ghil et al. (2002) note that an automatic time varying adjustment of the mother's wavelet shape provides an optimal choice

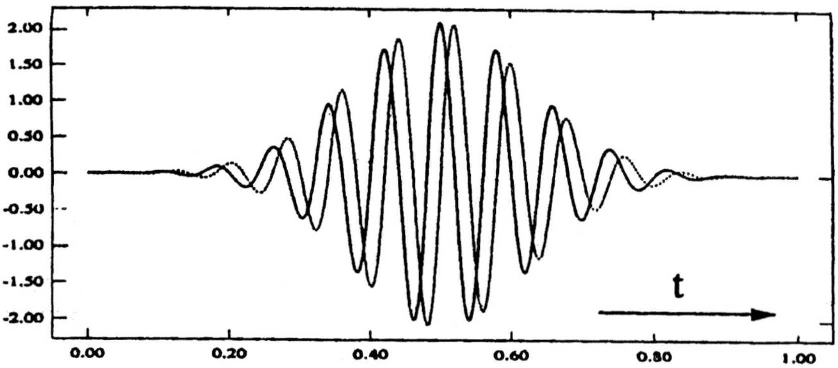

Fig. 2.8 Real (solid line) and Imaginary Part (dashed line) of Morlet Wavelet

(*Source*: A. Prokoph and F. Barthelms, 1996)

for multiscale decomposition of a given signal. Figure 2.8 shows the Morlet wavelet, which has been invariably used in geophysical data processing.

The complex valued Morlet wavelet can be represented by a periodic sinusoidal function. which is defined as (Morlet et al., 1983):

$$\psi(t) = \pi^{-1/4} e^{i\omega_o t} e^{-t/2}$$

where ω_0 is the frequency parameter that allows one to shift the frequency range for investigation. Wavelet spectrum will allow us to define with a greater precision the energy flow over the time interval adopted for a search of underlying cyclicity and hence temporal persistence of periodicities (Torrence and Compo, 1998). This method decomposes a given time series into scale components for identifying oscillations occurring at short period and others at a long period.

Hubbard (1996) and Weedon (2003) discussed the computational procedure for fast wavelet transform. Accordingly, a wavelet can be used for multiplication with a time series to produce a wavelet transform, analogous to the sine and cosine waves for the windowed Fourier transform in evolutionary spectra. First the time series is divided by filtering into low and high frequency components. The high frequency component is then multiplied by the smallest version of the mother wavelet after successive shifts in the wavelet position along the length of data.The remaining low frequency component series is then divided again into low and high frequency part and the wavelet transform is re-applied to the high frequency component , but using a wavelet wavelength which is double the previous scale (Weedon, 2003). The procedure is repeated until the very lowest frequency components of the original time series have been performed (Hubbard, 1996; Weedon, 2003). The following Fig. 2.9 illustrates the tiling procedure as described above.

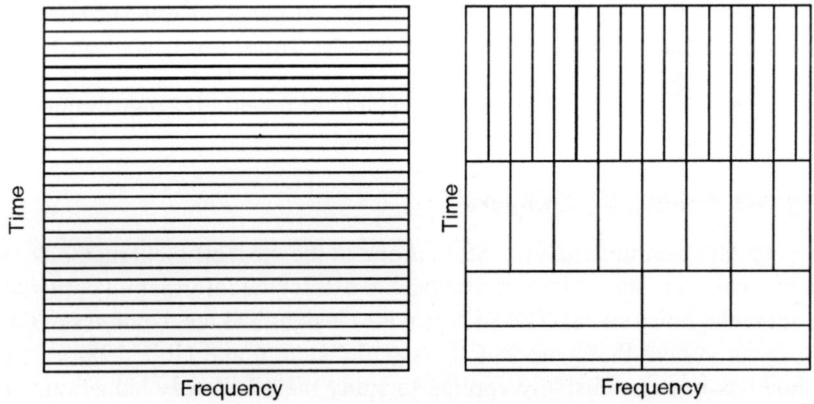

Fig. 2.9 Shows Tiling Procedure of Wavelet Analysis

In wavelet tiling, long wavelength wavelet covers a large portion of the data. Consequently, the frequency resolution is good, but ability to monitor non-stationary behaviour would be poor. On the other hand, however, short wavelength wavelets require a fewer data which will provde low frequency resolution but better time resolution e.g more information on the stationarity behaviour.This trade-off between time resolution and frequency resolution is described as an uncertainty principle (Mallat, 1998; Weedon, 2003). A detailed discussion on the statistical reliability test of wavelet coefficients and other related problems are discussed by Torrence and Compo (1998).

An example of wavelet analysis for different types of signal is displayed in Fig. 2.10.

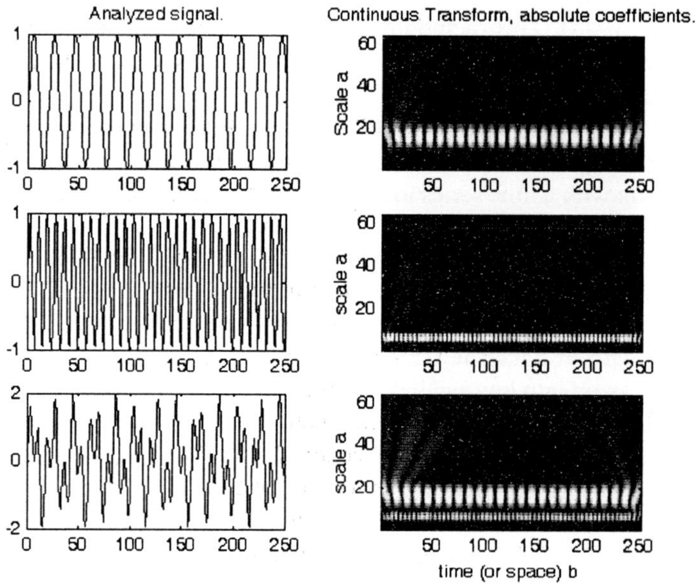

Fig. 2.10 An Example of Wavelet Analysis (left panel). Three Different Types of Signals and Corresponding Wavelet Transform (right panel)

As clearly apparent wavelet spectra is able to reveal different frequency component at signal in time frequency domain.

Singular Spectrum Analysis

The singular spectrum analysis (SSA) is one of the most efficient methods of spectral analysis. The method is designed to extract the information containd in a short and noisy signal. The SSA provides insight into the dynamics of the underlying system that generates the signal (Vautard and Ghil, 1989). The method has been extensively applied to study the oscillatory behaviour of palaeoclimate and other geophysical time series analysis (Ghil et al., 2002; Ghil and Vautard, 1991; Yiou et al., 1995 and several others).

The computational procedure of SSA starts with presenting the one dimensional time series in two dimensional phase space (trajectory) using the concept of "embedding" or delay operator with appropriate lag. For instance, a time series $X(t)$: $t = 1, 2, \ldots, N$ in a vector space of dimension M can be represented as $X(t) = X(t), X(t + 1), \ldots, X(t + M - 1)$. Here the vectors $X(t)$ are indexed by $t = 1, \ldots, N'$, where $N' = N - M + 1$. The method revolves around calculating the principle directions of extension of the sequence of augmented vectors in phase space. The SSA is based on computing of empirical orthogonal functions in the time domain, which uses data adaptive filters, that help separate the time series into components in the augmented vector space of interest (Ghil et al., 2002). These components are statistically independent at zero lag and can be essentially classified into trend, oscillatory pattern and noise from a complex non-statationary and amplitude and phase modulated signal. The $M' M$ covariance matrix Cx can be estimated directly from the data as a Toeplitz matrix with constant diagonals ; that is its entries cij depend only on the lag $i-j$ (Vautard and Ghil, 1989):

Table- A General Flow Chart of Spectral Methods

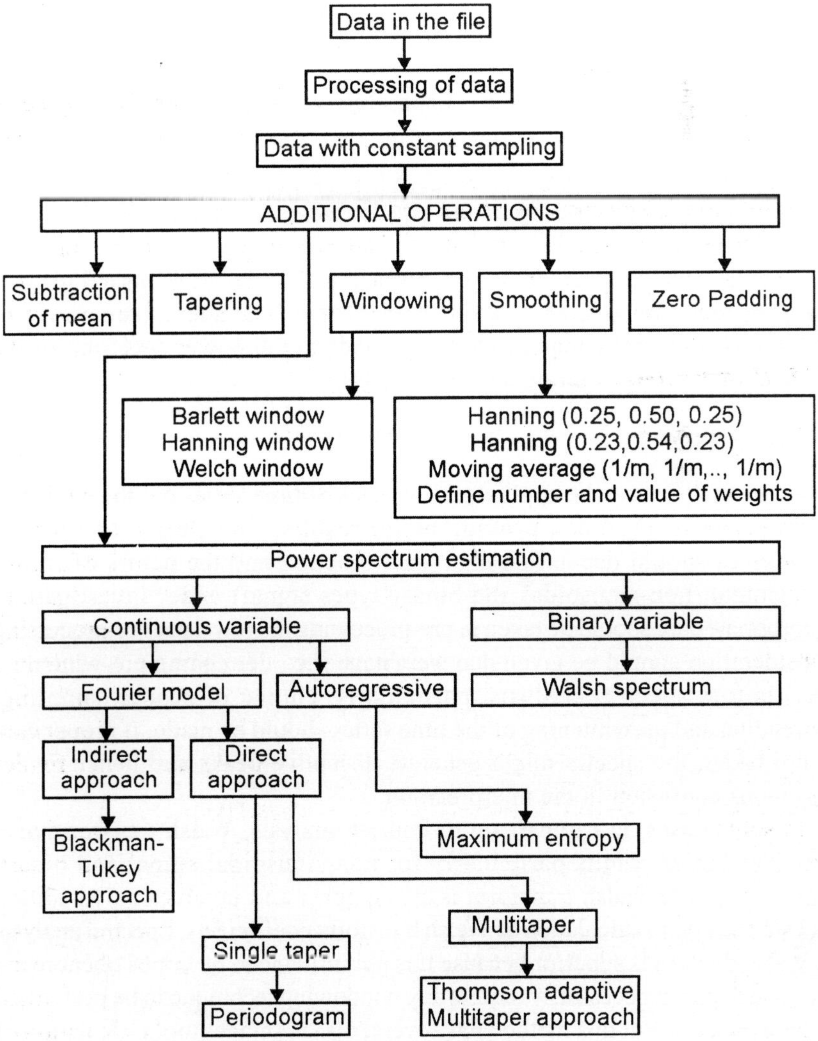

(Flow chart modified E. Pardo-Iguzquiza et al. 1994)

$$C_{ij} = \frac{1}{N-|i-j|} \sum_{t=1}^{N-|i-j|} X(t)X(t+|i-j|)$$

The eigenelements $\{(\lambda_k, \rho_k) : k = 1, ..., M\}$ of C_x are obtained by solving

$$C_x \rho_k = \lambda \rho_k$$

The eigenvalue λ_k equals the partial variance in the direction ρ_k, and the sum of λ_k, i.e., the trace of C_x, gives the total variance of the original time series *X*(*t*) (Ghil et al., 2002).

An equivalent formulation of above equation is also given by forming the $M X M$ matrix *Ex* that has the eigenvectors ρ_k as its columns and the diagonal matrix Λ_x whose elements are the eigen-values λ_k, in decreasing order:

$$E_X^t C_X E_X = \Lambda_x;$$

here E_X^t is the transpose of E_x. Each eigenvalue λ_k gives the degree of extension, and hence the variance, of the time series in the direction of the orthogonal eigenvector ρ_k.

Ranking the square root of the eigenvalues according to size is called 'singular spectrum'. The eigenvalues and eigenvectors define empirical orthogonal functions (EOFs) in phase space which can be projected to yield the principal components as a function of time. The sum of the power of individual principal components corresponds to the power spectrum of the original time series (Vaurard and Ghil, 1989).

CONCLUSIONS

Spectral analyses can be applied to any kinds of time series but the choice of appropriate method can provide better results. The choice of spectral techniques should depend on analytical function and the nature of record (sinusoidal, non-sinusoidal and binary types signal) under investigation. Appropriate care should be taken in pre-processing the raw data. Pre-processing consideration should be given due weightage (i.e., detrending pre-whitening etc.) before spectral analysis, particularly, choice of proper sampling, detrending and prewhitening of the time series should be made. If proper care is not taken, the spectra might generate deceptive peaks and hence render enormous confusion in the interpretation.

In some cases, in camparison to Fourier analysis, Walsh transform also provides better results particularly for non-sinusoidal signal and binary waveforms. The Walsh transform is also quicker and involves only addition and subtraction in calculation of Walsh transform coefficients. Spectral analysis in Walsh domain is superior because it is not influenced by Gibbs phenomena and also does not necessarily require any windowing technique to be performed in time domain. But this method has several disadvantages too. Calculation of correlations and convolutions in Walsh domain, however, is somewhat complicated where it requires modulo–2 statistics. The spectral analysis using autocorrelation function in Walsh domain is, therefore, more time consuming. Like the Fourier analysis, appropriate care has to be taken in selecting sampling intervals. There is no optimum criteria but one can appropriately decide sampling interval by visual inspection of given time history or of test of spectra with trial sampling interval, preferably choosing small sampling interval avoids aliasing in the spectral peaks.

Multitaper spectral method is a robust tool for analysing all kinds of time series. The method provides efficient means to identify periodic/harmonic components from the noisy time series. Modern spectral method based on singular decomposition provide robust tool for handling non-linear signals. Geophysical signals are often found to be associated with narrow band variability that is not purely periodic. Such strict harmonic components are rarely detected in geophysical and environmental records. One of the reasons for this is that the period, amplitude and phase are functional of slowly changing parameters (Ghil et al., 2002). Such signals are well captured by the SSA's pair of eigen elements (Ghil et al., 2002) and MTM method.

Stationarity is one of the problems which create confusion in interpretation of periodicty. In some cases, it is sufficient to divide a given time series into a consecutive number of shorter segments, while one is interested in small order periodicity, so that one can assume that the data is stationary or nearly so, over a short time range. However, resolution is a problem here. Spectral analysis using the wavelet transform has some advantages over the Fourier analysis in the sense that one can judge the stationary behaviour of the spectral peaks. Wavelet spectral method is another efficient technique which has the advantage for analysing nonstationary signal. However this method also suffers resolution due to constraint posed by choice of controlling parameters.

3

Earth's Astronomical Environment

INTRODUCTION

Earth is a cosmic object and a member of our solar system. Earth is also subjected to a variety of interactions exerted by the elements of galaxy through which the solar system moves. Several quasi-periodic phenomena including long-term cosmic cycles exist due to variety of influences exerted by the interactions of the solar system and the galaxies. Geological evidence indicates that exogenic cycles cause periodic stresses in long and short-term earth processes (Williams, 1981). For instance, there appears to be enormous tidal disturbances due to gravitational forces during the passage of solar system through the Oort clouds in the galactic region, which cause cometary impacts on the surface of the earth. Geological evidences also suggest that in some of the cases, the dynamical interactions might represent significant quasi-periodic influence leading to considerable hazards on the various processes of the earth and its environment. On a relatively shorter time scale, the changing position of the earth in its varying orbit around the sun significally influences the processes of geo–bio–ocean and atmospheric system. In particular, the periodic changes in the position of the earth's orbit causes insolation changes impinging on the earth's atmosphere leading to climate changes. These components are identified as orbital cycles of eccentricity, obliquity and precession. Such quasi-periodic variations occur due to the interaction of gravitational forces in the rotating Sun–Earth–Moon system. On a still shorter time scale, there are several other external cyclic forces on the earth's processes that arise due to the interactions of planetary motions, tidal forces and solar radiation with terrestrial phenomena. For instance, during the planetary motions, Jupiter and Saturn transmit momentum by gravitational torques, thereby causing changes in planetary orbits and spin rates to successive planets and the sun itself (Fairbridge, 1984). These changes are almost cyclic, governed by gravitational forces. The short-term tidal cycles also arise due to gravitational torques of the sun and the moon on the earth and are identified in many terrestrial processes including climate, earthquake, volcanic events and earth's rotation etc. Interactions of solar–terrestrial phenomena induce cycles, which are quite important for understanding the earth's processes on relatively shorter time scales. The tidal, orbital and cosmic cycles commensurate well with many earth phenomena on the respective time scales. Therefore, it is interesting to understand the origin of these astronomical cycles which could be useful for understanding the earth's processes.

CHARACTERISTIC TIME SCALES IN ASTRONOMY

There could be many sub-divisions of characteristic time scales in the astronomy. Here, we divide this scale mainly into four parts, as per our convenience to interpret geo–bio–ocean–atmospheric records — which are well known and their impact has been realised significally on respective scales of the various earth's system processes — cosmic cycle, orbital variations cycles, planetary conjunction and tidal cycles, and solar and tidal cycles. We briefly discuss the possible origin of these four groups of exogenic cycles.

COSMIC CYCLES

The Universe and Galaxies

The majority of astronomers strongly believe in the big-bang theory of the origin of the universe from a primordial nucleus some 15 to 16 billion years ago. It is also believed that the universe is expanding. Expanding universe consists of galaxies, cluster of galaxies and radiations. The galaxies are aggregation of stars together with interstellar gas and dust particles. There are billions of such galaxies in the universe (Bailey, 1983). These galaxies appear to be independent and hence may be regarded as a more or less closed system. The sun and collection of hundreds of billions of stars comprising our galaxy is known as Milky Way galaxy (Clube and Napier, 1986). There is bulk of diffused interstellar matter in the form of gas and dust. The interstellar medium is composed of a number of distinct phases which all seem to be in pressure equilibrium except for the densest condensation which are self-gravitating and collapsing (Torbett, 1989). Further, according to Torbett, the interstellar medium exhibits a hierarchical structure in that, generally, the hot, dilute inter-cloud phase has embedded within it clumps of atomic hydrogen clouds which, in turn, have embedded within them clustered clumps of denser molecular clouds filled with clumpy structure down to pc = parsec = 3.258 light year scales (Scalo, 1985). The dense molecular clouds, which have dynamical consequences, attain huge sizes and enough masses to exert gravitational influence on the solar system. These massive molecular clouds seem to exhibit a preference for residing in the density waves responsible for the spiral structure as well as being concentrated to the plane of the galaxy with scale heights of approximately 70–80 pc at the solar position (Scoville and Sanders, 1986). The galactic movements define one of the important conditions of evolution in our solar system. It is believed that our solar system is greatly influenced, at least, in its 4.6 billion years of life. Since the earth is a member of our solar system, it might be affected by these activities. The earth's evolution might, therefore, hide the historical events of our Milky Way galaxy.

The Milky Way Galaxy and the Solar System

The Milky Way galaxy is a spiral stellar system approximately 30–50 kilo parsec (kpc) in diameter. It is undergoing differential clockwise revolution (Fig. 3.1). The Milky Way galaxy is shaped like a flat disc with a ratio of thickness to diameter of approximately 1:20 and contains over 100 billion stars (Williams, 1975). It is believed that there is a gigantic black hole at its centre. The galactic plane is central to the disc — stars either orbit above or below it and because of complex gravitational conditions, they also oscillate about it (Kitz, 1996). The solar system is part of the Milky Way galaxy. It consists of the sun, nine planets, moon, several number of asteroids and millions of comets and meteorites and interplanetary dust and gases (Fig. 3.2). All planets revolve around the sun in the same direction in approximately the same plane, called the ecliptic plane, with nearly circular orbit. The Pluto, however, has highly elliptical orbit with reference to the orbital plane of the rest of the planets. Several studies suggest that the sun orbits with an elliptical orbit, $e = .07$ around the galactic centre with mean velocity $v = 225$ km/second (Williams, 1975 and references

Fig. 3.1 A View of Milky Way Spiral Galaxy

(*Source*: Courtesy the Observatories of the Carnegi Institution: Napier, 1989)

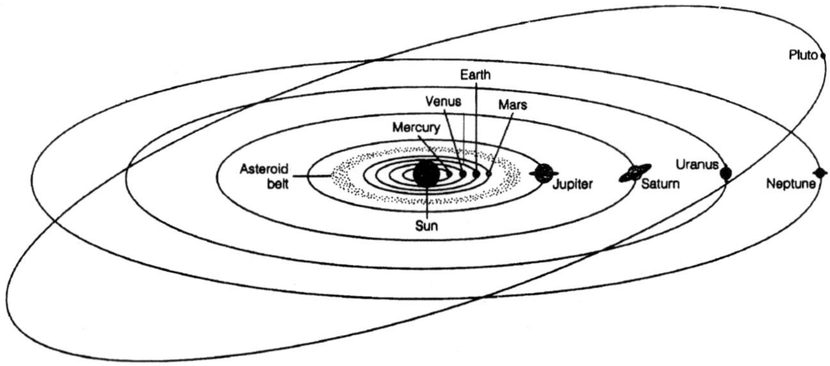

Fig. 3.2 Schematic Diagram Showing Solar System with Its Nine Planets and Structure of Asteroid Belt

Source: http//www.nineplanet.org/overview.html

therein). Its distance from the galactic centre varies from 8100–9400 pc with present value being approximately 8200 pc (Bash, 1986) (Fig. 3.3). One

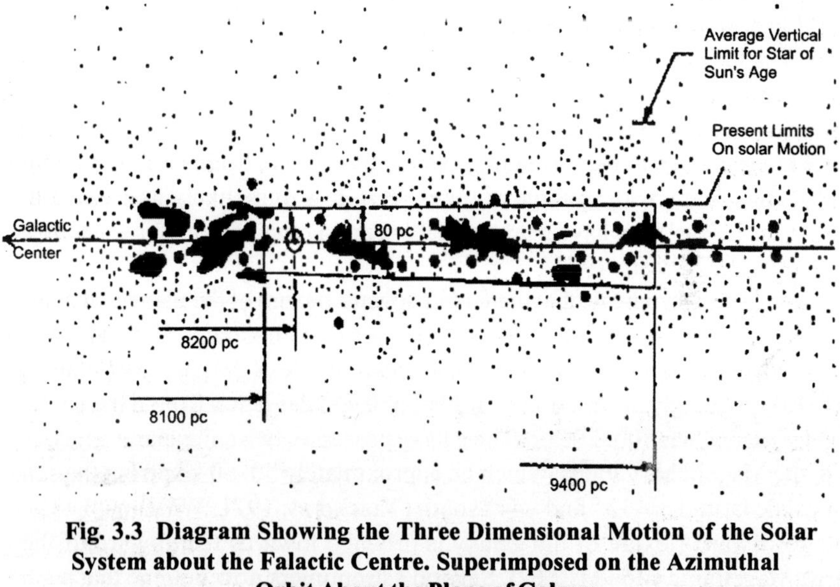

Fig. 3.3 Diagram Showing the Three Dimensional Motion of the Solar System about the Falactic Centre. Superimposed on the Azimuthal Orbit About the Plane of Galaxy

(*Source*: Torbett, 1989)

complete revolution of the solar system around the plane of galaxy takes nearly 250–300 m.y. and is called a 'Cosmic Year'. The earth has already completed almost over 20 circuits since it was born some 4.6 billion years ago (Steiner and Grilmair, 1973). According to the recent data, the sun is located above the plane of galaxy, 8 ± 12 pc and currently moving upward with a velocity of 7 km/sec. As shown schematically by Torbett (1989), (Fig. 3.4) superimposed

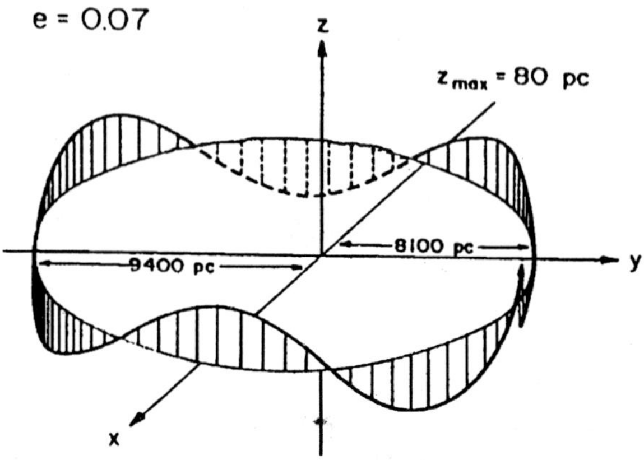

Fig. 3.4 Schematic Diagram Showing the Local Environment and Vertical Distribution of Galactic Components Compared to Present Limits and Radial and Vertical Motion of the Solar System. Figure Approximating to Scale

(*Source*: Torbett, 1989)

on this azimuthally motion is a vertical oscillation, with half period of about 26–37 m.y., about the galactic plane with present amplitude of 80 pc (Bahcall, 1986; Thaddeus, 1986). The full oscillation frequency (67 ± 3 m.y.) translates to 3–4 oscillations for every cosmic year (250–300 m.y.). The sun also is assumed to be spiralling inward with a radial contraction velocity component of 1.4 km/sec. Such inward spiralling eventually causes a decrease in the length of 'Cosmic Year', with passage of time. Thus, it is quite likely assumed that the 'Cosmic Year' may have been a variable period ranging from 200–400 m.y. (Steiner and Grillmair, 1973).

Several studies indicate that there is an upward flexure of the plane of neutral hydrogen in the outer spiral arms between galactic longitudes 20° and 140°, and a reciprocal downward flexure on the opposite side of the galaxy (Williams, 1975). Accordingly, near the sun's orbital path the deviation from the galactic plane is as much as 100–200 pc. The galaxy possesses two companion galaxies also, the Magellanic Clouds, which lie approximately 50–60 kpc from the sun at galactic latitudes ~32 ° and ~44 ° south (Thackeray, 1971; Williams, 1981). The downward flexure of the galaxy is oriented towards the upward of the Large Magellanic Cloud (LMC). Several astronomers now believe that such flexures are a tidal distortion caused by the clouds, particularly the LMC (Davies, 1972). It has been further suggested that the Large and Small Magellanic Clouds (LMC, SMC) are orbiting around the Milky Way and make close approach to the galaxy once in every 2500 m.y. (Williams, 1981). There is intriguing coincidence of some quasi-periodic earth processes, such as magmatism, with periods of about 600 to 800 m.y., which are fourth and third harmonics respectively, of this 2500 m.y. mega galactic rhythms (Williams, 1981; Prokoph et al., 2004).

Williams (1975) has suggested another periodicity, which arises due to flexure in galactic region caused by the gravitational attraction of the Oort cloud. Accordingly, for stationary flexed regions, the mean period of solar passage between regions of given diametrically opposite galactic flexure is 152 m.y. The absolute duration of such passages vary upto 10% from the mean due to the retrograde rotation of the major axis of the sun's orbit. Interestingly, this period is quite close to the second harmonic of the 'Cosmic Year Term'. The following table gives the major long-term cycles in the movement of the solar system in the galaxy.

Table 3.1 Main Periodicities of the Solar System in Galaxy

S.No.	*Name of the cycle*	*Approximate Period*	*Appearance in Geo-system*
1.	Mega cosmic cycle	2500 m.y.	Proposed obliquity change
2.	Second harmonic	1200 m.y.	
3.	Third and fourth harmonics	600–80 m.y.	Geo-tectonism/ Magmatism, climate
4.	Cosmic year	280–300 m.y.	Geotectonics, Magnetic Reversals Sea-level changes etc.
5.	Passage of solar system through Spiral Arm	150 ± 10 m.y.	Climate change
6.	Vertical oscillation of solar system (full cycle) perpendicular to the galactic plane	67 ± 5 m.y.	Not very prominent in geo-record
7.	Vertical oscillation of the solar system (half cycle) to the galactic plane	33 ± 3 m.y.	Prominent in almost all Geo–bio–ocean–atmospheric processes
8.	½ Harmonic of 33 m.y.	15 m.y.	Magnetic reversals and other processes

The table is based on various information as quoted in text.

COMETARY IMPACTS

Asteroids and Comets

Asteroids are large extra-terrestrial bodies probably formed as planetesimals. Fig. 3.5 shows aerial view of meteor crater from Arizona representing a typical fresh simple bowl shaped crater (Grieve, 1989). Astronomers have discovered a large number of asteroids of different sizes. These asteroidal bodies are present in asteroidal zone of the solar system as depicted in Fig. 3.2. A minor

Fig. 3.5 A Huge Impact Crater
(*Source*: Grieve, 1989)

fraction of these bodies are also present near the earth's orbit. There are about 750 asteroids whose orbits cross the earth's orbit (Torbett, 1989).

Comets are interplanetary bodies composed of rocky and icy materials. There are atleast half a dozen viable mechanisms dealing directly or indirectly with the perturbations of the Oort clouds to explain why comets might bombard the earth quasi-periodically (Napier and Clube, 1979). The Oort clouds were discovered by a Dutch astrophysicist, Jan Hendrik Oort, in 1950 from the motions of the comets that have been observed to pass by earth. The constituent comets of Oort cloud are very small and so far distant that it is not possible to detect them from the earth. It has been suggested that both short and long period comets are derived from the Oort cloud either by perturbations of the tidal field of the galaxy (Morris and Muller, 1986; Torbett, 1986a; Heisler and Tremaine, 1986 or by deflections from the randomly passing stars (Weissman, 1986; Hills, 1986 and Bailey et al., 1987). Once as an earth's crossing orbit, asteroid either collides with earth and other planets or is ejected from the solar system by close encounters (Shoemaker, 1984). Disruption of an asteroid of bigger size (~ larger than 100 km) may inject a sufficient number of fragments into a resonance on an average every 600 m.y. (Shoemaker et al., 1979 and Shoemaker, 1984). For objects larger than 10 km, the rate is roughly once every 30–50 m.y. (Torbett, 1989). The mean impact speed of earth crossing asteroid is $\simeq 20\ km^{-1}$ sec (Torbett, 1989). Hence, impacts of the objects of this size can have profound effects on the earth and its environment. There would be 'tidal compressional shock' at the time when the Sun and Oort cloud cross the galactic plane. It would disturb the Oort cloud enough to cause comet shower (Rampino and Stothers, 1984). This can provide at most a 15% modulation to comet influx. During the mid-plane crossing, the probability of encountering the effect of dense molecular clouds is higher since they tend to cluster towards the plane (Torbett, 1989). The resultant gravitational perturbations of the orbital configuration of Oort cloud will lead to comet showers inside the planetary system (Rampino and Stothers, 1984). The present amplitude of the Sun's motion is only of the order of the scale height of the clouds themselves and hence, dense clouds would be encountered more or less randomly with at the most a 50% modulation (Thaddeus and Chanan, 1985). The scale height of the clouds is only 60 pc, which is so small enough that one should see at least a 50% modulation in encounter rate (Stothers, 1985).

Evidence for Periodic Impacts on the Earth

As discussed in the previous section, history of various movements of the solar system and its interaction with Oort cloud lead to enhanced collisions with molecular clouds or other matters and provide a viable mechanism for the evolutionary history of the earth's processes. The modulation may have been stronger in the past if a recent deflection has reduced the amplitude of the vertical motion of the solar system to its present smaller value (Torbett and Smoluchowski, 1980). Several workers believe that the galactic periodicities and cratering records might be produced by purely stochastic influences acting on a complex but open system. For instance, it may be that some of the processes do not respond significantly to continuing random impacts or gravitational influences. It is likely that stochastic variations in the rate of asteroidal bombardment can be expected by virtue of the random nature of the collisions responsible for delivery to the inner solar system (Torbett, 1986b).

Shoemaker (1984), however, estimated a 600 m.y. periodicity for disruption of an asteroid larger than 100 km, which may inject a sufficient number of fragments into a resonance. Furthermore, the statistics of the ages of impact structures indicate an increase in cratering between 300 and 400 m.y. in the past (Grieve, 1982; Weissman, 1982). Spectral analysis of impact cratering record for the Phanerozoic reveals a quasi-periodicity of the order of 33 m.y. (Rampino and Stothers, 1984; Alvarez and Muller, 1984). The evidence of periodicity in the cratering record is, however, debated from the statistical point of view (Grieve, 1982) but more detailed analysis based on precise impact record (Table 3.2) combined with mass extinction and other geological data does reveal a 26–30 m.y. periodicity (Napier, 1998). Further, a systematic non-parametric analysis of impact cratering record (Table 3.2) also reveals a 30 ± 3 m.y. cycle suggesting that there exists a link between terrestrial and extra-terrestrial processes (Tiwari and Rao, 2003). It is also maintained that there is a possibility that astronomical forces may always create instantaneous catastrophic impact events leading to glacial/inter-glacial, tectonic and biological upheaval (Barlow, 1990; Yubushita, 1991; Rampino and Caldeira, 1993).

Table 3.2 (Napier, 1998) Gives Record of Impact Craters. Repetition Events and <10 km Diameter Craters are not Considered for the Analysis, Reported Above in the Text

Crater	*Age (m.y.)*	*Diameter (km)*
Zhamanshin	0.9 ± 0.1	13.5
El'gygytgyn	3.5 ± 0.5	18
Bigach	6.0 ± 3.0	7
Ries	15 ± 1	24
Haughton	23 ± 1	24
Popgun	35 ± 5	100
Chesapeake Bay	35.5 ± 0.6	85
Wanapetei	37 ± 0.2	7.5
Mistastin	38 ± 4	28
Logoisk	40 ± 5	17
Kamensk	49 ± 0.2	25
Montagnais	50.5 ± 0.76	45
Ragozinka	55 ± 5	9
Marquez	58 ± 2	13
Chicxulub	64.98 ± 0.05	170
Kara	73 ± 3	65
Ust-Kara	73 ± 3	25
Manson	73.8 ± 0.3	35
Lappajarvi	77.3 ± 0.4	23
Boltish	88 ± 3	24
Dellen	89.0 ± 2.7	19
Steen River	95 ± 7	25
Carswell	115 ± 10	39
Mien	121.0 ± 2.3	9
Tookoonooka	128 ± 5	55
Gosses Bluff	142.5 ± 0.8	22
Puchezh-Katunki	175 ± 3	80
Rochechouart	186 ± 8	23
Manicougan	214 ± 1	100
Araguainha Dome	247.5 ± 5.5	40
Siljan	368.0 ± 1.1	52
Kaluga	380 ± 10	15

MECHANISMS

Spiral Arm Crossings

As mentioned above, our Milky Way galaxy is a spiral steller system, where some kind of dynamical interactions during the movement of the solar system through the spiral arms passages takes place. It has been suggested that spiral arm passage can generate comet showers with long periods ~ 250–300 m.y. (Napier and Clube, 1979; Clube and Napier, 1982a, b). It is possible for the spiral arm passage to generate some comet influx. The average stellar density inside a spiral arm is ~ 10% larger than between the arms. Thus, increased passage of stars through or near the Oort cloud could give a small modulation (Torbett, 1989). However, Scoville and Sanders (1986) noted that the clouds are distributed so widely that the probability of passing through or near enough to any cloud during arm passage is only 10%. Thus, this mechanism can be expected to provide a minimal degree of modulation in comet flux. Torbett, (1986) however, suggests that the radial distribution of molecular clouds may provide a modest degree of modulation. The solar orbit lies just outside an enhancement in the density of molecular clouds known as "molecular ring" (Scoville and Sanders, 1986). Accordingly, the radial motion of the solar system carries it between 8100 and 9400 pc. The density of molecular clouds varies by a factor of approximately 2 over this range. Thus, according to Napier and Clube, 1979 and Clube and Napier, 1982, increased probability of collision at times of closest approach of the solar system to the galactic centre could give rise to an approximately 250 m.y. periodicity in the geological record, which seems too quite believable.

Solar Orbit

Innanen et al. (1978), adopted an alternative spheroidal density distribution model in the galaxy in which the sun moves eccentrically between about 8900 –1030 pc from the centre and takes about 233 m.y. for a complete circuit round the centre in space, or about 460 m.y. for a complete circuit of the galaxy, as defined by its spiral structure. In this model, the effective half-thickness of the gas-dust layer is quoted as 125 pc, and the authors find that the sun remains within 70 pc of the central plane. They find an average period of about 67 m.y. for a complete oscillation through the plane. The model calculation also shows that there are six intervals in the past 960 m.y. during which the sun was immersed in material about 80% of maximum density and the lengths of these intervals ranged from less than 30 m.y. to about 90 m.y. The range of quasi-periodicity may be interpreted as non-stationary behaviour of the phenomena. Interestingly, these periods compare well with the well-known six intervals of ice epochs (McCrea, 1975). Therefore, it is tempting to believe that ice epochs are somehow related to the broad features of the distribution of interstellar matter associated with the spiral structure of the Milky Way galaxy. Innanen et al. (1978) also noticed a near coincidence of a geological 'period' with the period of vertical oscillation of the sun through the galactic plane, over about 10 complete oscillations during its (Cosmic year $\simeq$ 280–300 m.y. period) revolution around the milky way galaxy.

Vertical Oscillations of the Solar System

The sun also vibrates upto~100 pc perpendicular to the galactic plane (complete vertical oscillation) with a mean period of about 67 m.y. (Fig. 3.6). The half period of the solar system oscillations perpendicular to the galactic plane is about 33 m.y. according to the best astronomical estimates (Rampino and Stothers, 1984). Intriguing coincidences of this cycle (the plane crossings period, also known as Z oscillation) (Fig. 3.6) with the boundaries of geological eras rendered considerable motivation for astronomers to propose viable hypotheses to explain earth's cyclic evolution. The powerful gravitational pull exerted by the galaxy's massive clouds of gas and dust renders required force to the solar system to bob up and down through the galactic plane much the same way as the earth's gravity keeps a pendulum swinging (Rampino and Stothers, 1984). Accordingly, when the solar system is at a point above the plane, gravitation pulls it back towards the galactic midline, but with no friction to stop it, the solar system continues moving through to the other side. Then, when it reaches a point below the plane, it is pulled back in the other direction. The solar system weaves back and forth atleast eight to ten times during its two hundred and fifty million years revolution around the Milky Way galaxy. Rampino and Stothers (1984) have suggested that there is high probability of encountering of the solar system with dense molecular clouds during midplane crossing, once in every 33 $\pm$ 3 m.y., or so, for a stretch of a few million years, depending on the density of interstellar clouds near the point at which the solar system crosses the galaxy's midline. Since the molecular clouds tend to cluster towards the plane of galaxy, the resultant gravitational perturbations of the Oort cloud lead to comet showers. Thaddeus and Chanan (1985) challenged the above idea on the grounds that the present amplitude of the sun's motion is only on the order of the scale height of the clouds themselves and hence dense clouds would be encountered more or less randomly with at the most a 50% modulation. However, based on more detailed reanalysis, Stothers (1985) rebuts this and upholds the conclusion periodic comet encounter and clarifies that even for strict modulation, one does not

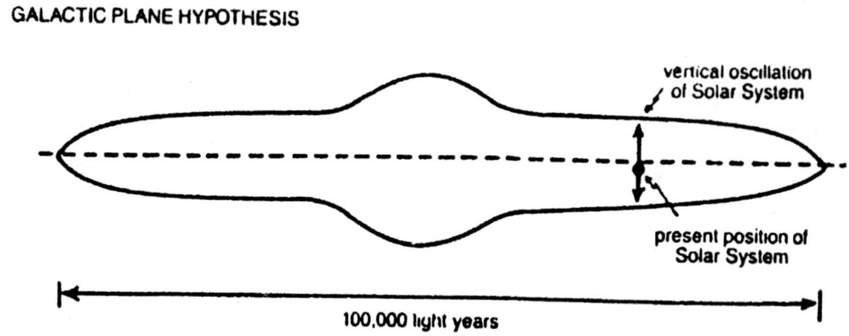

Fig. 3.6 Schematic Diagram of Milky Way Galaxy, Galactic Plane and Position of Solar System

expect to see perfect periodicity since phase jitter is caused by deflection of the solar path by gravitational encounters with stars and massive clouds. Furthermore, there are several other suggestions and arguments in favour and against of this mechanism and it appears that future discussions will provide more fruitful and reconciled view.

Nemesis Cycle

Nemesis is a distant dim star and as yet an undiscovered companion of the sun. It has been speculated that it periodically exerts the gravitational force with 26 m.y. to perturb the sun and comets. Raup and Sepkoski, 1984 found the relevance of this 26 m.y. cycle in marine mass extinction record. According to the Nemesis theory, this hypothetical object would have to be a low mass star or brown dwarf (Torbett, 1989). It is on an eccentric orbit and is believed to be periodically approaching the inner Oort cloud (Davis et al., 1984 and Whitmire and Jackson, 1984) (Fig. 3.7). Researchers note that the existence

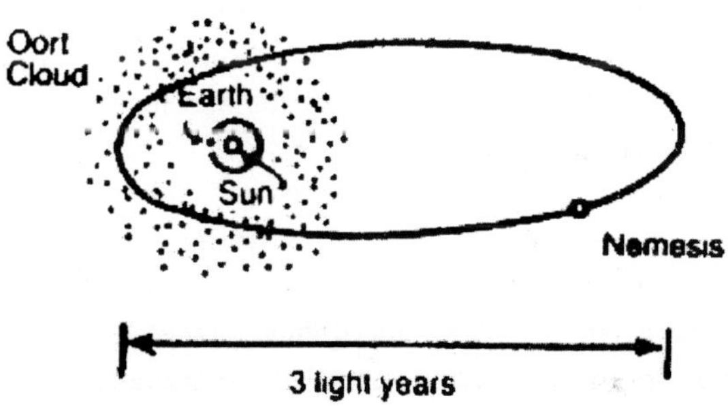

COMPANION STAR HYPOTHESIS

Fig. 3.7 The Nemesis Orbit Around the Sun

(*Source*: Sepkoski, 1988)

of a dense inner Oort cloud is crucial to this model since perturbations of normal Oort cloud comets will not induce sufficient modulation (Torbett, 1986). The required mass of such a companion is estimated to range from 0.01 m – 0.1 m (Solarmass) depending on the eccentricity of the orbit and the exact distribution of the comets in the inner Oort cloud (Hills, 1985; Torbett, 1989). According to the researches, the close position of the elliptical orbits disturbs the planetary system and comets leading to cometary bombardments. The required mean distance of Nemesis from the Sun for its periodic orbital evolution (26–30 m.y.) is, however, debated. (Smaluchowski and Torbett, 1984; Hut, 1984). A serious problem for the success of this model, therefore, is the difficulty of keeping objects this distant from the sun in a bound orbit with the 26–30 m.y. period (Torbett, 1989). The Oort cloud seems to be truncated at ~ 40,000 AU. It is roughly half its expected tidal limit and is thought to be due to the action of gravitational perturbations of molecular clouds (Torbett, 1986b). Theoretical estimates (Bailey, 1986) and numerical simulations

(Shoemaker and Wolfe, 1984 a, b; Weinberg et al., 1987) of the perturbing effects from realistic distributions of molecular clouds and stars agree that the probability of such a companion surviving to the present day is less than 1%. It has been estimated that there is a 15% chance of a solar companion in a Nemesis-like orbit being perturbed by passing stars to make an excursion inside the orbit of Pluto (Hills, 1985).

The astronomers, however, put atleast three immediate objections to the Nemesis theory (Torbett, 1989). Firstly, it is difficult to keep objects this distant from the sun in a bound orbit with the proper period. Secondly, the orbit is unstable, and thirdly a companion star should reveal itself in the current cloud. These present serious difficulties for the success of Nemesis hypothesis.

X Planet Hypothesis

Whitmire and Matese (1985) have suggested an alternative mechanism to generate 30 m.y. year periodicities of comet influx. Accordingly, in order to generate periodic showers, they employ a hypothetical undiscovered tenth planet of roughly 1–5 earth masses circling the sun located beyond the orbit of Neptune periodically which upsets the innermost part of the Oort clouds (Torbett, 1989). However, for generating periodic shower, the condition is that this object would need to have a rather strange trajectory (an elliptical orbit with moderate inclination whose perihelion is slowly advancing with a 60 m.y. period). From a theoretical point of view, planet X could have a shifting orbit always tilted at an angle to the orbit of the planets and the innermost comets of the Oort clouds so that it would intersect the comet belt only once every 30 m.y. (Torbett, 1989). During this encounter planet X disturbs and sends earthwards a shower of comets (Torbett, 1986). This proposal is also not without criticism (Tremaine, 1986; Bachal et al., 1985; Shoemaker and Wolfe, 1986 a, b) and hence did not receive much support.

Radiations

Schwartz and James (1984) provided an alternative suggestion particularly for the terrestrial mass extinction, employing the above idea of the Z oscillation. Accordingly, at times when the solar system is at maximum height above the galactic plane, increased fluxes of X-rays or cosmic rays could imperil the biosphere leading to deleterious effect on the evolutionary processes. It has been noted, however, that for the irregularly cloudy nature of the interstellar medium, the modulation provided by this mechanism would also be small. Furthermore, this mechanism possesses the wrong phase when plane-crossings are compared to extinction events. The last several extinction events seem to occur during plane-crossing rather than at the extreme of the solar vertical motion (Torbett, 1989).

Cosmic rays are generally believed to be produced primarily at the shock fronts generated by supernovae explosion (Torbett, 1989). The level of cosmic rays have changed significantly on many occasions during the geological time, causing severe increases in the genetic loading that affects all living things on the planet (Kitz, 1996). It may be that the actual distribution of cosmic rays within the galaxy is not known, but it has been estimated that their concentration is higher at the galactic plane than anywhere above or below it (Schwartz and James, 1984). Therefore, in crossing the galactic plane, the earth experiences a severe increase in background radiation.

Shklovsky (1968) noted that over the history of the solar system there have occurred roughly three to four supernovae within 10 pc of the sun and thirty within 20 pc. He has further noted that the result of such a nearby explosion would result in much larger transient increase in the cosmic rays flux, may be by a factor of 100, which would increase the ionisation rate in the atmosphere leading to catalytic reduction of ozone layer by 90% for periods ranging from few years to a century (Torbett, 1989). Such effects are certainly deleterious for the evolution of life and the ecosystem. As reviewed by Torbett, 1989, there exists a variety of radiations within the galaxy that could conceivably attack the stability of evolution and other processes.

In summary, we can say that the sun and its planets, asteroids and comets move slowly up and down across the galactic plane. A complete cycle of movement of solar system from below the galactic plane to above the plane and back to an original position below the plane is 62–67 m.y. The sun crosses the galactic plane twice in the complete oscillation and, therefore, the plane crossing period is 31–33 m.y. This particular cycle of 30 m.y. has been found in almost all long-term geological records and impact craters and hence need emphasis for formulating appropriate physical mechanism. According to William (1981), the well-known period of 280–300 m.y., 140–155 m.y., 70–80 m.y., 40–45 m.y. may be equated respectively with cosmic year's half, quarter and eighth periods. These various quasi-periodic oscillations arising due to multiple interactions of the sun with cometary objects in Milky Way galaxy might lead to cometary impacts on the earth. These quasi-cyclic periodicities coincide well with various long-term cycles of the earth's evolution. Hence, they have implications for modulating the earth's evolutionary processes. Cometary impacts on the earth can create several hazards to the evolution and stability of earth. Torbett (1989) has summarised the effects of either cometary impact or an asteroid on the earth. Accordingly, objects of the size of 10 km or more present the most serious hazards. The main consequence is the production of craters, triggering of massive earthquakes and the injection into the atmosphere of dust lofted by the impact. Secondly, severe thermal effects result from suspending ~19 × 10^{-2}cm of micro size dust in the upper atmosphere where it could effectively block the bulk of the sun's radiation and lower continental temperatures by a few degrees (Toon et al., 1982). The material injected into the upper levels of the atmosphere can remain for long time with severe effects on climate.

Although the possible implications of cosmic 'periodicities' in the geological record is a subject of hot discussion, it does appear that striking correlations exist among various earth's evolutionary processes including mass extinctions, geo-chemical phenomena and cratering events (Raup and Sepkoski, 1984; Shoemaker and Wolfe, 1986; Rampino and Caldeira, 1993 and reference therein; Negi et al., 1996; Tiwari and Rao, 1998, 1999a, 2003). The question of how periodic are the data largely miss the point since most mechanisms proposed to account for the 'periodicities' are expected to exhibit a substantial degree of phase jitter (Rampino and Stothers, 1984). The available data do, however, appear to be qausi-periodic although the reason for the physical significance of periodicity has yet to be clearly identified (Stothers, 1986). The model employing vertical oscillations of the solar system about the plane of the galaxy with a periodicity of 30 ± 5 m.y. in particular, which leads to enhanced collisions with molecular clouds, or other matter seems to be comparatively more viable mechanism for answering some of the questions in earth's system processes.

EARTH'S ORBITAL VARIATIONS

The planetary system seems to have been stable for the age of the solar system. However, the planetary orbits have been subjected to considerable variation caused by the mutual perturbations between the planets (Torbett and Smoluchowski, 1984). The researchers have tried to understand, in detail, the variations of orbital elements, departures from strict Keplarian motions of the sun, moon and other planets. Earth moves around the sun. The planet earth experiences several dynamical mode of action due to gravitational attraction of sun. The most important among these modes are the transits of the earth about the sun, the rotation (axial and polar motion) of the earth upon its axis, and a very slow wobbling motion of rotational axis.

The earth rotates on an axis, which is not perpendicular to the ecliptic plane. It is tilted towards the earth's equatorial plane and is inclined by about 23.5° with respect to the perpendicular of the ecliptic (Fig. 3.8). This angle is called 'tilt' or the obliquity of the ecliptic. The earth is also subject to a number of perturbations of its orbital element by the interaction of gravitational forces in the rotating Sun–Earth–Moon system. This leads to variations in the distance from the sun as well as differences in timing between perihelion passages and on sets of winter or summer (DeBoer and Smith, 1994). Thus, astronomical forcing of periodical orbital behaviour is quite important for earth system processes. The orbit of the earth changes considerably both in its shape (eccentricity) and orientations (direction of perihelion) in quite complicated ways. Berger (1988) has provided detailed account of variations of orbital parameters and their link with climate changes.

Accordingly, at any latitude on the earth, the insolation available at the 'top of the atmosphere' is a single-valued function of the solar constants S_0 of

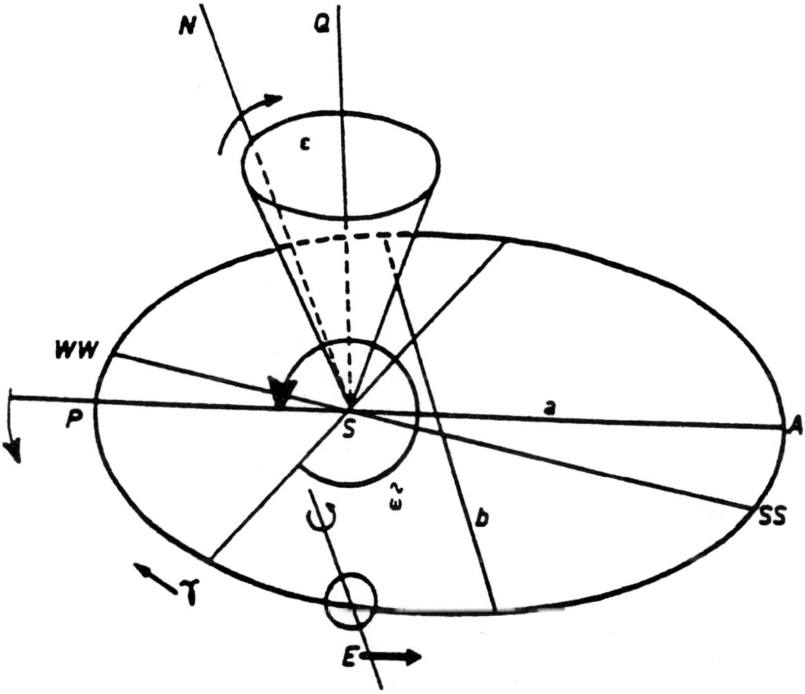

Fig. 3.8 Elements of the Sun's Orbit.
Source: Berger, 1988. For details see the text

(i) the semi-major axis of the earth's orbit 'a' (related to the length of the year); (ii) the orbital eccentricity 'e' and its obliquity; '∈' (iii) the longitude of the perihelion, ' ω ', measured from the moving vernal equinox; and (iv) the rotational angular velocity of the earth (related to the length of the day and to the general circulation). While '∈' and 'e sin ω ' only act to redistribute insolation among latitudes and seasons, the energy influx integrated over all latitudes and over an entire year is found to depend only on 'e' (a' semi-major axis of the earth's orbit and S_0 being assumed constant here) (Berger, 1988).

Eccentricity Cycles

Orbital model is linked with changes in the shape of the earth's orbit around the sun. This is caused by the shifting interplay of gravitational forces in the solar system; as a result, the Earth's orbit stretches slightly from circular to elliptical and back again with period of roughly 100,000 years. For a circle, $e = 0$ and for ellipses $0 < e > 1$ (Berger, 1988). Due to elliptical orbit, the solar flux varies with maximum at perihelion and minimum at aphelion. The relative amplitude (maximum-minimum)/median of this variation can be computed from the inverse square law for flux, and is $4e/(1-e^2)^2$, which for small values of 'e' is roughly equal to 4e. In case of the earth, for which $e = 0.017$, the solar flux varies by nearly 7% between aphelion and perihelion, and it is on the average equal to the solar constant (Berger, 1988). When the orbit is circular, the earth receives the same amount of heat from the sun every day of the year. When the orbit is more elliptical, on some days of the year our planet is closer to the sun, and receives more heat, than on other days in the same year. According to Berger, however, the total amount of heat received by the whole planet over a whole year is always the same (as long as the sun's output is constant); the three astronomical cycles only redistribute heat between the seasons, and do that only by a very modest amount. As the eccentricity 'e' varies between 0 and 0.04 with its present value 0.016, the orbital flux modulation becomes more or less severe. The eccentricity varies with principal periods of 95,000 years, 123,000 years and 413,000 years. In addition to this, eccentricity variations also have several beat periods and higher order frequencies. (Berger et al., 1989 a, b). In addition, there are several higher order eccentricity cycles in expansion of 'e' term. The most prominent among them are 2.00 m.y., 1.3 m.y., 0.8 m.y. and 0.6 m.y. In addition to this, there are several harmonic and beat cycles.

The 41,000-years Obliquity Cycle

The tilt or obliquity varies between (21° – 24.5°) with its present value of 23° 27' and affects the seasonal contrast and latitudanal gradient of insolation. Obliquity executes harmonic motion with an average period of about 41,000 years. Berger (1977) has reported many obliquity cycles ranging from 30,000–54,000 years. The most prominent in pre-quarternary period are 41,000 and 54,000 years (Berger et al., 1989a). Some of these periodicities are also dominant in climate and other environmental changes record. The values of these cycles have been changing during the past geological times due to change of earth–moon distance. (Walker and Zahnle, 1986; Berger, 1976; Berger et al., 1992).

The 21,000-years Precession Cycle

The longitude of perihelion is a measure of the earth–sun distance at a particular season (summer solstice). Parameters 'e' and 'ω' combine to form the climate precessional parameter 'e sin ω'. This parameter plays an opposite role in each hemisphere and is a measure of the difference in length between half-yearly astronomical seasons and of the difference between the earth–sun distances at both solstices (Berger, 1988). Accordingly, at the present day the earth is about of the perihelion at the northern hemisphere (winter solstice). Sun's precession executes periodic motion with an average period of about 21,000 years. However, in geological/climate records this period is found as 23,000 years and 19,000 years.

It may be noted that the eccentricity cycles are the product of interplanetary gravitational forces and they have probably remained stable through atleast 600 m.y. (Walker and Zahnle, 1986). However, on the other hand, precession and obliquity cycles have changed due to continued evolution of the earth–moon system. Evidence suggests that the earth's rotational velocity has been decreased as a result of transfer of the angular momentum to the moon. The transfer of angular momentum also caused increase in the moon's orbital velocity. Consequently, the recession of the moon eventually attenuates the periods of the earth's precession and obliquity cycles (Walker and Zahnle, 1986; Berger et al., 1992). The orbital motions of the earth and moon also give rise to ripples on the precessional cone. The largest ripple, the principal nutation, is associated with the regression of lunar orbit's line of node with a period of 18.6 years (Walker and Zahnle, 1986). In addition to these principal astronomical cycles, there are a number of higher order and beat astronomical cycles, which arise due to non-linear effects (Berger, 1977 and Stothers, 1987). Table 3.3 gives the details of orbital cycles based on the literature quoted in the text.

Table 3.3 The Main Orbital Cycles

S.No	*Name of the Cycle*	*Period*	*Reference*
1	Higher order eccentricity cycles	3.47 m.y. 1.03 m.y. 0.60 m.y.	Berger, 1977
2	Eccentricity bands	413 k.y. 203 k.y. 113 k.y. 95 k.y.	Berger, 1977
3	Combined effect	59–64 k.y.	Berger, 1977
4	Obliquity bands	54 k.y. 41 k.y. 39 k.y. 28 k.y.	Berger, 1977
5	Precessional bands	24 k.y. 23 k.y. 22 k.y. – 19 k.y. 18 k.y. 16 k.y. 16 k.y.	Berger, 1977
6	Non-linear effect	12 k.y. 10 k.y. 8 k.y. 6 k.y.	Berger, 1977

TIDAL CYCLES

On a still higher frequency end of the spectrum of astronomical forcing, we are aware that there are cycles of alternation of day and night, tidal cycles, seasonal cycles, sunspot cycles and cycles related to the position of earth's axis and to the earth's rotation. Changes in day and night, period of earth's rotation on its axis within a period of 24 hours and period of earth's revolution around the sun is approximately 365 days. Here the impact of tidal cycles arising due to gravitational attraction of sun and moon assume greater importance for the understanding of the earth's short-term processes.

The tides on the earth are generated due to combined gravitational attraction of the sun and the moon. It has been suggested that the moon generates higher tides on the earth than does the sun. Understanding the combined tidal effects of the sun and moon on the earth system processes are intricate due to the elliptical orbits of the earth around the sun and moon around the earth (Smith, 1981). The height of the lunar tidal wave is about 70 cm and height of the solar tide is approximately 30 cm. These cyclic tidal forces widely affect the processes relating to the ocean-atmospheric and the earth's dynamics. These tidal waves ferry around the earth because of the earth's rotation. The tidal effects are more prevalent at the shores because of local topography or resonance. Tidal forces can deform solid body and planet. The tidal effects of moon in oceans are so high that it can cause large water to move in the opposite direction of the moon. Tide dissipate energy and act to decrease rotation of the earth. As a result, rotation is decelerated until it becomes synchronous with orbital motion. Hence, tidal frequencies are seen in the earth's rotation and various ocean-atmospheric phenomena (Smith, 1981).

Tidal Periodicities

Tidal cycles are well known in astronomy. Because of their impacts on earth-ocean atmospheric processing, they can be also decoded from geophysical record (Williams, 1981). Burroughs (1992) and Weedon (2003) have reviewed the possible impact of different tidal periodicities on the earth–ocean and atmospheric processes. According to them there are several short-term tidal periodicities which vary from days to several years and from decades to centuries. The earth's orbit is elliptical with the sun at one focus and has a period of 365.24 days (the plane of this orbit is known as the ecliptic). The moon's orbit is also an elliptic with the earth at one focus (Synodic month) and has period of 27.553 days. Once in every 29.531 days, during the motion of the earth around the sun and the motion of the moon around the earth, the earth–moon–sun approximately align in line. This is called a full moon period or 'conjunction cycle'. This is also called as anomalistic month. The moon orbit makes an angle of 5°9' to the earth equatorial plane. The moon moves above and below this plane each month. During this motion, it crosses the equator periodically, when its declination with respect to ecliptic is zero, with a period of 13.661 days. The cycle length between equators crossing in the same direction is 27.332 days and is known as 'tidal month'. Some of these tidal cycles are very dominant in earth's axial rotation record.

The 18.6-years Earth–Moon Tidal Cycle

A well-recognised tidal cycle of 18.61 years arises in the earth–moon system. The regression of node, the line joining the points where the moon's orbit crosses the ecliptic, defines how the angle of moon's orbit to the earth's equatorial plane combines with or partially cancels out the tilt of the earth's axis. This has the effect of altering the variation in the maximum declination of the moon from the ecliptic (Burroughs, 1992). According to Burroughs (1992) the tilt of the earth's axis, the equatorial plane is at an angle of 23° 27' from the ecliptic, this combines with the angle of moon's orbit to the equatorial plane so that the maximum declination to the ecliptic is (28°40') N and S. But this extreme value occurs only every 18.6 years; at the opposite extreme. Halfway between, when the maximum declination is the difference between the tilt of the earth's axis and the angle of the moon's orbit from the equatorial plane, the value ranges only between 18° 21' N and S (Burroughs, 1992). The significance of this variation is that when the declination is greatest, the tidal forces at high latitudes are greatest. Recent peaks in these forces have occurred in 1913, 1931, 1950, 1969 and mid-1988. The relevance of this cycle is also discussed in the context of weather (Burroughs, 1992).

It is interesting to note, that the tidal forcing could also influence the release of tectonic energy in the form of volcanic activity and earthquakes. The tidal effects of planetary conjuction are quite modest. However, periodic tidal variations are evident in geological records due to short-term effects of fluctuations in the basic energy flow from the sun due to changes in orbital angular momenta and disturbances in orbital configuration of all planets. There seems to be some intriguing coincidences of association of tidal forcing peaks as mentioned above and major Indian earthquakes. The major Indian earthquakes occurred in July 8, 1918 Assam (7.6); in July 1930 Dhubari, Assam (7.1); in August 1950 Assam (8.5); in 1967 Koyna, Maharashtra (6.5); in 1988 Bihar–Nepal border (6.4). Although strict cyclicity is not possible due to complexities involved in the system process, however, it appears that there is striking coincidences between tidal forcing and earthquake process. Tidal forcing might not trigger directly the earthquake but if the dynamics in a particular region is in the state of criticality, it is possible to happen.

Longer Tidal Cycles

The principal cycle generated by the planet's orbital system is the Saturn–Jupiter Lap (19.857 yr), where the two planets possess 86% of the angular momentum involved (Fairbridge, 1984). According to Fairbridge (1984) all of the outer planets realign themselves in the same pattern ('OPR') once every 4448 years (224 × 19.857), and the 'all planets restart' ('APR') cycle is 93,408 years, which is apparently similar to the earth's eccentricity period.

The changing positions of the planets during the motions also exert tidal forces on the earth. The planetary tidal cycles of order of 567 years, 1133 years and 1650 years are known to occur as a result of earth/moon orbital motions. They are widely known as 'Planetary Conjunction Cycles'. The geophysical significance of these cycles are discussed by Fairbridge and Hillare (1977). The maximum period of spring tides repeat approximately after 1668 years when moon is in perigee and earth on perihelion and have significant impact on earth's ocean-atmosperic processes. Another cycle of the order of 434 years is approximately 1/4 harmonic of the 1668 years period. These cycles are evident in many geophysical records such as sea-level fluctuations, temperatures and climate records (Negi and Tiwari, 1988).

The following table, which is based on the various literatures quoted in the text, gives the summary of all tidal periodicities.

Table 3.4 Well-known Tidal Cycles on Years to Millennial Time Scale Based on Literature as Quoted in the Text.

S.No	*Cycle Name*	*Period*
1.	In the diurnal system spring tides occur at maximum declination periodically while neap types occur when the moon passes the earth's equatorial plane.	13.6 days
2.	Semidiurnal spring tides.	14.8 days
3.	The angle of the moon related to earth's equator is known as the lunar declination. In the nodical or tropical month it takes 27.2 days for the moon to oscillate through a complete declination cycle.	27.2 days
4.	Maximum spring tides in semi-diurinal regimes occur periodically (the period of anomalistic month) during the closest approach of the moon or perigee.	27.6 days
5.	The time taken for the moon to orbit the earth is called the sideral month.	27.3 days
6.	Period of lunar or synodic month.	29.5 days
7.	Conjunction/opposition coincides with perigee during the March or September equinox.	4.5 years
8.	The projection of the long axis of the moon's orbit rotate in a space periodically (this cycle is connected with the advance of the moon's perigee which determined the times of the alignment of the perigee with the earth's perihelion).	8.85 years

Contd . . .

9.	Periodic regression of the longitude of the node- the line joining the points where the moon's orbit crosses the ecliptic.	18.61 year
10.	The changing position of the planets during the motion also exerts tidal forces.	567 years
11.	Planetary conjunction cycle.	1133 years
12.	Perigee and conjunction/opposition coincide with perihelion.	1600 years

There are also evidences that the tidal and orbital periods have not been stationary during the past geological time. A comparison of modern and the Late Proterozoic tidal periodicities are compared in the following table.

Table 3.5 Late Proterozoic (~ 650 Ma) and Modern Tidal and Rotational Parameters

Parameters	*Late Proterozoic**	*Modern*
Solar days/lunar month	30.5 ± 0.5	29.53
Solar days/sidereal month**	28.4 ± 0.5	27.32
Lunar months/year	13.1 ± 0.1	12.37
Sidereal months/year**	14.1 ± 0.1	13.37
Lunar apsides cycle (years)	9.7 ± 0.1	8.85
Lunar nodal cycle (years)	19.5 ± 0.5	18.61
Solar days/year	400 ± 7	365.24
Sidereal days/year**	401 ± 7	366.24
Length of solar day (h)	21.9 ± 0.4	24.00
Earth–Moon distance (RE)	58.28 ± 0.30	60.27
Lunar retreat rate (cm/year)	1.95 ± 0.29 (650-0 Ma)	3.7 ± 0.2 (Laser range)

* Values indicated by tidal rhythmites of the Late Proterozoic Elatina formation and Reynela Siltstone, South Australia (Williams, 1984, 1988, 1981b; Williams and Sonette, 1985).

** Units in invariable time required for dynamical calculations.

SOLAR CYCLES

The sun's surface is gaseous and in a state of continuous turbulence. There is hidden deep chronometer in the sun (Dicke, 1978). The solar photosphere evolves every second and lasts for hours rendering short-term transient perturbations and emitting solar flares and other eruptive phenomena. The close observations of sun's photosphere have revealed dark patches in lower latitude between 30° N and 30° S (Burroughs, 1992). These dark spots, known as 'sunspot' vary in number, size and duration and have been observed for thousands of years (Fig. 3.9). The observed data indicate that the average number of sunspots vary quasi-periodically in a period range of 9–13 years

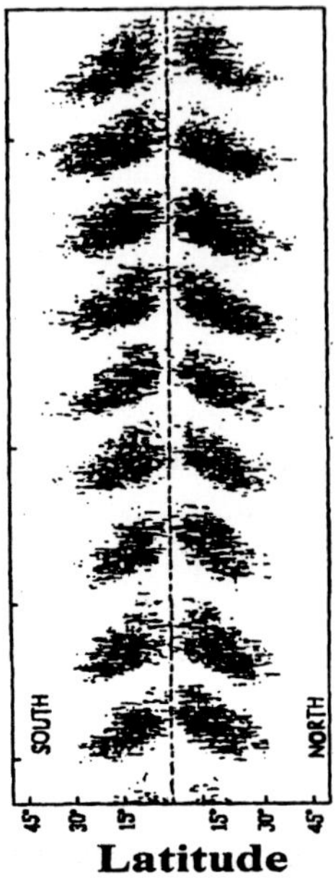

Fig. 3.9 The Observed Variations in the Number of Sunspots from 1875–1975
(*Source: Glovanelli, 1984; Burroughs, 1992*)

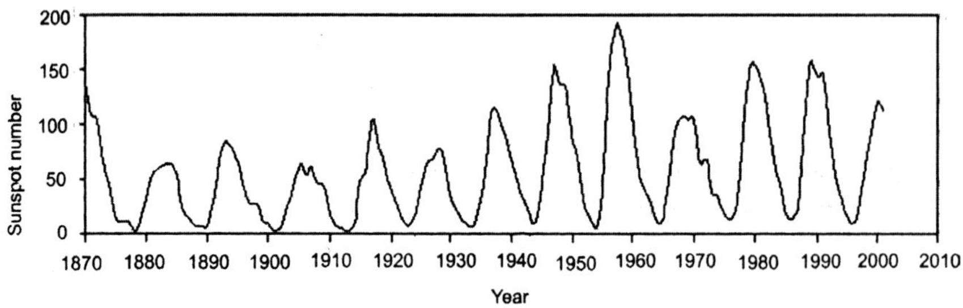

Fig. 3.10 A Record of Variations of Sunspot Numbers

with a mean period of about 11 years (Fig. 3.10). Cycle begins with appearance of some spots in both the northern and southern hemispheres some 35° away from the solar equator. During the development of cycle, the older spots slowly disappear and more new ones appear at lower latitude (Burroughs, 1992). By the end of completion of each cycle, the number decreases and the spots are concentrated at latitude some 5° from the equator.

The generation mechanism of solar cycle, recognised in Sunspot Numbers, is generally attributed to a magneto-hydrodynamic process (Burroughs, 1992). Accordingly, this dynamo model involves the interaction of toroidal and poloidal magnetic field generated by the sun's surface differential solar rotation. According to this model, the non-uniform differential rotation of the sun's surface affect the sun's magnetic field leading to buoyant instabilities leading to rash sunspots and solar flares. At the end of each cycle, the magnetic field dies away, as discussed in detail by Burroughs (1992) and the dipolar magnetic field is recreated with reverse polarity peak repeating after 22 years, known as Hale cycle (Hale, 1924; Dicke, 1979).

The 11-year Solar Cycle

The 11-year solar cycle (in occurrence of sunspot) first was recognised in 1843 by H. Schwabe and since then there have been several efforts to demonstrate that the amount of energy reaching the earth from the sun is affected by the incidence of sunspot. It is now widely recognised that the sun's radiance varies with sunspot number and may have some connection with terrestrial record of climate and weather. However, if solar cycle has indeed a role in modulating terrestrial climate and weather, the evidence of 22 years Hale cycle must also be sought. The duration of 10–11 year's sunspot cycle is, however, variable which might indicate that mechanism at work might be of stochastic/chaotic nature.

The 22-year Magnetic (Hale) Cycle

Evidence of 22-year Hale cycle detected in the sunspot number provides evidence for global solar change. Measurements of the recent part of the Hale cycle indicate solar constant changes upto 0.1%. Apparently, the changes that result in the sunspot and cosmic rays modulation also induce changes in the solar constant (Stuiver and Quay, 1980). The Hale cycle is less variable than the 11-year Schwabe cycle. The magnetic polarity of sunspot alternates between positive and negative successive cycles. Evidence indicates that the 22-year Hale cycle also appears to be responsible for a weak cycle of 20–24 years in the pressure temperature El Nino parameters and associated drought rainfall fluctuations. Infact, solar cycles are numerous—they are real, demonstrably cyclic and reproducible (Burroughs, 1992). When such a peak appears in the spectral analysis of a terrestrial time series, it should be considered as being valid and not written off as some kind of average between other cycles such as those having periods of 18.6 years and 22.4 years. The forcing mechanisms of solar cycles are related to planetary motions. They are both periodic and slightly irregular. They may reinforce one another and/or they may cancel out. Searches for signals from such cycles in smoothed terrestrial time series should not obscure these characteristics.

The 80-year 'Gleissberg Cycle'

A longer solar cycle of the order of 80–90 years in variation of the peaks in successive cycle is also reported (Gleissberg, 1958). This cycle is suggested to be associated with the change in the period between successive peaks of solar activity. In the past two-and-a-half centuries, both maxima and minima have followed a cycle of strictly over 90 years. However, this cycle is not sufficiently regular to attain much significance in geological record.

Table 3.6 Main Solar Cycles (based on literatures quoted in the text)

S.No	Name of the Cycles/ Name of the Solar Cycle	Period
1.	Schwabe – sunspot	11 years
2.	Double sunspot – Hale cycle	20–22 years
3.	Gleissberg cycle	80–90 years
4.	King Hele cycle	180–190 years ≅ 200 years
5.	Long-term solar cycle	420 years
6.	Quasi-biennial oscillation -a cycle of 2.2 years (26 months). There is regular alteration of west-east equatorial stratospheric winds. The length of this cycle is one-fifth of the solar cycle.	2.2–2.8 years

The King Hale Cycle of 180 years

The existence of a longer period, ~180 yr cycle has been proposed (Cohen and Lintz, 1974; Okal and Anderson, 1975) which matched with a comparable, length cycle that some investigators claim to have found within sunspot numbers. However, it is explained that it arises from the beat of the 11 and 9.8 yr cycles and is not an intrinsic periodicity. There are some evidence of this periodicity in geological and geophysical records.

Higher Order Solar Cycles

The solar-wind plasma in our interplanetary space deflects part of the cosmic-ray flux (very high energy particles originating outside the solar system) through its magnetic interaction with the electrically-charged incoming particles, for the reason that changes in the magnetic properties of the plasma originate at the sun's surface. The cosmic rays flux arriving at the earth therefore depend on the changing surface conditions of the sun (Stuiver and Braziunas, 1989). For this reason, the measurement of variable production rate of the cosmogenic isotopes (such as ^{14}C) in the atmosphere provide a time history of changing conditions of the sun's surface. The concentration of ^{14}C and ^{10}Be in the atmosphere has been found to be inversely proportional to sunspot numbers. These two isotopes, therefore, have been used as possible indicator of solar cycle.

This link provided an alternative way to find the long-term solar cycles. Trees, through carbon dioxide assimilation, lay down a ^{14}C record, which provides clues towards the causes underlying ^{14}C production rate changes (Stuiver and Braziunas, 1989). The ^{14}C and ^{10}Be have been decoded in the tree ring record and ice core respectively and are used to decipher long-term cycle in the solar activity over periods of decades to thousands of years. Stuiver and Braziunas, 1989 analysed a 9,600 years-long high precision ^{14}C chronology using the maximum entropy method of spectral analysis and identified a basic long term period of about 420 years along with several other harmonics centred around 218, 143, 85, 67, 57, 52 yrs. There is also a strong evidence for periodicities of 200 years, and 114 years in a tree ring width record. Stuiver and Braziunas (1989) link this periodicity with cyclic changes in the solar constant. The following table 3.7 gives the summary of main periodicities of solar cycles.

In several investigations Sonett et al., 1985, 1988 and Sonett and Finney (1990) performed spectral analysis of long-term ^{14}C time record and detected number of peaks including some primary periodicities and several other beat frequency. They found major peaks at around 2400 years and 1000 years along with several beat cycles, which are also listed in the table 3.7.

Short Term Cycles

Evidence indicate that sun undergoes periodic changes on many different time scales. Hence, periodicities are also observed over wavelength from several years in the solar irradiance, X-ray to radio flux, sunspot area and flares activity. (Szabo et al., 1995 and references therein). Bai and Sturrock (1991) found evidence of shorter periodicities of order of 51, 78, 104 and 129 days in addition to 154 days in solar flares and sunspot records. They also suggest that these periods are close to intergral multiples (by factors of 2, 3, 4, 5 and 6) of 25.8 days fundamental periodicity. The above periodicities then might be interpreted as subharmonic of this fundamental periodicity. Oliver et al. (1998) have suggested a 158-days periodicity in sunspot area due to emergence of magnetic flux of the sun. Further a 1.3 year periodicity has been reported in magnetic field observation which nicely match with similar cycle of solar wind (Szabo et al., 1995).

Table 3.7 Plausible Periodicities in ^{14}C Record Obtained by Spectral Analysis

Symbol	*Period*	*Source*
$f_{c,}1$	2311	primary
$f_{c,}2$	965	primary
$f_{c,}3$	753	$3 f_{c,1}$
$f_{c,}4$	717	$f_{c,2} + f_{c,1}$
$f_{c,}5$	493	$2 f_{c,1}$
$f_{c,}6$	413	$2 f_{c,2} + f_{c,1}$
$f_{c,}7$	357	$2 f_{c,2} + 2 f_{c,1}$
$f_{c,}8$	229	primary
$f_{c,}9$	208	$f_{c,8} + f_{c,1}$

Source: Sonnet and Finney (1990)

CONCLUSIONS

External forcing mechanisms for the earth's processes work on different time scales ranging from very short time to millions of years in quasi-periodic manner. In this chapter, the basic idea was to provide overall information of external forcing agents on all time scales which have direct or indirect influence on the various earth–ocean–atmospheric processes. These are the quasi-periodic cycles that arise due to the movements of the solar system around the galaxy, planetary and orbital cycles and tidal and solar cycles. Quasi-periodic forcing displays periodic stresses—through different mechanisms on different time scales—on the earth's various processes and on its environments. In some of the cases, these forcing effects are not very clearly realised but in many other cases it has been seen that external forcings have considerable effect on the earth processes. The planetary and orbital changes exert close control or modulate to a considerable extent the physical and evolutionary processes of the earth's climate. However, the role of the eccentricity cycles (95 k.y. to 400 k.y.) on climate is still a debated issue. Although there are many ambiguities, it appears that external forcing mechanism of earth processes have several advantages and implications. Discovery of a common 30 m.y. periodicity in various geological and geophysical records identical to one of the galactic cycles is quite encouraging from the prediction and modelling point of view. Similarly, the well-established orbital cycles (e.g. eccentricity, obliquity and precession) are already providing several implications for tuning the climate and geological records. On a still shorter time scale, although the mechanism/ mechanisms of the solar, lunar and tidal cycles are not well explored in the various ocean – atmospheric processes of the earth, its appearance calls for searching appropriate models which might be of immense value for understanding the earth's climate and related phenomena.

4

Long-Term Cycles in Geo-Tectonics and Solid Earth's Processes

INTRODUCTION

The origin of the earth is believed to be some 4.5 billion years ago. Since then it has undergone distinct phases of evolution. For a long time, the earth scientists have been deeply concerned with understanding the forces that drove the earth's various interior processes and shaped the surface of the globe. The researchers invariably analysed the crucial data pertaining to the various geo-tectonic activities stemming from the interior processes of the earth, geomagnetic field reversals and ocean-atmospheric processes including the global carbonate silicate cycle to unravel the evolutionary mysteries of the earth. Attempts were also made to visualise a direct/indirect physical link through common external/internal forcing mechanisms. Consequently, there have been several speculations regarding the nature of these records as well as their physical linkages. Several studies based on visual inspections and time series analysis lead to considerable confusion about the nature of these records. As a result, several diverge views emerged that suggested that various processes of the earth's interior can be expressed as mixed response of random, catastrophic, chaotic and cyclic rhythms in time and space. The view of 'randomness' of long-term earth processes dominated for a long time and became the central dogmas in the earth's science. Most of the above conclusions, however, were based on old geophysical/geologic observations available at that time, which were incomplete and far more complicated to understand the earth's evolutionary history. The modern view, however, does not reconcile with the theory of randomness of the earth's processes but provides crucial evidence based on the new data, which holds the view that most of the earth's processes are chaotic and/or quasi-cyclic in nature.

The past few decades have witnessed major developments in the field of geo-sciences. The remarkable advent of the plate tectonics theory, a major paradigm in the earth sciences, provided huge amount of new data on the spatio-temporal variations of geological, geo-chemical, geomagnetic, geophysical and geo-morphological records. Most of these data are based on modern multiple dating techniques and are tuned with the geomagnetic reversals and micro-palaeontological records. Hence, these records are naturally more precise and accurate for understanding the evolutionary history of the earth processes. These new data have also facilitated more detailed studies to understand their coupling behaviour and physical linkages, thus, rendering a unifying concept to link directly or indirectly the geo-bio-hydro-atmospheric processes.

During the past two decades, several studies were carried out using the new data and modern time series analysis techniques. These various studies have provided substantial ground to propose more usable physical mechanisms of earth's dynamical processes. In this chapter, we present and discuss the available results of those records where atleast new data are analysed quantitatively using the modern time series techniques and discuss the possible physical mechanisms and also concentrate on only those records which are related to the solid earth processes. Some other results pertaining to the long-term ocean- atmosphere and biological extinction processes are presented and discussed in subsequent chapters. It is worth mentioning here that even some of these new records have drawbacks and cavity. However, the recent results have used appropriate computer-based time series analysis techniques to handle these data that, will enable us to examine the robustness and statistical reliability of cycles. Further, these cycles can be compared with those well-known cycles that have already been identified in the astronomical records. The rigorous analysis of the new records would naturally eliminate and/or provide some of the most intriguing and convincing examples of cyclic behaviour in earth processes. We would also mention here that the precise very long-term geological records for detailed quantitative studies are limited and may therefore permit only to decipher authentically shorter periodicities in a period range of upto approximately 300 m.y. Many of these records, however, extend over much beyond this period and there is evidence for longer cycles of 500–700 m.y. also. However many of these cycles have only limited impact due to the fact that longer records lack the precision of age dating. Hence, it may also not be possible to argue with certainty on the stability of very long-term cycles and about their physical correlation.

STRUCTURE OF THE EARTH

In order to understand the mega-cycles of solid earth processes, it is essential to know the processes and structure of the earth. Hence, first we briefly describe here the first order structure of the earth and then proceed to understand its dynamical behaviour. The earth can be broadly divided into three main concentric layers: the crust, the mantle and the core (Fig. 4.1).

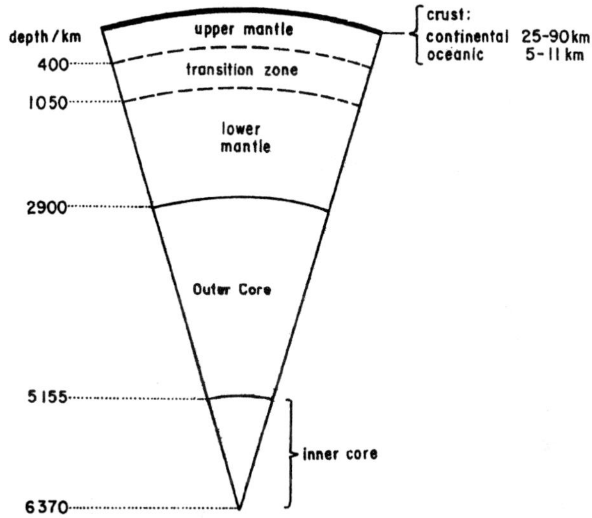

Fig. 4.1 A Major Division of the Earth's Interior Depicting the Crust, Mantle and Core

Crust

The outermost layer of the earth is known as the crust. Thickness of the crust in continent and ocean varies. Continental crust is thicker and it varies in thickness ranging from 20–40 kms extending as far as 80–90 kms beneath the biggest mountain region like Himalayas. The average thickness of the crust in ocean is estimated to be 10–18 kms. The crust is almost 4 billion years old. According to the plate tectonic theory to be discussed in the next section, the outermost crustal layer is broken in eight major and several minor tectonic plates called 'lithosphere'. The average thickness of lithosphere is about 100 kms.

Mantle

The earth's second major division is the mantle. The mantle, 'the middle part' of the earth, just beneath the crust is about 2,900 kms thick, including sub-crustal portions of lithosphere. The main rock type is peridotite, which is dark, dense igneous rock containing largely iron and magnesium. The mantle comprises about 80–85% of the earth's volume. It is denser than the crust (density ranges from 3.15–5.8 g/cm^3). It can further be divided into three distinct zones based on physical characteristics. The outer mantle is solid and surrounds asthenosphere. The asthenosphere, with thickness 200–400 kms, surrounds lower mantle. The crustal plates move over the asthenosphere as a result of underlying convection cells in the lower mantle (Condie, 1976).

Lower mantle has plastic behaviour and is slowly churning flows in partial melt within the asthenosphere and generates magma of viscous material. The main rock content is periodotite. The melting rocks due to high temperatures are called 'magma'. Magma rises to the earth's surface under pressure because it is less dense than the rock from which it is derived.

The temperature of the earth's interior increases with depth. The temperature beyond 100 km depth (below lithosphere) increases from 1300°C–1500°C. This is enough to melt some of the silicate materials in the mantle rock. As a result, the rock is able to flow to create a soft layer at about 200 kms. on which the rigid lithosphere floats (Condie, 1976). There is yet another layer below the asthenosphere called 'mesosphere' or 'D-layer' at the base of the mantle. The continents are embedded with oceanic plates, which make up much of the sea-floor. The dominant force drives the tectonic plates in the form of convective motions, powered by heat from the decay of radioactive elements. Mantle rocks are constantly driven up towards the surface by the enormous temperature pressure below. It is, however, believed that the radioactive decay is a continuous process whose rate has declined smoothly with time. Thus, the production of heat cannot, by itself, account for the continued power generation for driving the plates.

Core

The lower mantle surrounds the core, the third major division of the earth. It is the interior most part of the earth occupying about 15–17% of earth's total volume. Seismological data indicate that the core can further be sub-divided into small solid inner core (5100–6400 km) and apparently a large liquid outer core (2900–5100 km). The outer core is so hot that the metal is always molten but in the inner core pressures are so high that melt solidifies even at high temperature of 3700°C. Both parts consist of high density material (largely iron and nickel) of density of order of 10–13 g/cm^3. It is believed that the earth's magnetic field originates in outer core due to spiralling currents and has presence of primordial magnet there (Jacobs, 1994).

PLATE TECTONIC THEORY

During the sixties, a remarkable discovery was made about the dynamical behaviour of the earth and its interior processes. Hence, before dwelling on the long-term cyclic processes of the earth, it is prudent to describe briefly the main features of the above development that are mainly related to the plate tectonic theory. Historically, it was Prof. Alfred Wegner, a noted German meteorologist, who first developed the novel concept of the continental dynamics during 1910–1920, based on a study of palaeoclimatology, continental geometry, palaeontological provinces and structural correlation. He suggested that the continents had moved relative to each other during the geological past.

He suggested that 'Super continent Pangaea' existed before the continental drift broke up due to forces associated with the rotation of the earth. However, his idea was criticised (Jeffreys, 1924) due to insufficient forces for driving the plates and also lack of adequate driving mechanism for continental movements. Holmes (1929) demonstrated that thermal convection was capable of driving mantle convection and continental drift. During the late-forties, however, extensive studies based on palaeomagnetic data, marine magnetic and seismic investigation provided considerable evidence to convince the earth science community that continents had, indeed, drifted away from one another.

During 1960, palaeomagnetic studies provided crucial evidence for relative continental motion from apparent polar wandering curve. Study of rock samples in North America and Europe gave different position for the magnetic pole leading to the important conclusion that the differences observed were the result of continental drift between the two continents. Similarly, marine geologists discovered that the ocean-floors were youngest at the ridges and became gradually older as one traversed normal to the ridges. Study of magnetic field over the ocean indicated a regular striped pattern of magnetic anomalies lying parallel to the ocean ridge. Hess (1962) hypothesised that the ocean-floor may be moving in response to convective motion in the mantle. He suggested that the sea-floor was created at the axis of a ridge and moved away from the ridge to form an ocean basin. At the ocean ridge, adjacent plates diverge from each other in a process known as 'sea-floor spreading'. Vine and Mathews (1963) provided convincing evidence to support Hess and showed that the magnetic striping of either side of a mid-oceanic ridge are symmetrically arranged. Morgan (1971) set a basic framework for understanding the plate tectonic theory. He realised that relative motion of lithospheric plates with respect to one another was a natural consequence of the thermal convection in the mantle. The oceanographers established the existence of mid-oceanic ridge, a continuous chain of ridges and mountains, which is an area of pronounced seismic activity. These various extensive studies of ocean-floor characteristics, its surface features and magnetic and seismic properties, as discussed above, provided a basic framework for comprehensive understanding of the geological and geodynamic process of continental drift which finally gave birth to the plate tectonic theory.

In essence, according to this theory, a portion of rigid upper mantle and overlying crust constitute the earth's rigid shell called 'lithosphere' which is split up into number of plates (twelve large individual pieces). These plates are being continually created as hot mantle material rises along mid-oceanic ridges and is consumed at oceanic trenches where plates bend and descend into the interior of the earth, a process known as 'subduction'. The various elements of tectonic processes are depicted in Fig. 4.2 (Wicander and Munroe, 2002). The plates diverge and move apart in response to underlying convection cells and are destroyed in subduction zones where one of the converging plates plunges into the hotter mantle. Where plates slide by one another, large

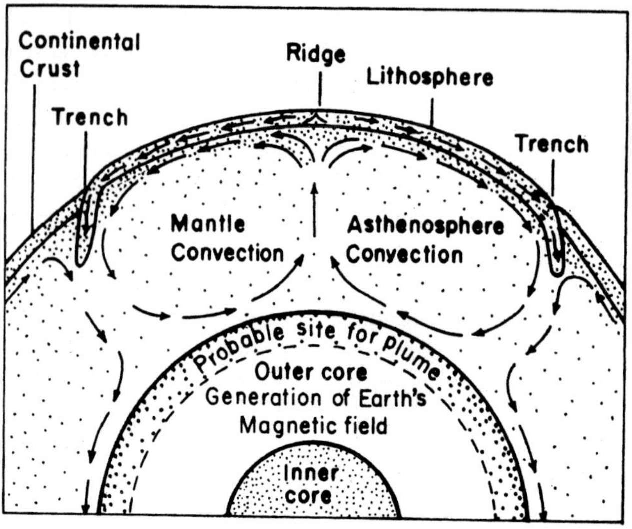

Fig. 4.2 Schematic Diagram Depicting Various Processes of Earth's Interiors e.g. Oceanic Ridge, Subduction Zone, Trench and Mantle Convection

Source: http//pub.uggs.gov/publications/text/unanswered.html

transform fault occurs. Interactions of these plates are responsible for such phenomenon as earthquakes, volcanic eruption and formation of mountain ranges and ocean basins (Wicander and Munroe, 2002). The oceanic lithosphere is continually recycled as it is created at the oceanic ridges and subducted at the ocean trenches. Because of this cycling, the average age of the ocean-floor is about 90–100 m.y. The movement of the tectonic plates and their breakup and reassembly and interactions/collisions would explain the cyclic phenomena of mountain-building processes, volcanic eruptions, formation of magma chain and evolution of ocean basins etc. These megacyclic processes would also accelerate sea-floor spreading and thereby influence sea-level changes. It would promote the processes of cyclic continental rifting, magmatism and episodes of mafic-dykes swarm kimberlite production.

LONG-TERM GEOTECTONIC CYCLES

Wilson Cycle

The theory of plate tectonics has provided a unifying concept for studying the global earth dynamics and also provided very neat explanation and understanding for many geological and geophysical phenomena. During the 1960s, the plate tectonic paradigm provided a new concept of cyclic evolution of mega-geotectonic processes. A noted Canadian geologist, J. Tuzo Wilson (1966), who was also one of the major contributors in the discovery of the plate tectonic theory, proposed a fundamental long-term cycle of continental drift. He suggested that oceans open and close, continents fragment and reassemble cyclically with a period of approximately 400–500 m.y. This super-continental cycle, also known as 'Wilson cycle' was based on evidence of opening and closing of the Atlantic ocean. The Wilson cycle explains many long-term geophysical phenomena that are linked together and evolve in a cyclic manner. The cyclic chain of evolutionary processes start from the breakup of continents that occur on continental rifting zones as a result of fracture and culminate into a rift valley leading to the formation of ocean-floor spreading centre or ocean ridge. Nance et al. (1988) suggested a mechanism to explain how the continents assemble and break apart cyclically. According to these workers, upwelling hot rock material from deep mantle partially melts to form new ocean crust at first stages of an ocean ridge. As sea-floor spreading continues at the spreading centre, an ocean is formed. Ocean-floor at an oceanic ridge is very nearly a symmetric process. As the sea-floor at continental margin grow older, the lithosphere become thicker and denser. Eventually, the lithosphere becomes sufficiently unstable leading to the formation of an ocean trench and subduction zones. The cycle repeats, the authors argue. Geological evidence suggests that all the continents had come together by the end of Palaeozoic era to form super continent 'Pangea'. Pangea began breaking apart during the Triassic period leading to subsequent movement of plates resulting in present distribution of continents and ocean basins (Condie, 1976). Evidence also suggest that super continent consisting of most of the land masses form breakup and reform in a cycle spanning about 400–500 m.y. Thus, super continental is one more example of how various earth's system and sub-system are interrelated and integrated. During this process, the continents disperse and collide leading to the opening and closing of Tethys Ocean and formation mountains (Wicander and Munroe, 2002).

According to Nance et al. (1988), continental crust is only half as efficient as oceanic crust in the process of conducting heat. The mantle heat would accumulate under super continent causing it to dome upward and eventually break apart. As fragments of the super continent disperse, heat can be transferred through the new ocean basins created between them. After a certain amount of heat below the continental fragments escapes, it may be driven back together over millions of years and the process continues. They also suggested a megacycle of the order of 400–500 m.y. similar to Wilson cycle in plate motions and its consequences on various earth processes including long-term climate and sea-evel changes. The super continent cycle (Nance et al., 1988 and Worsely et al., 1985) notices surprising regularity of the timing of various episodes of mountain building and major episodes of rifting through geological time. Accordingly, major tectonic episodes are broadly centred around 2600 m.y., 2100 m.y. 1600 m.y., 1100 m.y., 650 m.y. or 250 m.y. ago. These dates are globally distributed and indicate the signatures of intense tectonic episodes and related activities. Indeed, there are some more inferences of periodicity from these temporal geologic records. Several geologists have also identified five cratonic sequences within the plate-form sediments of North America with duration of the order of 50–150 m.y. A cycle of 50–60 m.y. has also been suggested by Vogt (1972, 1975) in flood basalt volcanism and later by Seyfert and Sirkin, 1979 in tectonic record.

Recently, Cooper (1990) suggested a more precise periodicity in complete tectonic records of South Africa. He inferred a common cycle of the order of 350 ± 50 m.y. in various tectonic records. On the basis of available geo-chronological data, he demonstrated a periodicity of 320 ± 25 m.y. in major phases of continental volcanic events, magmatism and carbonatite intrusions of South Africa (Fig. 4.3). The data have also shown an evidence of relatively shorter periodicity of the order of 260 m.y. during early Archaean i.e., pre-3200 m.y. In addition to this, another rock succession displays identically striking episodicity in anorthosite sequences. According to Cooper (1990), the mid-Proterozoic anorthesite episode also displays an approximate cyclicity of 300 m.y. Several other tectonic phases have shown long-term cyclicity of the order of 320 ± 25 m.y. including major granite forming events. The major granite forming events from South Africa also suggests 640 m.y. mega periodicity for certain type of granite-forming events, which is similar to the 'Chelogenic cycle', proposed earlier by Sutton in 1963. Major orogeny phases

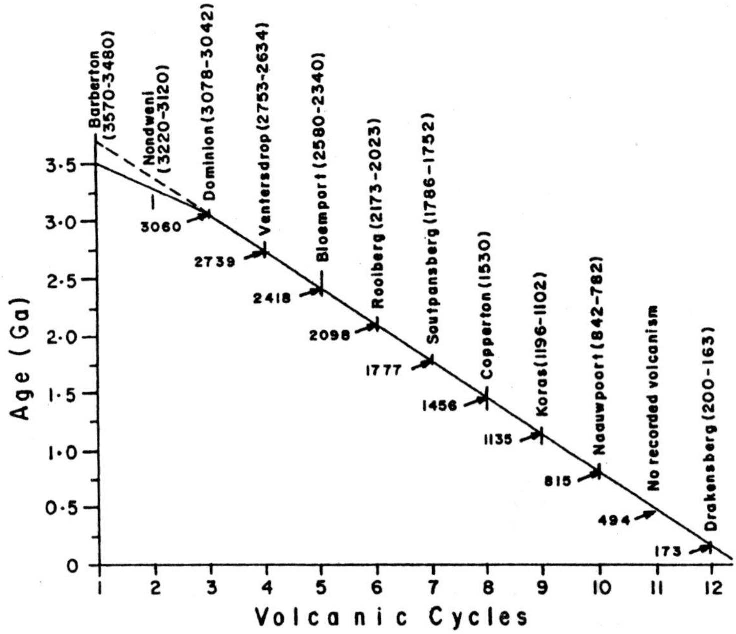

Fig. 4.3 Ages of Selected Volcanic Cycles from Southern Africa with Most Precise Age Range in Brackets. The Bold Figures Indicate Approximately Uniform Interval Suggesting a Periodicity of 320 ± 25 m.y.

(*Source*: Cooper, 1990)

have also shown striking regularity with higher order cyclicity of the order of 640 m.y.

A better quantitative analysis of tectonic records of Cooper using the method of Stothers (1979) reveal strong periodicities at around 160 m.y., 350–400 m.y. and 650 m.y. (Fig. 4.4). Interestingly, the later cycle of 650 m.y. also matches with the recently reported cycle of Abbot and Isley (2002) in the global record of plume volcanic activities. The other two cycles also match with earlier suggested cycles in the pulsation tectonics. The important point to note here is that the super continental cycle is linked to the model of pulsation tectonics of Sheridan (1983), which explains the break-up and assembly of super continent and links various processes of earth's interior. The process obviously would have affected other processes such as climate and sea-level

(a)

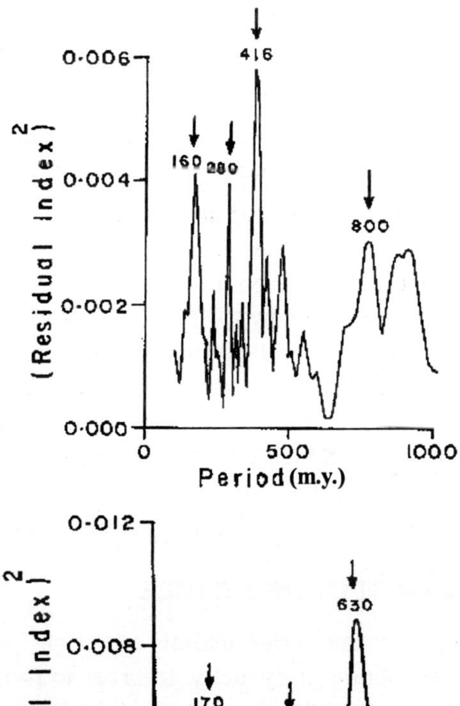

(b)

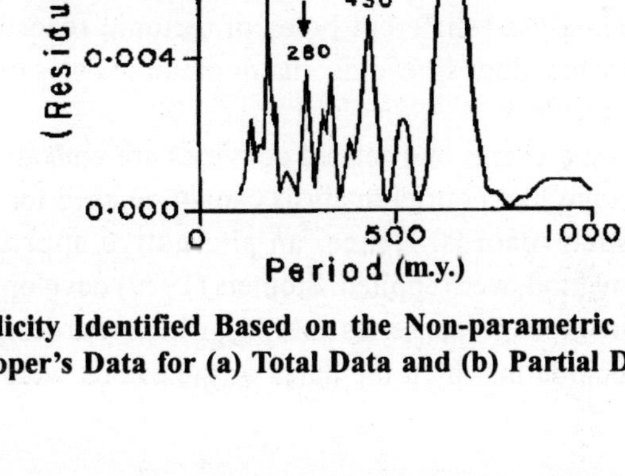

Fig. 4.4 Periodicity Identified Based on the Non-parametric Analysis Using Cooper's Data for (a) Total Data and (b) Partial Data

in a similar cyclic manner. A shorter cycle, in comparison to supercycle, of the order of ~ 65–100 m.y. (Sheridan, 1983) has also been suggested. The gross system analysis of the earth provides a working model to understand the long period cycles of the entire environment (Sheridan, 1983).

It is interesting to note that a recent study by Prokoph et al. (2004) applied a wavelet based spectral and cross-spectral techniques to better data set of igneous province for the past 3500 m.y. Their analysis revealed non-stationary mega-cycles of order of approximately 500–800 m.y., 300–400 m.y., 230–270 m.y., 160–170 m.y. and 30–35 m.y. They have also claimed that these periodicities are statistically significant against red background noise level and would have physical implications. Interestingly, all these periodicities fall in cosmic and galactic frequency bands as discussed in chapter 3.

Episodicity and Crustal Evolution

Glikson (1999), in his several other works, has presented detailed account of quasi-periodic episodes of crustal evolution and their relation to asteroid/comet mega-impact events on the earth for the Precambrian period. According to these workers, the global peaks, based on K-Ar, Rb-Sr and U-Th-Pb ages, appear at 2.6, 2.1, 1.8, 1.4, 1.0, 0.48, 0.35 and 0.105 G.y. Several other studies also outlined peak events at ~2.6, 1.8, 1.1 and 0.6 G.y. based on K-Ar and Rb-Sr ages. Achaean age concentration maxima at 3.5, 3.0, and 2.7 G.y. are well documented and are supported by U-Pb isotopic data, which indicate an episodic recurrence cycle of about 350–500 m.y. in major crustal events. These researchers further suggest that such age anomalies can be correlated with several types of events, including the development of intracratonic rifts, major periods of dyke formation and high crustal movement rates and reversals represented by apparent polar wander paths. The following Table 4.1 is based on Glikson data and references therein, and its interpretation.

A 30–MILLION YEAR TECTONIC CYCLE

New data and appropriate time series analysis of various geologic records provided a more clear picture of cyclicity. In order to produce statistical significance and hence to strengthen the physical correlations, Rampino and Stothers (1984) analysed different types of tectonic records of mountain building episodes, sea-floor spreading, major plutonic dates of carbonate and intrusion Kimberlite etc.

Since the tectonic events and related activities are episodic in nature, the traditional technique like Fourier analysis cannot be used for the analysis of periodicity in such records. Hence, an alternative approach, based on non-parametric method, were applied. Stothers (1979) developed an objective non-parametric method of time series analysis, which was designed especially for those time records in which the dates of the events were recorded only

Table 4.1 Major Global Tectonic Episodes and Impact Events in Earth's History (Data from *Glikson*)

Major Peaks	*Mega Tectonic Events*	*Evidence for Impacts and Geological Interpretation*	*Locations*
~3.2 G.y.	Rifting events	The iridium rich spherule coincide with onset of major felsic volcaniclastic and greywacke sedimentation	Western Australia
~2.45 G.y.	Dyking events	A vestige of contemporaneous impacts, signifying either a single or clustered event represented by spherule and diamictite unit intercalated with Wittenoom dolomite and Dales Gorge member of the Hamersley Group Pilbara block	Zimbabwe
~2.1–2.0 G.y.	Rifting events	The Bushveld complex, Transvaal, comprising an area of 65,000 km² of a layered mafic/ultramafic lopolith, coeval granite, and felsites has been interpreted in terms of a mega-impact	Western Africa And Guyana
~1.85–1.8 G.y.	'Hudsonian' mobile belt events	The Sudbury impact structure (d=200 km) for several plutonic and thrusting episodes	North Central Australia
~1.3–1.05 G.y.	'Grenville' rifting and mobile belt events	Geophysical structure and breccia occurrence provide clues to a bombardment period that triggered rifting, basic plutonic and dyking events	Sweden
~0.7–0.6 G.y.	'Lapetus Ocean' and igneous events	Evidence for clustered impacts structures and global rifting continental splitting events.	South Australia

and not the amplitudes. This method involves fitting a linear model of the type $t = t_0 + np$ (where p is a trial phase for most recent epoch and n is an integer) for a given time series of type t_i (i = 1,2, . . ., n). The resulting sum of the squares of the residual are minimised at each trial period to obtain a best fit for selected trial period; dividing it by the period and then subtracting the result from a numerical constant yields the residual index. Using the above method Rampino and Stothers (1984) analysed tectonic and several other episodic records for the past 600 m.y. (Phanerozoic) which revealed two dominant cycles of the order of 33 ± 3 m.y. and 260 ± 40 m.y. suggesting that earth's tectonic processes are periodically punctuated with atleast two different characteristic time constants (Fig. 4.5). They have also suggested that statistical significance of the above periodicity is reasonably high to propose physical linkages. It is interesting to note that these two periodicities match with cosmic

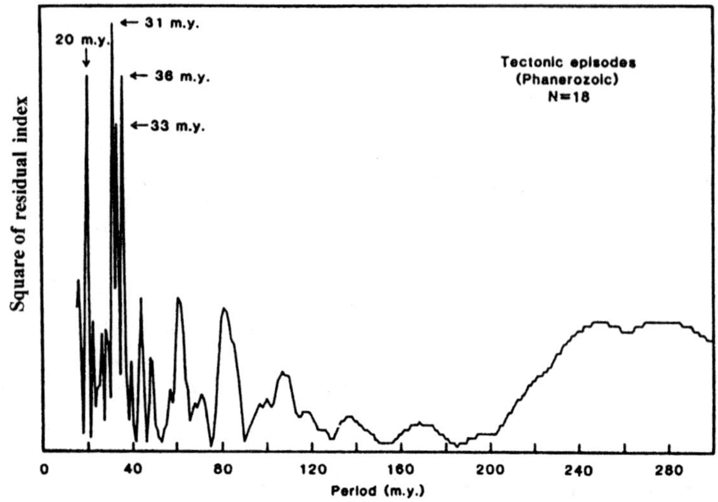

Fig. 4.5 Depicts Periodicity of Major Tectonic Episodes During the Phanerozoic Eon. The Square of Residual Index (abscissa) and the Period (base). The Best Fit is Seen for 31 m.y.

(*Source*: Rampino and Stothers, 1984)

cycles and periodic impact record. Rampino and Stothers (1984) have discussed the importance of these cycles in detail.

PROCESSES OF MANTLE DYNAMICS

Mantle Plumes

Mantle plumes are the expression of the upward limb of convection within the lower mantle. Figures 4.6(a) and (b) show schematic diagrams of plume coming from below the deep mantle. Morgan (1971) has suggested the existence of approximately 20 such thermal plumes. According to recent studies, the number has gone up to 300 and even more all over the world extending from the deep mantle to the asthenosphere. Mantle plume most likely arises from a hot thermal boundary layer at the base of mantle. The mantle plumes are believed to originate below 700 km or even from the base of mantle e.g. near the D-layer. The mantle plumes are stationary features and independent of plate motion. Morgan (1971) and Deffeys (1972) were perhaps the first to conceive the idea of deep mantle plumes. They suggested that these plumes carry heat from the lower mantle to the surface. However, later Loper and Eltayeb (1985) suggested that plumes do carry heat from the core to the surface but the convective pattern, which cools the lower mantle, is not directly related to plumes. The plume generates a thermal halo, which is buoyant and will tend to rise upward. Plume rises at much slower rate due to the strong variation of viscosity with temperature. However, it is capable of exerting larger tractions because it is more viscous and is driven

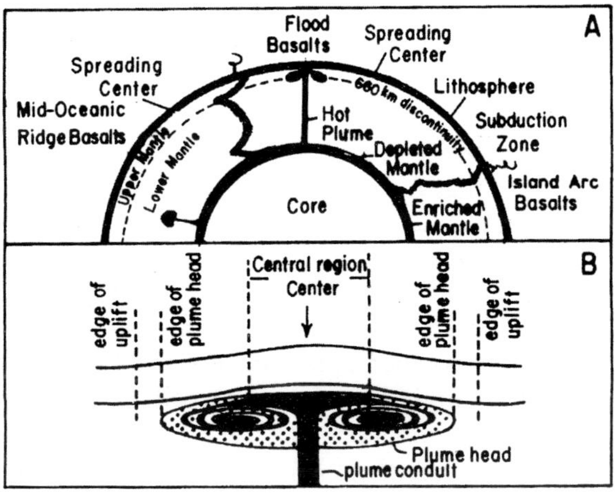

Fig. 4.6 (a) Schematic Diagram of Mantle Plume and (b) Its Probable Origin (*Source*: Earnst and Buchan (2003).

by a broader thermal signature (Loper, 1991). It is believed that the core is cooling (Stacey and Loper, 1983) and that the heat it casts off sustains a layer of heated material at the base of the mantle called D-layer. The thermally-induced structure of this layer is dynamically similar to that of a plume. It has low viscosity to act as a conduit for localised horizontal fluid flow (Stacey and Loper, 1983). Horizontal convergent flow to the bases of the plumes may supply them with the buoyant, low-viscosity material necessary for their existence. According to Loper and his co-workers, due to the large Raleigh number of the mantle, the heat from the core must be transferred to the surface of the earth via convection. The D-plume system is the manifestation of this convection. The hot and low-viscosity material, which flows up the plume chimney diverge, when it encounters the cold lithosphere and forms the hot-spot swells seen at the surface (Crough, 1978; Olson and Nam, 1986). A small percentage of the plume material reaches the surface via hot spot volcanism (Loper and Stacey, 1983). The remainder contributes to the low-viscosity asthenosphere just below the lithosphere or it under plates the lithosphere (White and McKenzie, 1989). The ensuing discussion and various studies amount to say that mantle dynamics is a physially driven turbulent process and may possess order in midst of these complex chaotic dynamics (Loper and McCarteny, 1986).

Plume Cycle

Generally, most tectonic activities are found to be located only along the plate margin. However, there are several volcanic centres, which are away from plate boundaries. These areas have been referred to as hotspots. Hotspots are surface manifestation of long narrow columns of hot mantle materials called mantle 'plume'. Several researchers (Jones, 1977; McFadden and Merrill, 1984) noted that significant change in temperature (100–200°C) in few tens of million years of the core mantle boundary might explain the observational changes in the long-term reversals frequency of the magnetic field. The rate of temperature change controlled conduction and radiogenic heat. However, the conductive heat transfer could cause rapid and significant temperature changes in a few million years if the velocity of mantle plume convection is great. Sheridan (1987) has suggested that for vertical thermal convection in a spherically symmetrical earth, the circulation in the convective cell is oscillatory in nature. Accordingly, the oscillatory nature of these scales reflect the circulation of the convection cell first by heating then cooling, and then cooling again as flow of hot mantle completes its circulation from one cycle to the next. The period of the cycle is determined by the equation $t_0 = L/V$, where L is the path length of material moving in the convection cell, and V is the velocity of convection. According to him, assuming L = 3000 km and the velocity V of convection (Yuen and Schubert, 1976) is about 10 cm/yr., the characteristic time scale for convection of the lower mantle comes to about 30 m.y., which augurs well with the other observations. Loper and McCartney (1986) proposed a 30 m.y. periodicity in the coupled core-mantle processes and linked it with magnetic reversals periodicity. The physical relevance of plume cycle will be discussed in the next sections.

Characteristic Time Scale of Mantle Flow and Volcanic Cycle

Based on seismic topographic and gravity data, Anderson (1998) suggested that there are important characteristic time scales of mantle convection. Accordingly, these time scales may be associated with spherical harmonic of degree (l = 1 to 6). Accordingly, convection of degree 6 shows up in the spectrum of hot spots and upper mantle topography. He suggested that cratons control the l = 6 patterns of convection and upper mantle convection may reorganise roughly every 30 m.y. as cratons move about.

Worldwide distribution of flood basalt provinces that erupted in the last 250 m.y. and associated hotspots are shown in Fig. 4.7 (Duncan and Richards, 1991). White and McKenzie (1989) argued that plumes are responsible for a continental breakup and triggering rifting process. The rifted continental margins are underplated with large volume of basalts due to magmatic activity. The available worldwide geochronological record of magmatism (including intrusive and carbonatitic activity) has been compiled for the past 250 m.y. These events exhibit a definite global enhancement in volcanic activity around 28, 42, 61, 83, 101 and 132 m.y. BP. The clustering of global magmatic activity enhancement showed an average quasi-periodicity of 33 m.y. (Pandey and Negi, 1987). Baksi (1990) argued against evidence of 33 m.y. periodicity in mafic volcanism. His main criticism was based on the inferior and incomplete magmatic record. Rampino and Stothers (1988) presented a new chronology of the initiation dates of major continental flood basalt volcanism from published literature of Potassium-Argon (K-Ar) and Argon-Argon (Ar-Ar) of basaltic rocks and related basic intrusions (Table 4.2). They have noted that

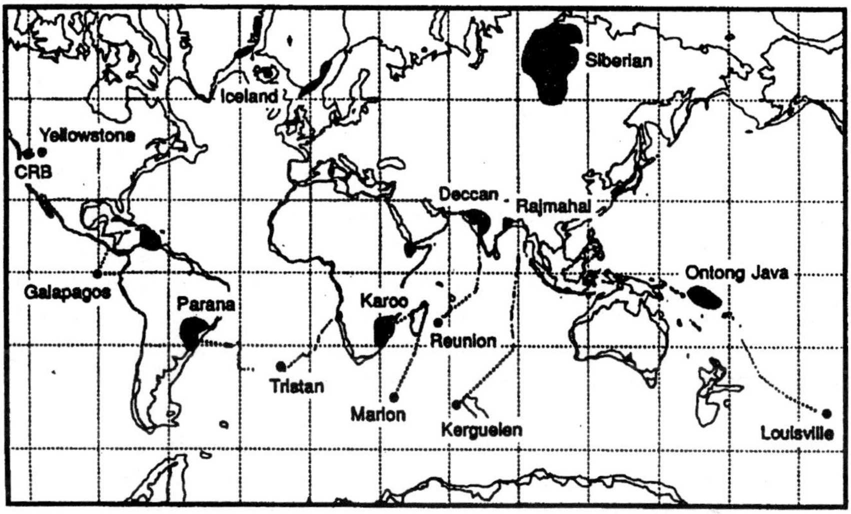

Fig. 4.7 Worldwide Distribution of Flood Basalt Provinces Erupted in the Last 250 m.y. and Associated Hotspots.

(*Source*: Duncan and Richards, 1991)

Table 4.2 Ages of Flood Basalt Initations

Episode	*Age (10^6 years)*
Columbia River (US)	17 ± 1
Ethiopian	35 ± 2
Brito-Arctic	62 ± 3
Deccan (India)	66 ± 2
Rajmahal (India)	110 ± 5
Serra Geral (S.America)	130 ± 5
South-West African	135 ± 5
Antarctic	170 ± 5
South African	190 ± 5
Eastern North American	200 ± 5
Siberian	250 ± 10

Source: Rampino and Stothers, 1988

these datings are independent of the biostratigraphic and paleomagnetic time-scales. They have also mentioned that estimated errors of initiation dates of volcanic episodes determined from the distribution of the radiometric ages are approximately plus or minus 4 per cent. Table 4.2 shows 11 distinct episodes during the past 250 m.y.

Rampino and Stothers (1988) performed time series analysis using improved and complete global record of volcanism (Table 4.2) and suggested a possible evidence for 32 m.y. periodicity. For this analysis, they assumed a periodic delta-function response to the unknown input signal and applied the nonparametric method as discussed earlier, to search for significant periods in the series of these dates. They note that the sensitivity of the absolute heights of the spectral peaks to modest errors in the dates are taken care by maintaining only a simple count of the number of times the highest spectral peak occurred at the same period for 1000 time series. The date of each episode was randomly selected from the age range indicated for that particular episode. They sampled only the period range of 18 to 100 m.y. to avoid spurious periods that may arise from both the high-frequency noise in the data and the low-frequency cut-off length. Rampino and Stothers (1988) presented a histogram of their results based on all 11 flood basalt episodes (Fig. 4.8a). The figure depicts frequency diagram of period at which the highest peak occurs, for 1000 pseudo-random time series of flood basalt initiation dates. If there is any periodicity in this record, the authors argue, the most likely period is either around 22–26 m.y. or around 31 m.y. They further emphasised that even after including an additional basalt at 90+10 m.y. or omitting a less well dated Siberian flood basalt from the analysis, the histogram is essentially unchanged thus revealing a remarkable stability of 32 m.y. periodicity. Further, after removing Colombia river basalts, which seem to have erupted in other than a continental rift related setting and do not belong to the class of the data presented in Table 4.2, the analysis reveals a possible periodicity of 32 m.y. Figure 4.8b shows

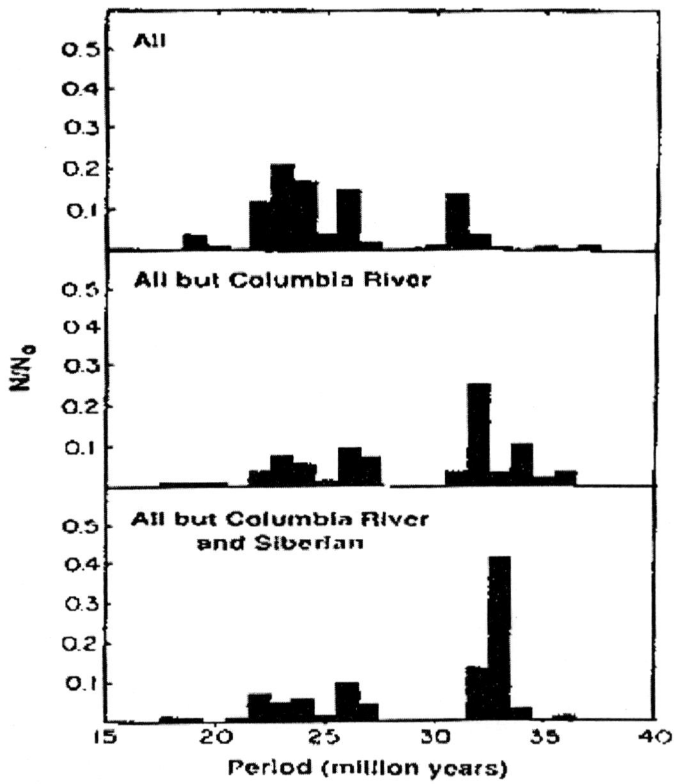

Fig. 4.8(a)

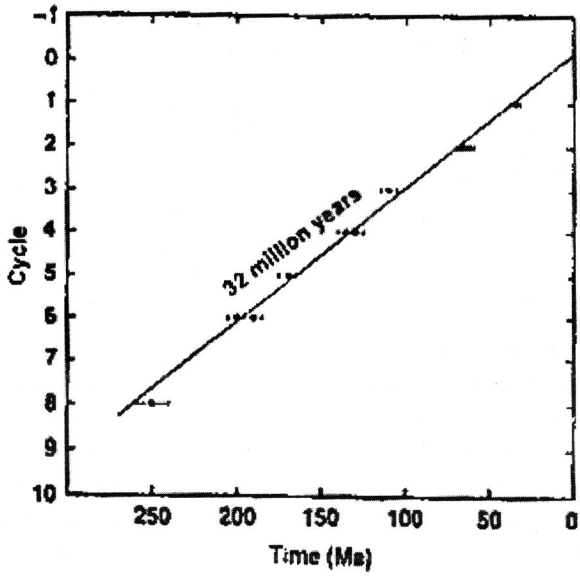

(b)

Fig. 4.8 (a) Frequency Diagram of Periods at which the Highest Peak Occur for 1000 Pseudo Random Time Series of Flood Basalt Initiation Dates (Table 4.3) (Rampino and Stothers, 1988) (b) Time- Cycle Diagram for Flood Basalt Episodes, with a Line Showing the Best Fit Period of 32 m.y.

(*Source*: Rampino and Stothers, 1988).

time-cycle diagram for flood basalt episodes, with a line showing the best fit period of 32 m.y. A bar indicates the probable range of accidental and systematic errors for the initiation date of the episode (Rampino and Stothers, 1988). It is interesting to note that independently dated sets of other types of tectonic activities e.g. active tectonism and alkalic and silicic volcanism also show identical periodicities. Various independent studies of new data pertaining to crustal and mantle processes, thus suggest that periodic behaviour exhibited in these data/processes cannot be due to chance, it rather emphasise true signal of earth's internal dynamics.

CORE DYNAMICS

Origin of Geomagnetic Field

The origin and mechanism of generation of earth's magnetic field have been studied extensively for several decades (Elsasser, 1946; Bullard et al., 1950). Accordingly, the geomagnetic fields are believed to originate from magnetohydrodynamic (MHD) processes in the earth's fluid and electrically conducting outer core (Gubbins and Bloxham, 1987). The earth's main magnetic field is generally thought to be created by dynamo action in the earth's fluid outer core. The theory requires the electromagnetic induction processes, flow of electric currents in the earth's fluid and electrically conductive core. The outer liquid core, extending from a depth of 2900 to 5150 km, consists primarily of Fe alloyed with small amount of Ni and lighter elements (S, O and H) (Merrill and McFadden, 1990). The liquid outer core is bounded on one side by the solid/viscous mantle and on the other side by solid nickel-iron boundary of the inner core. The geomagnetic field originates due to spiralling motion of the molten iron-rich, and electrically conducting material in the outer core. Perhaps currents were initiated from chemical irregularities that separated charges and set up battery's action to generate weak current. It is believed that large currents are churning and rotating in the molten outer core, which eventually gives rise to the earth's magnetic field (Jacobs, 1994). However, any system of electric currents flowing through a bounded and electrically conducting body will decay. One of the questions, therefore, which bother geomagnetist is, how can a magnetic field be maintained indefinitely by the motion of an electrically conducting fluid, despite the continual drain of magnetic energy by the electrical resistance of the fluid (Roberts, 1988; Bloxham and Gubbins, 1987). This question encapsulates the fundamental competition between two processes (McFadden and Merrill, 1995). Accordingly, the field is altered and magnified by the shearing motions of the fluid in the outer core. This occurs because fluid is a good conductor and so carries the field line along with it. In effect, this is conversion of mechanical energy into electromagnetic energy via Lenz's law. Some initial field is of course needed. However, on the other hand the core is not a perfect conductor, so the field diffuses and therefore decays by dissipation due to electrical resistance.

The next basic requirement is knowledge of how the geodynamo is powered, i. e., what is the process that provides the mechanical energy for convection. Alternative idea has recently emerged that proposes that the fluid movement within the core is caused by the energy released when the inner and outer cores are separated. The convection might occur due to the cooling of the earth core. There is growing consensus that the inner-outer core is freezing boundary (Buffett, 1992) so that cooling of the earth leads to growth of the solid inner core with time. It is believed that upon freezing, the less dense elements are released and rise (Poirier, 1992; McFadden and Merrill, 1995). As the earth cools, growth of inner core by freezing preferentially extracts Fe from the outer core. As a consequence, liquid enriched in the less dense elements is released from the inner-outer core boundary. The chemical buoyancy of this liquid is currently thought to be the most likely energy source for driving convection and hence the geodynamo (Merrill and McFadden, 1990). Several others believe however that the dominant cause of outer core convection is compositional (or chemical) buoyancy derived from gravitational energy. Because of lack of knowledge of density contrast across the inner-outer core boundary and chemistry of the core, considerable uncertainties exist concerning the primary energy source.

Another view is that the energy produced through the radioactive decay of such elements as U, Th and K (which are thought to be contained initially to a certain extent in the core) may drive thermal convection in the metallic core. It is also suggested that the core was in molten state at the time of its formation consisting mainly of Fe and Ni as well as 20% of light elements (Jacobs, 1994 and reference therein). Later, Fe and Ni condensed and settled down to the centre of the core where they solidified and gradually formed an inner nucleus. The following electric current is free to decay with a time constant governed by the electromagnetic properties of the core. Therefore, a mechanism of electric current regeneration is required for the magnetic field to be maintained in spite of the free decay. Such a regeneration mechanism is demonstrated by Bullard using a simple disk dynamo model.

There are several other issues, which are discussed extensively by Merrill and McFadden (1990) and Gubbin (1981, 1987). For instance, whether the direction of rotation of the outer core changes or the current reverses its direction due to some reason or the other, the poles should interchange their positions. Whether the inner earth becomes hotter or cooler, as discussed above, in a quasi-periodic manner, creating abnormal convection currents in the outer molten core and whether the rotation of the earth becomes faster or slower causing disturbances in the outer core, or do large comets and asteroids strike the earth occasionally sending shockwave down to its interior to shake up the fluid core or tilting its axis, are the issues to be explored. But what causes these changes is still an enigma and matter of conjecture. Whatever may be the reason, it is widely recognised that reversals of geomagnetic field is most probably a result of physical processes occurring in the fluid core of the earth. The polarity reversals of magnetic field have very wide applications in understanding basic questions such as evolution of history of the earth. It is infact an important time marker of earth's evolutionary history and assumes great importance for understanding all the geo-processes of different time scales.

TIME SCALES OF GEOMAGNETIC FIELD VARIATIONS

The geomagnetic field varies on widely different time scales ranging from a second to millions of years. The full spectrum of geomagnetic field can be classified and recognised by different recording processes on different time scales. Here in this section, we discuss cyclic behaviour of only long-term geomagnetic reversal processes.

Geomagnetic Observations

It is known that roughly 90% of the magnetism at the earth's surface can be described by a geocentric dipole (essentially a bar magnet) tilted 11° with respect to the rotation axis (Merrill and McFadden, 1990). The remaining 10% of the magnetic field is non-dipolar at the earth surface. The resulting residual magnetic field (that is when the dipolar part of the field is subtracted from the main field) is a complex field consisting of several magnetic foci (where the field is vertical). For entirely dipolar field, geomagnetic pole (the extensions of the best fitting dipole) and the magnetic poles (where the inclination of the magnetic field is vertical) would coincide, and they would be 11° away from the geographic poles (Merrill and McFadden, 1990). The dipole field also changes with time. The recent studies, however, indicate that this rate has been much slower than the non-dipole field. Accordingly, during the past 150 years there has been a westward drift of about 0.05 degrees per year in longitude but no progressive motion in latitude of the overall dipole orientation. The moment of the geocentric axial dipole has been decreasing at about 0.5% per year. This decrease should not necessarily, however, be taken as an indication of an impending reversal. (Merrill and McFadden, 1990).

Although it is known that geomagnetic field has existed for more than three billion years, the historical record, which gives the valuable information on the nature of geomagnetic field, is barely three centuries long. Hence, study of palaeomagnetism is essential to elucidate the detailed nature of secular variations, excursions and reversals. Such studies will play a critical role for simulating and testing geodynamo models and are important since geomagnetic field reversal process has also provided an important time-marker of the recent geological history and the concepts of plate tectonics have used global palaeomagnetic reversals as crucial observations. When a rock forms, it typically acquires a primary magnetisation that is parallel to the external field at the time of formation. If the rock can be dated and primary magnetisation can be distinguished from secondary magnetisation (magnetisation acquired after the rock formed), valuable information can be obtained on the direction and on the intensity of the ancient magnetic field (Merrill and McFadden, 1990). There are several problems due to recording medium and rock magnetic problems discussed elsewhere (Jacobs, 1994) that can lead to considerable errors in construction of reversals chronology. However, marine magnetic anomalies have provided the best possible source of information about magnetic reversals for the past 165 m.y., i.e., the age of the ocean-floor that preserves a record of reversals. The volcanic lava flows and sea-floor sediments have recorded ambient geomagnetic field signatures at the time of their formations, like a tape-recorder. Recent advances made during the past decades provide opportunity to reassess geomagnetic field records and look for new paradigms to explain the records and link these to numerical formulations of the geodynamo (Bloxham and Gubbins, 1985). Past behaviour of geomagnetic field can be studied by palaeomagnetic time series, which arise from:

(i) the magnetic polarity time scale,
(ii) fluctuations in polarity bias and in the frequency of magnetic reversals,
(iii) records of the geomagnetic secular variation obtained from archaeomagnetic measurements on baked clays, lacustrine and rapidly deposited marine sediments, and
(iv) direct observation of the field elements.

The following points merit discussion:

(a) Long-term changes of the geomagnetic field: reversals frequency and palaeointensity through the earth's history.

(b) short-term changes in the geomagnetic field: high-resolution record and the relation to geodynamo simulation.

Geomagnetic Excursion and Secular Variation

There is evidence of irregular movements of the poles with even shorter durations and unestablished opposite polarities called excursions. Excursion occurs when the magnetic field deviates significantly, that is, 45° or more, from that of a geocentric axial dipole field. These features probably indicate the apparent local behaviour of reversals of the magnetic field. Although, a lot of work has been done on excursions, it is not yet clear whether they are aborted reversals or related to some other core phenomena (Coe et al., 1984; Merrill and McFadden, 1990).

There is also evidence for change in the magnetic field over time intervals of a few years to few thousand years, which is generally referred to the secular variations. There is evidence that the present magnetic field has been declining at about 5% per century (Merrill and McFadden, 1990 and references therein). This does not, however, mean that free decay is primarily responsible for that decline (McFadden and Merrill, 1995). Further, it may be noted that the field intensity has increased and decreased several times during the past 10,000 years most likely due to dynamical processes. This is also an intimate part of geodynamo process. The data on secular variations can be obtained by the direct, archeomagnetic and paleomagnetic observations. During the past decades, considerable efforts have been made to characterise the nature of long-term secular variation. Substantial theoretical as well as data analysis work has been carried out to understand the phenomena. (Jacobs, 1994) and references therein. It is not yet clear, however, whether the changes present in the short historical record are typical or not, since similar changes are also clearly apparent in the palaeomagnetic data. This study also requires good database, which is now rapidly reaching a stage where significant inferences can be drawn about the geodynamo.

Polarity Reversals of Geomagnetic Field

The magnetic field reverses polarity (i.e., geomagnetic north and south poles change places). The palaeomagnetic studies of the global rock samples covering a large span of geological time has established that periods of either polarity are systematically of longer or shorter durations. Palaeomagnetic records obtained from volcanic lava flows and deep sea sediments when decoded, revealed that the engraved geomagnetic field can be represented by a dominant component of an axial geocentric dipole, which exhibits two stable polarity stages—normal (present) and reversed polarity. The direction of the geo-magnetic field swings (during a polarity change) through about 180°, demonstrates polarity changes in a flip-flop pattern while transit duration is

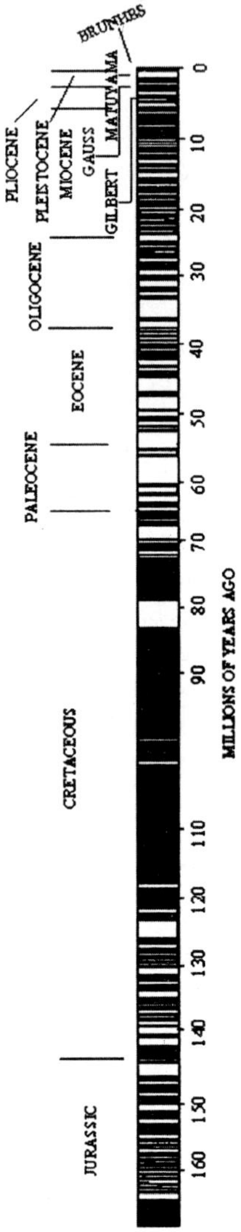

(a)

Eon	Era	Period	Epoch	Time Span	Duration (My)
Phanerozoic	Cenozoic	Quaternary	Holocene	- 0.01	0.01
			Pleistocene	0.01 - 1.6	1.59
		Tertiary	Pliocene	1.6 - 5.3	3.7
			Miocene	5.3 - 23.7	18.4
			Oligocene	23.7 - 36.6	12.9
			Eocene	36.6 - 57.8	21.2
			Paleocene	57.8 - 65	7.2
	Mesozoic	Cretaceous		65 - 140	75
		Jurassic		140 - 205	65
		Triassic		205 - 250	45
	Paleozoic	Permian		250 - 290	40
		Carboniferous		290 - 355	65
		Devonian		355 - 410	55
		Silurian		410 - 438	28
		Ordovician		438 - 510	72
		Cambrian		510 - 540	30

(b)

Fig. 4.9 (a) Polarity Reversal Time Scale (After Hoffman, 1988) for the Past 165 m.y. Black (normal) and White (reversal) Polarity (b) Geological Time Scale

(*Source*: http//www.geol.ucsb.edn/outreach/timescale)

short enough to be assumed as instantaneous on geologic scale, and shows highly variable stable-state-survival timings, ranging from less than 4×10^4 years to more than 10^7 years. The Palaeomagnetic observations thus indicate that the process of geomagnetic polarity reversal is instantaneous on geological time scale although it takes place gradually over thousands of years. There are major normal and reversed polarities of geomagnetic field. These are called 'geomagnetic polarity epochs'. Superimposed on these epochs are short-term fluctuations(events) in magnetic polarity called 'reversals wiggles'. The duration of these sub-chrones are as short as 20,000 years. A reasonably well-established geomagnetic reversal chronology for the past 165 m.y. is now available, which is based on the marine magnetic anomalies. These data reveal a very fine structure of undetected short polarity intervals (Hoffman, 1988).

There are conflicting views on changes in intensity during reversals of magnetic field. Van Zijl et al. (1962) and Kaponovich et al. (1966) have shown a marked reduction (about 10% of initial field) in intensity during the reversal. Dodson et al. (1977) have also found evidence of considerable fluctuations in intensity inside the core with little changes in direction. Kawai et al. (1976) found evidence of the onset of directional changes to be coincident within a sharp drop in intensity. In a palaeomagnetic study, a marked change in intensity associated with change in direction has been observed. According to Hammond et al. (1979), there is evidence of a decrease in intensity associated with the change in direction. This in turn supports the view of existence of non-dipole transition model. They have found a 10% difference (reduction) of intensity value in core before and after the transition. The finding agrees with their earlier works where they have reported a simultaneous change in intensity and direction.

Spectrum of the Geomagnetic Reversals Time Records

During the past several years, considerable efforts have been made to examine the nature of earth's magnetic field. Cox (1968) concluded that the process of polarity reversals is random. Crain et al. (1969), however, attempted to explore evidence for long-term cyclic pattern by using the Fourier Power Spectrum analysis of proportion of available normal polarity measurements during the Phanerozoic. They have proposed two periodicities of 75 m.y. and 300 m.y. and associated these cycles to movement of the solar system around the plane of galaxy. McElhinny (1971) compiled the worldwide palaeomagnetic data representing the overall proportions of samples exhibiting normal and reversed polarity during different time intervals. Using this, he inferred a cyclicity of 350 m.y. in the normal polarity history. Later, Ulrych (1972) constructed the normal polarity history curve using the data of McElhinny and performed maximum entropy spectral analysis to examine for evidence of periodicity. He suggested two periods of approximately 250 m.y. and 700 m.y. but did not find 75 m.y. cycle of Crain et al. (1969). Irving and Pullaiah (1976) have examined long-term palaeomagnetic data (land-based geomagnetic polarity bias data) for different geological periods in

Phanerozoic. The presentation of normal to reversal polarity bias indicate a marked difference between different geological periods. Irving and Pullaiah (1976) also performed identical MEM analysis to obtain periodicities of 300, 113 and 58 m.y. Their analysis have shown that all peaks are relatively insensitive to sampling and averaging intervals. The evidence for a period of about 300 m.y. is strong.

Evidently, the above mentioned analysis were based on the assumption that polarity reversals obey a sinusoidal process. It may, however, be mentioned that the polarity reversal record ideally resemble binary telegraphic wave-form with numerous discontinuities and abrupt changes. The physical behaviour as well as the signal characteristics of such an instantaneous process seems to be quite unrelated to that of sinusoidal signals. Obviously, the choice of a more appropriate technique for data analysis should match with the dominant signal characteristics. Keeping this in view, Negi and Tiwari (1983) examined the data presented by McElhinny assuming that a telegraph-signal type model is more realistic for presenting the behaviour of the reversal data and application of binary transform technique (based on Walsh functions) matching with the flip-flop reversal characteristics are more appropriate than sinusoidal based classical methods. The same logic and scheme as used earlier by McElhinny (1971) and Ulrych (1972) was followed to construct signal-wave form, which represents the duration of normal or reversed polarity bias corresponding to the proportion of time taken during each interval. The telegraphic signal model combines the full history of normal and reversed polarities in contrast to Ulrych (1972) and McElhinny (1971) where only normal behaviour is taken into consideration for analysis. Negi and Tiwari (1983) applied a more appropriate technique based on Walsh functions to the Phanerozoic geomagnetic reversal sequences. Walsh power spectra of telegraphic signal revealed integrated spectral structure, particularly five distinct peaks centred around 285, 114, 64, 47 and 34 m.y. It is interesting to note that these periodicities do include approximately the earlier periods obtained by Crain et al. (1969), McElhinny (1971) and Ulrych (1972) directly or as harmonics. The two periods of 34 m.y. and 285 m.y. were found above 99% confidence level according to Fisher's statistical significance test and also passed other peaks above 90% confidence level. The stationarity behaviour of the reversal sequences was examined by dividing the whole record into two sub-groups and calculated mean and variance. Similar spectral analysis by dividing the entire time sequence into two subsequences revealed more or less identical peaks. Evidence of periodicities was highly debated (Lutz, 1985) and hence alternative idea was explored to examine the reversal record.

Characteristics of Mean Rate of Polarity Reversals

Alternatively, the mean frequency rate of geomagnetic reversal is shown in Fig. 4.10 (Stothers, 1985). From this figure, it is clear that the present reversal rate is quite high, about 4–5 reversals per million years. From 83–12 m.y., the mean reversals rate increased steadily but may be decreasing now again (Fig. 4.10).

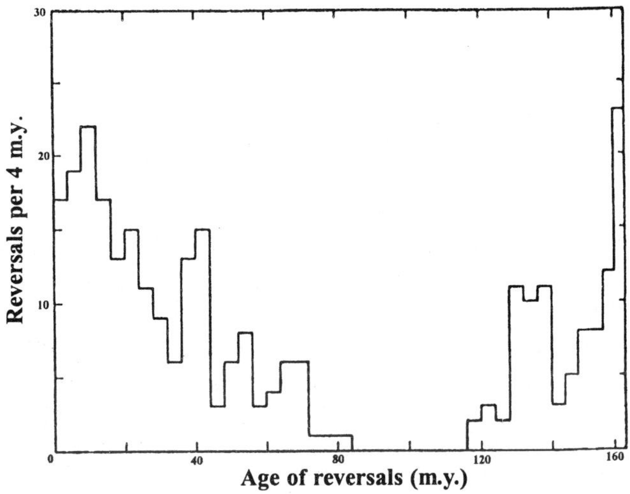

Fig. 4.10 Reversal Frequency Per 4 Million Years vs Age for 160 m.y.

Source: Stothers, 1985

It is interesting to note that this has not always been the case. For instance, the Cretaceous superchron from 83–118 m.y. represents one of the two well-documented long quiet intervals during which no reversals occurred. As can be seen from the above figure, the increase in reversals rate after the Cretaceous superchron is nearly mirrored by a steady decrease before the superchron (Merrill and McFadden, 1990). Observations also indicate that there is a larger interval of a reversed polarity of about 50–60 m.y. known as 'Kiaman reverse superchron' about 200 m.y. before the Cretaceous superchron.

Non-Parametric Method of Spectral Analysis

Raup (1985a) performed the non-parametric analysis of more updated reversal rate record for the last 165 m.y. and suggested evidence for a stationary 30 m.y. periodicity supporting the findings of Negi and Tiwari (1983). He analysed the pulses of increased reversal activity using a dispersion technique developed by Stothers (1979). He assumed magnetic reversals as isolated events, ignoring the bi-modality behaviour (from normal to reversal or vice versa). The residuals index is calculated using the formula: $(P\sqrt{(n^2-1)/12n^2}-\sigma)/P$, where 'P' is the period length, 'n' is the number of observations and 'σ' is the standard deviation of departure of the analytical procedure proposed by Stothers (1979). Lutz (1985), however, showed that the result of Raup (1985a) analysis is sensitive to the length of record and hence argued that there is no evidence of periodicity. He also suggested that periodicity observed in Raup's analysis is due to length of the data record. Lutz (1985) demonstrated, by progressively eliminating the most recent events, that the spectrum changes showing 30 m.y. periodicity are an accident of the record length. Further, Lutz and Watson (1988) demonstrated that long-term variation alone could account for most of the spectral evidences that have been interpreted wrongly as an evidence of 30 m.y. periodicity in the geomagnetic reversals record. They noted that a trend in time series analysis can seriously affect spectral estimates and that the reversal record also contains a quiet superchron (such as the Cretaceous normal superchron), roughly one cycle in length and unexplained by the periodicity models. One can separate the data into two parts, the longest of which would contain fewer than three cycles. Consequently, Raup retracted his claim (Raup 1985b). Stothers (1986) re-examined the reversal record in more details. He considered the criticism of Lutz (1985) and upholds the finding of Negi and Tiwari (1983) and Raup (1985a). Stothers performed a separate analysis for the two intervals (0–83) m.y. and 118–165 m.y. and explained 35 m.y. quiet interval (the Cretaceous normal superchron). He maintained that 30 m.y is statistically significant basic period term in geomagnetic reversals. In fact, Pal and Creer also used the same data only for 80 m.y. (Fig. 4.11(a)) and detrended it by fitting an appropriate procedure (Fig. 4.11(b)) and then used the detrended time series (Fig. 4.11(c)) for MEM analysis.

MEM is capable of resolving hidden cycle even in shorter time record. MEM

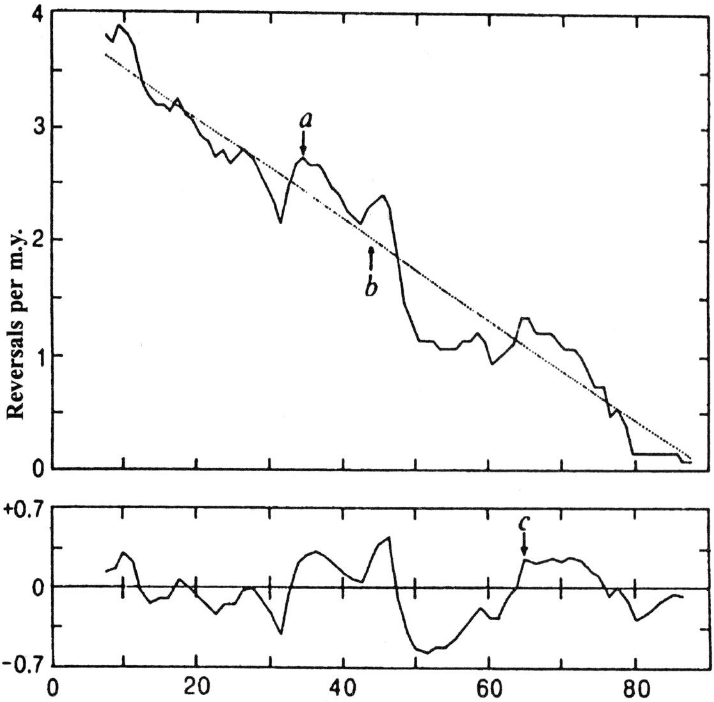

Fig. 4.11 Magnetic Reversal Time Series for the Past 80 m.y. (a) Number of Reversal Per Million Year (original data) (b) Best Fitting Line to Remove Trend (c) Detrended Time Series

(*Source*: Pal and Creer, 1986)

spectral analysis of reversal time series displayed in Fig. 4.11(c) yielded a 32 m.y. cycle (Fig. 4.12).

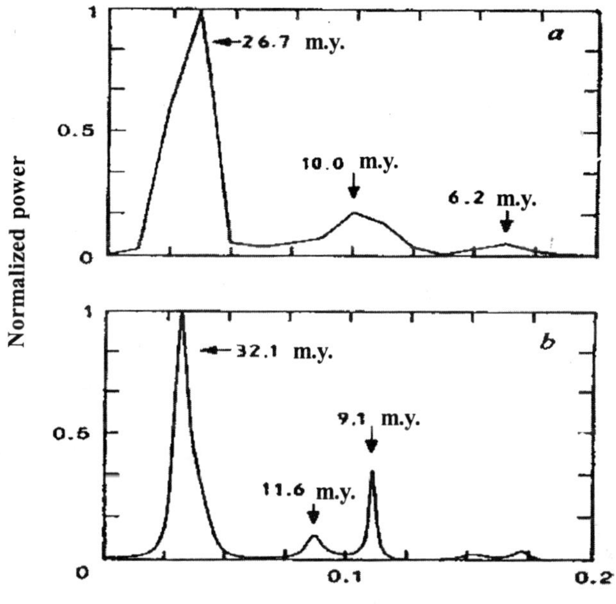

Fig. 4.12 Spectrum of Geomagnetic Polarity Reversal Time Scale for 80 m.y.
(*Source*: Pal and Creer, 1986)

It is interesting to note that the listed dates of known magnetic reversals (Harland et al., 1982) since the Cretaceous quiet intervals (83 m.y. ago) show a progressive increase in the reversal (high frequency) towards the present time. Mazaud et al. (1983) proposed a periodicity of 15 m.y. superimposed on a monotonic function of the rate of geomagnetic polarity reversals (Fig. 4.13). McFadden (1984, 1987), however, criticised the result of Mazaud et al. (1983) and argued that there is no such periodicity. In a more recent analysis, Marzocchi and Mulargia (1990a,b), Mazaud and Laj (1991) showed that there was a periodicity of 13–16 m.y. in the rate of reversals occurrence. They have analysed three most commonly used geomagnetic polarity time scales of LaBrecque et al. (1977); Lowrie and Alvarez (1981) and Berggren et al. (1985) covering a time span of the past 100 m.y. The authenticity of this periodicity was tested by comparing these polarity time scales with a large number of synthetic sequences produced by random process. They have concluded that detected periodicity is presumably not a simple statistical fluctuation of an aperiodic generator. It rather represents the long-term 'memory' of the geodynamo. However, the 15 m.y. periodicity found by

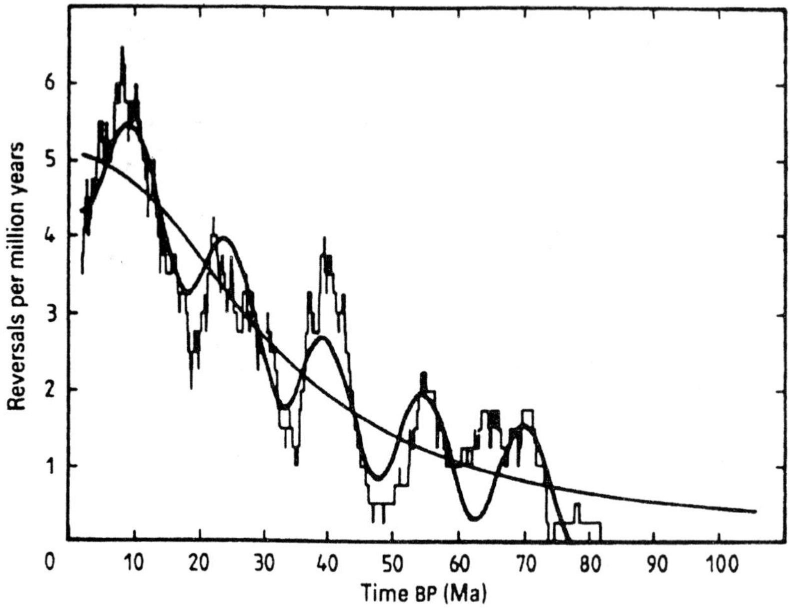

Fig. 4.13 Direct Least Square Fit of the Frequency of Reversals (in the Lavreque et al., 1977, Time Scale), Using a Lorenzian Function

(*Source*: Mazaud et al., 1983)

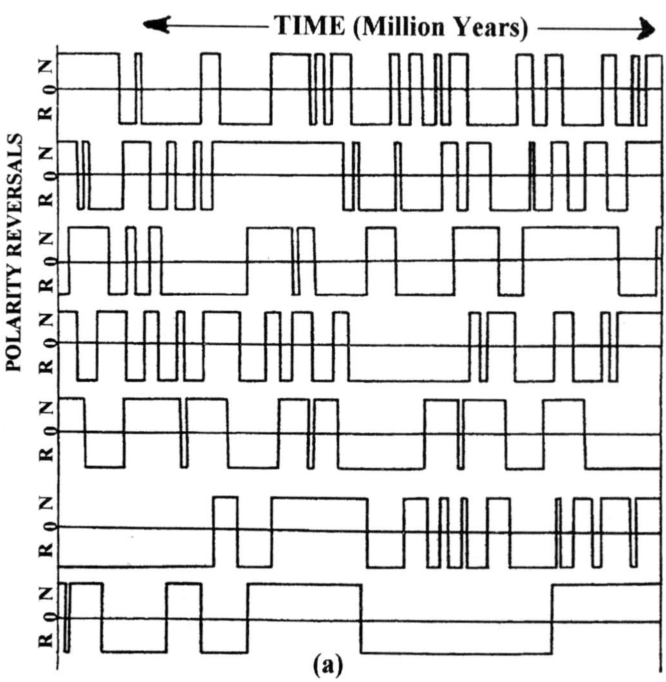

Fig. 4.14(a)

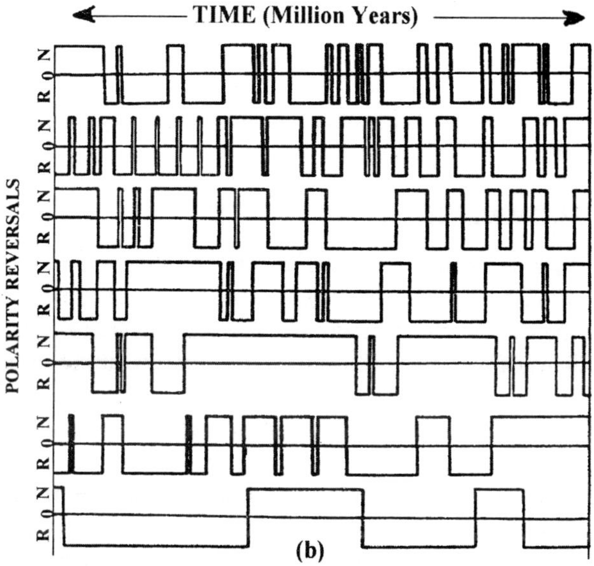

Fig. 4.14 Telegraphic Model of Polarity Reversal Time Scale for the Last 45 m.y.

Mazaud et al. (1983) in the record of the last 83 m.y. is probably a harmonic of 30 m.y. periodicity rather than a fundamental period.

Further, Negi et al. (1993) constructed telegraphic signal model of detailed record of reversals for the past 50 m.y. and used Walsh spectral analysis to show evidence of smaller periodicities of the order of 1.3 m.y., 0.60 m.y. and 0.41 m.y. (Fig. 4.15). Interestingly, these periodicities apparently seem to

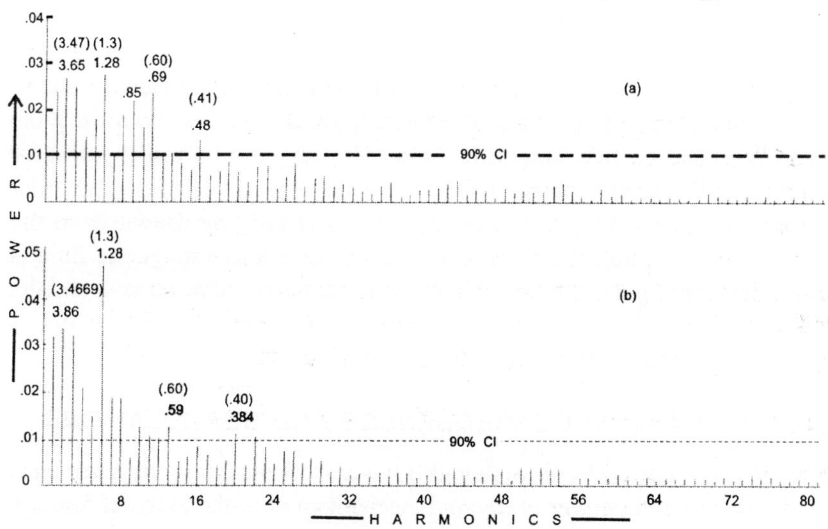

Fig. 4.15 Walsh Power Spectrum of Polarity Reversals Time Scale
(*Source*: Negi et al., 1993)

coincide with higher order eccentricity cycles. Some researchers performed a non-parametric analysis using the reversals frequency record, averaging the reversals per five million years, and conjectured the statistical reliability of these cycles. Such averaging, however, seems to be inappropriate because we are here searching not for the frequency rate but hidden frequency of individual reversal.

CHAOS IN GEOMAGNETIC REVERSALS

Rikitake (1958) was first to discover evidence of chaos in geomagnetic reversals. He presented a double disk dynamo model using ordinary differential equations. The model oscillates randomly between two states of fixed scales. A detailed analysis of model exhibits all characteristics similar to the chaotic attractor in non-linear forced dissipative hydrodynamic system (Cook and Roberts, 1970; Ito, 1980; Cortini and Barton, 1994; Kono, 1987). Further, additions of random noise and/or varieties of stochastic processes do not give clear picture as whether the physical system dynamics is governed by chaotic process or noise dynamics (Marzocchi et al., 1995). It is, however, favoured that system dynamics influence physically derived stochastic process for triggering reversals. There is suggestion that hydro-magnetic dynamo might exist in the core. According to this model, atleast two different acceptable velocity fields are presumed in which reversals take place and there is change from one wavelength field to another. Other models also suggest fractal nature (Consolini et al., 2000; Ivanov 1994 and Gaffin, 1989) and/or bifurcating solution that is evidenced in fluid dynamical system (Chandrasekhar, 1961; Robert, 1988). Robert (1988) proposed that reversals could be associated with catastrophic transitions between primarily geographic and magnetostatic states.

The Rikitake dynamo is a good tool for modelling reversals, however, the solution of the relevant system of differential equations can only be determined by numerical calculations. Chillingworth and Holmes (1980), considered the third order and have shown that there is evidence for a low-dimensional attractor in terms of catastrophic theory for modelling the reversals of the earth's magnetic field. The observed strange attractor behaviour of the earth's magnetic field indicates a system of atleast the order of three. Two important conclusions that can readily be drawn from the above analysis are that the reversal process of the earth's magnetic field is non-random, ruling out the earlier view of randomness of reversals, and the evidence of low dimensional chaos supports the view that the reversals sequences might possess order in the midst of chaos.

POSSIBLE CAUSES, CONSEQUENCES AND CORRELATIONS

There are mainly two schools of thoughts, which link the long-term geological periodicity/processes either to internal mechanism or to the external forcing. Some others also believe that the external forcing is a primary trigger to create earthly upheavals. We discuss here both of these mechanisms.

Internal Forcing Mechanisms

Plate Tectonic and Geo-magnetic Reversals

As we have seen above, there is possible evidence of atleast two quasi-periodicities of the order of 280 ± 40 m.y. and 33 ± 5 m.y. in long-term geological records. There have been several speculations on the possible linking of approximately 300 m.y. periodicity in reversal records with the time scale of plate tectonics. It was suggested that the magnetic field would reverse frequently during times of active convection and tectonics. Some researchers suggested that variation in the length of polarity originated in processes near the core-mantle interface (Courtillot et al., 1987). They correlated a 300 m.y. periodicity found in reversal record with the characteristic time scale of plate tectonics (Vogt, 1975; Irving and Pullaiah, 1976; Creer, 1975). Whyte (1977) suggested that discontinuities in the rate of change of earth's rotation may be the possible cause of geomagnetic reversals.

Several others also have suggested that small variations in core motion caused by changes in core mantle boundary condition could cause the polarity reversals of the dipole field. Some correlation between the rate of geomagnetic reversals, mantle convections and global tectonic activity exists (Courtillot and Besse, 1987). Hence, there is a suggestion that long-term changes in reversals can be used as possible indicator of time changes of core mantle boundary. There is also a suggestion that a strong correlation between the morphology of the transitional fields of the past 10 m.y. and the lateral structure of the core-mantle boundary exist (Mazaud and Laj, 1991). Larson (1991) estimated increase of 50–75% in formation of oceanic crust between 120 and 80 m.y. According to him, this increase both in the spreading rate from ocean ridges and in the age distribution of oceanic plateaus, which was concentrated in the Pacific ocean, coincides with the Cretaceous normal superchron during which earth's magnetic field showed no reversals. He notes that this could be a result of a super plume that originated about 125 m.y. ago near the core mantle boundary. It moved upwards by convection through the mantle and erupted beneath the Pacific. It is recognised by several workers that the core–mantle boundary region, in particular heterogeneous D-layer at the base of the mantle, is a likely site for plume generation (Stacey and Loper, 1983).

Larson and Olson (1991) argued that the rate of production of oceanic plateaus, seamount chain and continental flood basalt correlate inversely with magnetic reversals frequency (Fig. 4.16). They proposed deep mantle plume activity as causal link between these two phenomena. Figure 4.16 shows worldwide distribution of the rate of production of oceanic plateaus, seamount chain and continental flood basalts plotted with magnetic reversals time scale as a function of geological time for the past 150 m.y. However, there are contradictory statements regarding the physical understanding of Cretaceous superchron. One of the two groups suggests that superchrons correspond to inactive periods in the D-layer, when the heat transfer is relatively small

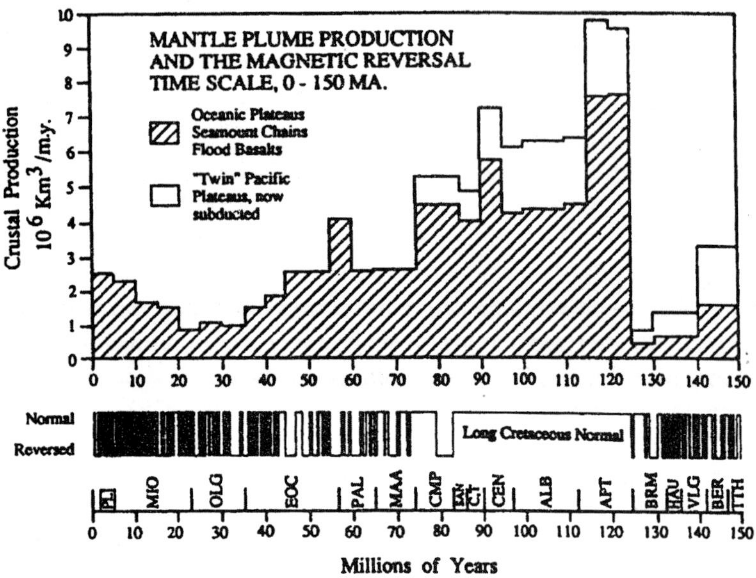

Fig. 4.16 Worldwide Distribution of Oceanic Plateaus, Seamount Chains and Continental Flood Basalts Plotted with Magnetic Reversals

(*Source*: Larson and Olson, 1991)

(Loper and McCartney, 1986; Courtillot and Besse, 1987). The other group however suggests that increasing the amplitude of convection in the outer core i.e., during the active period, increases heat flux from the core (Larson and Olson, 1991). In this model, superchron represent high energy states for the dynamo, and they occur when the heat transfer in D-layer is large. This is explained in Fig. 4.16, which demonstrates the response of core mantle boundary conductive heat flow and outer core convection to a super plume episode. The D-layer prime is thinned to fuel the plumes, raising its vertical temperature gradient. This raises conduction across the core mantle boundary which in turn raises convection intensity in the outer core. Convection intensity characterised by a magnetic Reynolds number, is inversely proportional to reversals frequency.

Core-Mantle Coupling and Geo-magnetic Reversals

An interesting new development in the study of mantle plumes is the probability of time dependent behaviour (Loper, 1991). An internal mechanism with the capability of triggering the initiation of plumes and having a relaxation period close to 30 m.y. has been suggested by Loper and McCarteny (1986) and Loper (1991) based on the stability of the D-layer by Loper and Eltayeb (1985). Evidence of a 30 m.y. periodicity is more prominent in almost all the records of geotectonic and geomagnetic reversals processes. Loper and his co-workers (1986, 1988) found a similar periodicity of the order of 30 m.y. in the mantle plume dynamics. Hence, they suggested that the apparent cyclic pattern observed in many of the earth's tectonic process could possibly be a result of internal core mantle dynamics. Different kinds of long and short-term quasi-periodic variations and, in particular, reversals of the earth's magnetic field may be the result of periodic core-mantle dynamics, in particular, mantle plume could lead to such correlated episodes of geo-tectonic activity (Loper, 1991). These workers have demonstrated their claim through the time dependent model studies of thermal convection in the earth's mantle. They have shown that at high Rayleigh number, the route to chaotic thermal convection in mantle may lead the process/and related processes periodically and/or quasi-periodically. In their model, Loper and McCartney (1986) proposed periodic nature of energy supply to geomagnetic dynamo. They suggested that the source of energy supply is directly related to the rate of cooling of the earth's core which in turn is controlled by the D-layer. According to this model for the thicker D-layer the temperature gradient across it is small and energy supply to the dynamo is low, causing a quiet state with few reversals. On the other hand, in the case of thinner D-layer, the temperature gradient across it is large leading to greater energy supply and resulting in a more disturbed turbulent state with more frequent reversals. They found a recurrence time of about 30 m.y. for this process, which coincides well with an identical reversals periodicity found in the geomagnetic reversals record. Sheridan (1983) also showed that reversals frequency is low during periods of plume activity and high when plumes are absent. However, an increase in the rate of reversals of the magnetic field is triggered by a thinning of the D-layer as heated material erupts away from it. This is just an opposite correlation to that of Loper and McCartney's observations.

A negative correlation between long-term eustatic sea-level change and geomagnetic reversal rate for the past 150 m.y. has been reported (Gaffin, 1987). It has been further noted that sea-floor creation rate and subduction rates are closely associated with the frequency of reversals. This model, however, does not explain some other abrupt phenomena, such as ridge jump and sea-floor spreading. This might then link to the sinking process of oceanic slab into the mantle. Another possibility that could also be visualised as a driving force for different processes is the changing configuration of the earth's plates (Worsely et al., 1985).

Forgoing discussions indicate an intriguing chain of quasi-periodic linkages between plate tectonics, mantle processes and outer core fluid dynamo (geomagnetic field). The plate tectonic and magnetic field changes are explained by 'mantle plume' theory. It is hypothesised that plume eruption from the deep mantle could affect activities of the core mantle boundary and overlying asthenosphere. Loper, 1991, discussed this in detail. Accordingly, the lithospheric plates slide freely on the low-viscosity asthenosphere. The plume eruption will speed up plate motions by possibly forcing plates apart. It will also decrease the asthenosphere viscosity to allow the plates to descend more quickly in subduction zones and to move rapidly in response to a given set of boundary forces on plate edges. The phase lag between initiation and cessation of plume eruptions in the lower mantle will quickly change the magnetic field. It is caused by the time it takes for plumes to rise through the whole lower mantle convection (Vogt, 1975; Sheridan, 1983). During the period of plume rise, the asthenosphere gets heated up while the outer core is being cooled down. In the return half-cycle, asthenosphere is cooled while the outer core is heated up.

Mantle Plume and Magnetic Reversal Frequency

It is widely known that the origin of the magnetic field of the earth and the processes of its reversal is found in liquid outer core. It is also known that origin of the plate tectonic motions and their spreading velocities are controlled by the properties and stresses in the lithosphere and asthenosphere. Larson and Olson (1991) demonstrated an excellent example of coupling of earth's interior processes including how does the mantle influence the outer core and how does the core influence the mantle. Figure 4.17 (Larson and Olson, 1991) shows response of core-mantle boundary conductive heat flow and outer core convection to a super plume episode. D-layer is thinned to fuel the plumes raising its vertical temperature gradient. This raises conduction across the

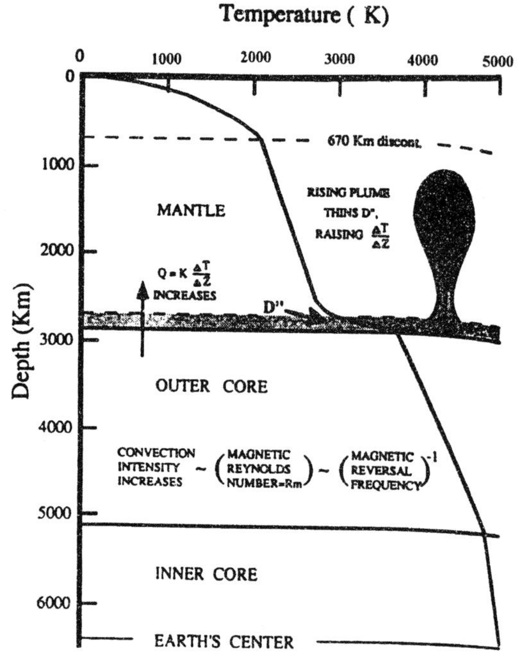

Fig. 4.17 A Schematic Diagram Showing Response of Core-mantle Boundary Conductive Heat Flow and Outer Core Convection to a Super Plume (See Text for Details)

(*Source:* Larson and Olson, 1991)

core-mantle boundary which in turn raises convection intensity in the outer core. Convection intensity, characterised by a magnetic Reynolds number, is inversely proportional to reversal frequency. (Larson and Olson, 1991).

EXTERNAL FORCING MECHANISMS

A wide range of exogenic factors also influence the earth's processes significantly. These are the movement of the solar system through the galaxy (Napier and Clube, 1979; Negi and Tiwari, 1983) and impact cratering on the earth induced by galactic forcing (Napier, 1989; Rampino and Stothers, 1984; Clube and Napier, 1982b). As has been discussed in Chapter 3, it has been proposed that the 'Cosmic Year Term' (cyclic revolution of the solar system around the plane of galaxy with a time period of around 285 m.y.) and half period (33 m.y.) of the oscillation of the solar system perpendicular to the galactic plane, both correlate well with earth's processes and have led to the major turning points in the earth's history (Negi and Tiwari, 1983; Rampino and Stothers, 1984; Alvarez and Muller 1984; Raup and Sepkoski, 1984). Further, it has also been suggested that the Oort clouds are strongly perturbed by the galaxy's molecular cloud system (Clube and Napier, 1998) and earth and all other inner planets are bombarded quasi-cyclically with a 33 ± 3 m.y. period by the galactic perturbation. (Rampino and Stothers, 1984).

Cyclic Impact and Geotectonic

Several lines of evidence suggest that major geological events are catastrophic in nature. A key question, however, is whether or not these catastrophies are extra terrestrially or internally driven. Specifically, meteorite and cometary impacts have caused or atleast triggered periodic recurrence of various geological processes (Napier and Clube, 1979; Clube and Napier, 1982) on the earth's episodes of mantle plume volcanism (Rampino, 1987). Several workers have suggested initiation of mantle-plumes by major impacts. Impact may serve as a possible mechanism to trigger plate motions (Seyfert and Sirkin, 1979; Jones et al., 2002; Ingle and Coffin, 2004). Several researchers have also suggested evidence of massive shock induced phase-transitions near the transition-zones. According to these researchers, the impact models do not contradict convection principles, on the contrary, they require them. Impacts are suggested to meet delicately balanced conditions, and to draw on latent mantle energies to initiate subsequent triggering (Burek and Wanke, 1988).

It has been claimed that large bolide impacts are the most energetic events that can perturb the earth's interior, convective circulatory pattern, geomagnetic field, break the crust and create major tectonic activity and atmospheric disturbances (Rampino, 1987; Melosh, 1989). So, as suggested by Rampino (1987), perhaps the major impacts on the surface of the earth might have been the principle trigger to tear the crust and subsequently to create the above said upheavals in history leading to various cataclysmic episodes.

Impacts might lead to basaltic volcanism directly by penetrating and deeply fracturing the earth's crust allowing magma to reach the surface. It can cause pressure-release melting in the upper mantle by the excavation and removal of crustal rocks, and thus triggering a rising plume of magma. (Rampino, 1987 and references therein). Very large impacts might also indirectly trigger volcanism at pre-existing volcanically active zones. O'Keefs and Ahrens (1987) show that for high acceleration and large scales, a deep transient bowl shaped crater is always produced. They conclude that the initial crater morphology is given by $d_c r = 0.002\ r_0^2$ (in eg. units, where 'd_c' is the depth of excavation, 'r' is the crater radius, for the earth's gravity and an impact velocity of 1 unit. It indicates that the large impact craters should have had deep initial penetration cavities (O' Keefs and Ahrens, 1987). According to these researchers, if the above relationship holds true, then the transient depth of excavation of 100–200 km diameter crater on the earth (as proposed for a 10-km diameter impactor at the K/T boundary) would be in the range of about 20–40 km; deep enough to penetrate easily through the 5–10 km thick ocean crust, and even deep enough to excavate much or all of the thicker continental crust. Hilderbrand and Boynton (1987) have demonstrated that the K/T eject layer has a substantial mantle component. Rare earth element abundance patterns in the boundary layer can be explained by a nature of ocean crust, continental sediments and oceanic mantle; the ejected mantle material indicates a minimum depth of cratering of about 40 km. These results, thus, suggest that the crater produced proposed K/T impactor could have been deep enough to generate direct volcanism in an oceanic impact and to create deep fractures in the lower continental crust (Rampino, 1987 and references therein).

Probable target areas for the largest K/T impactor lie beneath the Deccan Traps (India), the Iceland hotspot (the source of the Brito-Arctic basalt eruptions), or some other hotspot or oceans plateau (Alvarez, 1987 and references therein). Available geological records indicate that bolide impact, and flood-basalt volcanism have occurred repeatedly with specified intervals of ($\simeq$ 30 m.y.) over geologic time. There are few observations of impact-related features on the Indian craton where they are exposed near the traps. It seems that the mantle plume, possibly induced by large impact, must have triggered the eruption. However, it took several millions of years to ascend to the base of the lithosphere (e.g. Courtillot and Besse, 1987; Richards et al., 1989). Controversy surrounds, however, with regard to the timing of these events. It would have occurred more than several years before KTB time. Inevitable evidence of impact near KTB, however, suggest that it (impact) might have increased/accelerated the process of volcanism, even if such processes had already been initiated. It is also suggested that several impactors can hit the same area of the earth at different times over such a long time-span. Furthermore, the idea that large impacts can cause massive volcanism is based on the assumption that there is a layer of molten rock in the upper asthenosphere (Alvarez, 1987). The magma can be released by fracturing in the lithosphere and/or by removal of a volume of crust. The crater will lead to huge amounts of decompression melting. Although these ideas seem to be debated (White, and McKenzie, 1989), the effect of periodic volcanism and the coupling phenomenon of galactic forcing of the rhythms of the earth may introduce a very significant dimension in the understanding of the earth's history. The periodic cometary bursts during galactic crossings could be a primary cause of several geological rhythms of the earth (Rampino and Stothers, 1984). It is likely to be the direct and immediate cause for mass extinction boundaries.

Significant correlation has indeed been found between the major impacts and plate tectonics and its consequences on the palaeo-environmental activities. Modern version of catastrophism proposes that impact events significantly influence major earth processes. Most recently, Price (2001), in his monumental book *Major impacts and plate tectonics* suggested a model for the Phanerozoic evolution of the earth lithosphere through the astronomically induced impact events.

In a recent study, Isley and Abbott (2002) mentioned that both the rock records of mantle plume activity and terrestrial impact events are incomplete. However, in order to test the hypothesis that large impacts can strengthen existing mantle plumes or not, these researchers carefully examined the overall abundance of mantle plume activity versus the abundance of large impacts and also the temporal incidence of the strongest mantle plumes versus the temporal occurrence of large impacts. They found that large impacts do strengthen existing mantle plumes, more precisely, both the overall abundance of plume magmatism and abundance of very strong mantle plumes increase as large impacts increase. Further, Isley and Abbott (2002) performed cross-correlation analysis of all mantle plume volcanism and strong mantle plume volcanism versus the impact record. Their analysis reveal spectral peak in a range of 26 and 36 m.y. with their matching spectral characteristics for impacts (and mass extinctions) suggesting that the plumes (and mass extinctions) are being influenced by large impacts. Figure 4.18 Set-I—top shows spectrum of entire data and bottom time series of entire mg rock data set. Set-II—top shows spectrum of 256–0 m.y. data subset with age errors $\leq$ 7.5 m.y. and bottom time series of part of the data set spanning over 256–0 m.y. The dominant peak is at 34 m.y. Periodicity shows up particularly well in the youngest part of the data set. Set-III—top shows spectrum of mg data subset spanning over a period between 2856–2600 with age error $\leq$ 6 m.y. and bottom time series of 2856–2600 m.y. (Isley and Abbott, 2002). It may be mentioned that in Set I only spectral peak of 140 m.y. and higher are statistically significant whereas in Sets II and III spectral peaks at 26 m.y. or higher at 34 m.y. are significant.

They have concluded that the overall results of the cross-correlation and at spectral analysis strongly support a temporal relationship between large

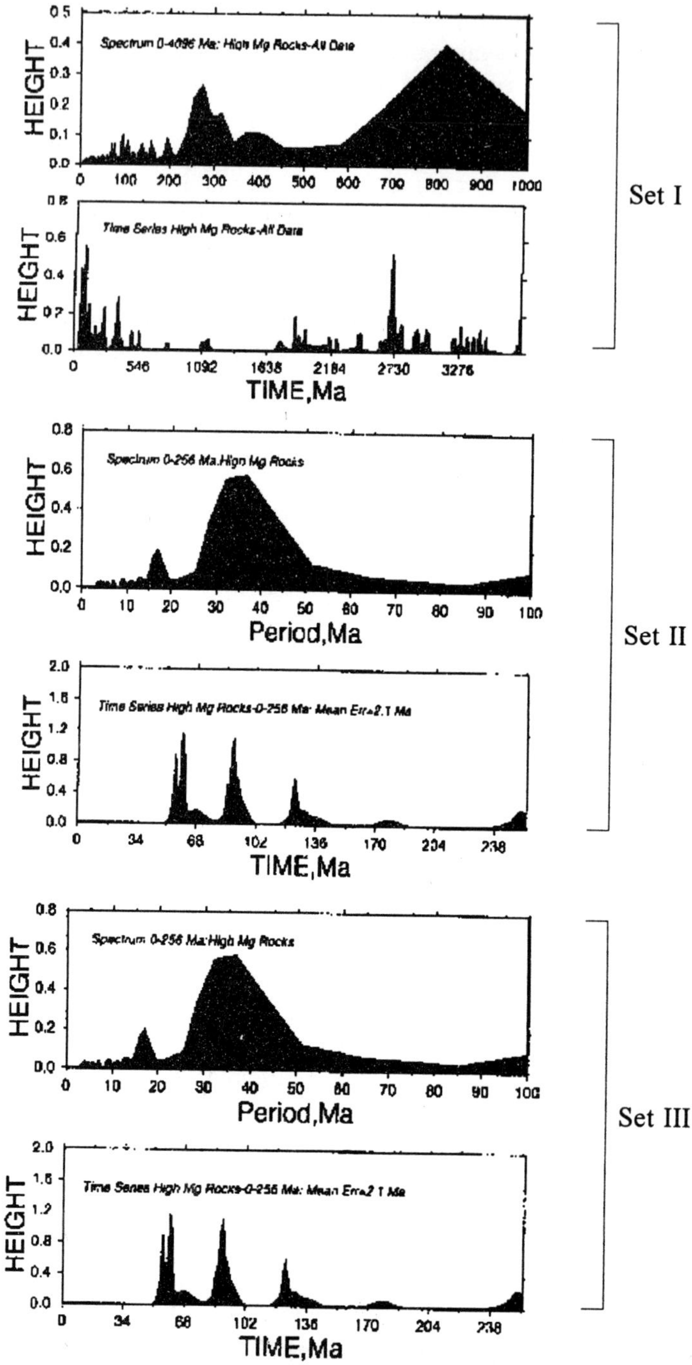

Fig. 4.18 Spectrum of the Entire and Subset of Data Series of Plume Activity. (See Text for Details)

(*Source*: Isley and Abbott, 2002)

impacts and resultant strengthening of mantle plumes. There is observational evidence also that supports a cause and effect relationship between large impacts and strengthening of existing mantle plumes (Isley and Abbott, 2002). According to these workers, the best documented is the temporal relationship between the K/T boundary impact at Chixculub and major phase of Deccan trap volcanism. The Deccan traps were active well before the K/T boundary impact. However, the most voluminous phase of Deccan trap volcanism occurred immediately after the impact, during chron 29R, which follows immediately after K/T boundary time. This volcanic phase produced the bulk of the Deccan traps and lasted less than 1 m.y. Thus, the Deccan plume was strongest immediately after the Chixculub impact (Isley and Abbott, 2002).

Isley and Abbott (2002) also proposed three non-exclusive mechanisms, but they do not rule out other mechanisms that may be equally plausible. Accordingly, impacts may cause cracking and distressing of the crust, allowing melts that have been trapped due to tectonic stress and/or impermeable boundaries to rise more easily to the surface. The impacts may also produce large cracks in the surface of the earth, allowing new plate boundaries to form with consequent thinner lithosphere and longer melt columns, and produce microdykes at the core mantle boundary.

There are several lines of evidence that indicate a temporal relationship between large meteorite and cometary impacts and stronger mantle plume activity (Rampino and Stothers, 1988; Isley and Abbott, 2002). Time series of mantle plumes and strong mantle plumes from 0 to 3.8 Ga correlate with a time series of large meteorite impacts at high confidence levels with a zero effective time lag (Isley and Abbott, 2002). Accordingly, there are four large-scale peaks in plume activity over the last 3.8 Ga that are directly correlated with the large-scale peaks in impact intensity. Spectral analyses of the data for all mantle plumes, for strong mantle plumes, and for meteorite impacts produce large spectral peaks at 32–36 m.y., coincident with the interval of vertical oscillation of solar system through the galactic plane. Glikson (2003), however, note that with existing data set it is somewhat difficult to propose extra-terrestrial link with mantle plume activity, but, Isley and Abbott, 2003 maintain that large body impacts do play major role in strengthening plume/magmatic and tectonic activities.

Finally, atleast one large impact, the 0.065 Ga Chixculub event, produced a consequent intensification of its associated mantle plume, producing the chron 29R portion of the Deccan traps. This latter evidence points to a direct cause-and-effect relationship between impacts of large meteorite or comets and a consequent strengthening of existing mantle plumes (Isley and Abbott, 2002).

Bolides Impact and Geomagnetic Reversals

A possible link between geomagnetic field reversals and steroidal impact events has been suggested (Muller and Morris, 1986). A steroidal impact on the earth's surface could physically disturb motions in the core and hence affect the earth's magnetic field. It has been pointed out that great field of tektites, covering Australia, Indonesia and a large part of the Indian ocean fell about 700,000 years ago at about the time of last magnetic reversals at the Brunhes-Matuyama boundary (Glass and Heezen, 1967). Merrill and McFadden (1988, 1990) however argued that an external origin for geomagnetic reversals seems to be unlikely although the possibility of such origin cannot definitely be ruled out. Muller and Morris (1986) suggested that dust from the impact crater and sooty particles from forest fires would change the climate and launch a little ice age. If the climate stayed cold for long enough time, sizable quantity of water in the form of snow and ice would be transferred from warm equatorial waters to the pole. This may cause difference in rotation rates to disrupt the convection flows and thereby the driving force of any dynamo. A similar idea has also been proposed by Doake (1977) who has noted that melting of ice on the surface of the earth may change the moment of inertia of earth's rotation resulting in the reversals of the earth's magnetic field. Glass and Heezen (1967) postulated that impacts could reverse the earth's magnetic field. It is indicated by the occurrences of tektites of the Australian Strewnfield at the base of the Brunhes/Matuyama reversal and tektites of the Ivory Coast Strewnfield at the base of the Jamarillo Event.

Possible link between impact and subsequent field reversal raise the question as how an impact might affect the motion of distant earth's core. Several researchers argued that impact-generated shock-discontinuity cannot reach the core without efficient chain-reactions from the higher mantle to the core. For this reason, Burek and Wanke (1988) suggested that phase-transitions near and/or at the transition zone are triggered by impact–shock associated with a density increase and a volume decrease. This effect could then cause a morphology disturbance (hill formation) at the core-mantle boundary leading to the earth's magnetic field reversals.

Currently, it is somewhat difficult to assign a definite link as to what causes the frequent reversals of the earth's magnetic field. There are two different conceptions: the popular endogenic-reversal hypothesis and the exogenic and impact-reversal hypothesis (Glass and Heezen, 1967). To exclude one of the possibilities and strengthen the cosmic theory of reversals, we need to find more examples of time-coincidence between impacts followed by a reversal (impact-triggered reversal hypothesis (Schwaarzschild, 1987).

CONCLUSIONS

Modern view based on the plate tectonic theory and recent periodicity analysis of long-term tectonic and earth's interior processes as discussed, emphasise that all geophysical and geological processes starting from the surface of the earth and down to the core possess common quasi-periodicities and hence emphasise the fact that these processes are orderly coupled and interrelated systems (Tiwari and Rao, 2003). In particular recent studies of various earth's processes have invariably shown a common cycle of about 33 ± 5 m.y. in the solid earth processes. Since evidence for a common 33 m.y. periodicity is based on various new updated geological records, which has also been analysed rigorously using the robust statistical techniques there is a need now for a unified mechanism to explain different physical processes. Although some controversy remains regarding the existence of long-term harmonic components in these records, the above studies using high quality of interdisciplinary data series and diverse scientific approaches establish the evidence for cyclicity and help understand the issues of external and internal forcing in the earth's evolutionary history. The quantitative analysis using high quality of data and sophisticated techniques considerably illuminated further knowledge of their physical linkages both with respect to external and internal forcing and thereby provide sufficient knowledge to address the question of periodicity in earth processes. In particular, strong support of possible link of impact cratering and geological processes add a new dimension to our understanding of earth and cosmic relationship. These recent results will provide far-reaching implications for fine-tuning of geophysical records and improved correlation studies (Rampino and Caldeira, 1993). Such quantitative analysis are crucial for addressing the underlying causes of evolutionary geological processes.

5

Long-Term Sea Level, Climate and Bio-Geo-Chemical Cycles

INTRODUCTION

Global sea level and climate changes and oceans, and geochemical processes are all coupled and intimately linked phenomena. Hence, the study of any one of these cannot be taken in isolation. In this chapter, we study the cyclic behaviour of these various processes and discuss their interlinkages. Understanding the behaviour of sea level fluctuations has important implications for organic productivity of the oceans and sediments distribution pattern along the continental margins and in the interior basins. Therefore, the study of periodic behaviour of these changes is of primary interest to hydrocarbon exploration (Haq et al., 1987). Long-term fluctuations of sea level changes have significantly influenced the sedimentary environment and ocean-chemistry. These changes have also influenced the fluctuations of global atmospheric CO_2 level and climate, significantly. Evidence suggests that increase in atmospheric carbon dioxide may have produced oceanic anoxia and global warming. Hence, a common cycle may be anticipated in these processes. Further, the changes in sedimentary environment and chemical composition of seawater have direct/ indirect bearing on the formation of carbonates including dolomite (Lumsden, 1985) and other rock types. Long-term global CO_2 variation has been linked to the earth's interior (lithospheric dynamics, mantle rheology etc.), oceanic (sea level changes) and biospheric (mass extinction) processes. Long-term global CO_2 variation has also significantly modified the earth's climate. A meaningful quantitative analysis of long-term ocean atmospheric and geochemical processes therefore may provide some clues to understand the coupling behaviour of various earth's processes and evolutionary climatic change and other related external/internal forcing mechanisms.

SCALES OF SEA LEVEL CHANGES

Global sea level changes have occurred on different time scales ranging from a few years to a few hundred million years. The amplitudes of sea level changes have also varied from hundreds of metres to few centimetres depending upon the time intervals. For instance, very long-term sea level changes on the time scale of $0 - 10^7$ m.y. occurred upto about hundred metres amplitude (Vail et al., 1977). During the past 2 m.y., short-term changes in sea level (Pleistocene and Holocene period) occurred upto about tens of centimetres in amplitude (Haq et al., 1987). Historical changes at time scale of few months/days to a few hundreds of years occurred in millimetre only (Pirazzoli, 1989; Meier 1990). Scientists have associated these changes to a vast set of physical forces, including both endogenic and/or exogenic origins. However, due to multiple forcing involved on different time scales, the precise forcing mechanism of these changes are not known (Cartwright, 1985). In this chapter, we confine ourselves to long-term cycles i.e., of the order of million years. The other scales of changes and variability will be discussed in the subsequent chapters.

LONG-TERM SEA LEVEL CHANGES RECORD

The records of long-term sea level changes are decoded indirectly through analysis of the past remains of sedimentary and bio-facies distributions. Long-term variations are also identified from evidence of raised beaches, emerged coral reefs, facies boundaries and stratigraphic unconformities. Variations in the facies and movement of the strand line are also possible indicators of relative sea level changes (Coleman and Roberts, 1988). A reliable construction of the global sea level curves requires correlations of facies changes over large areas located in regions of different tectonic settings. Mapping of the succession of transgressive and regressive facies is a common picture of sea level rise and fall (Vail et al., 1977). The depth of deposition of sediments and thereby succession of transgressive and regressive facies patterns are, influenced by mainly three factors — sea level changes, sediments supply and local tectonics leading to subsidence or uplift (Haq et al., 1987). Thus, the mapping of the transgressive and regressive pattern cannot provide absolute measure of sea level rise and fall unless these influences are appropriately removed. Further, the migration of continents through the equatorial zones, changing geoid, the vertical movements of large crustal areas, orogeny and global tectonics are some of the major problems in the reliable construction of sea level changes curve (Haq et al., 1987). Despite these various difficulties, tremendous efforts have been made to produce fairly reliable data on long-term scales. A major breakthrough came during the seventies when a research team led by Vail and his colleagues (1977) recognised that there is apparent parallelism between stratal surface and unconformities in geological sections. Taking this as a basis, these researchers proposed that depositional sequences bounded by unconformities and their correlative conformities represent primary units with chronostratigraphic significance. They applied sophisticated technique based on concept of seismic stratigraphy for determining relative changes in sea level. In particular, they have used stratal geometries and patterns of onlap, down lap, truncation and basinward shifts of coastal onlap to interpret sea level histories along various continental margins. They have constructed a series of charts showing global cycles of relative changes in sea level based on the apparent synchroneity of sea level falls in widely separated basins. First, they generated regional chart of landward extent of coastal onlap by mapping onlap and off-lap patterns within seismic sequences. A large number of regional onlap charts were averaged to estimate worldwide sea level. Vail et al. (1977) presented the longest sea level changes record covering the period of 0–600 m.y. The record of Vail et al. although good, has been subject of criticism (Christine Blick 1988; Christine-Blick et al. (1988); Mathew, 1988) on the following grounds:

(i) potential error in estimating the magnitude of sea level rises and falls,

(ii) the timing and global synchronicity of some of the major events and their significance in the deep sea, and

(iii) the need for updating the sea level curve in view of recent developments of time scale.

In view of the above, Haq et al. (1987) constructed a relatively better synchronised sea level curve taking into consideration the above problems. The resulting sea level curve has a better time resolution as it is based on multiple dating techniques such as magnetostratigraphy, biostratigraphy and chronostratigraphy (Fig. 5.1). The sequence-stratigraphic depositional models,

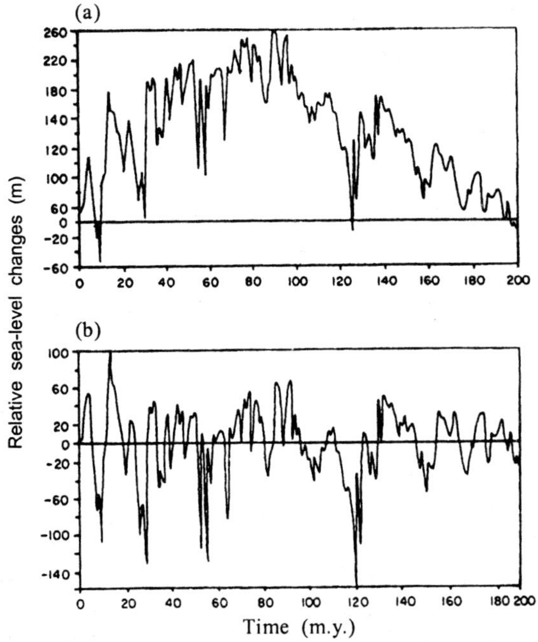

Fig. 5.1 (a) Relative Sea Level Changes Record for the Past 200 m.y. (b) Detrended Record of the Figure (a)
(*Source:* Haq et al., 1988.

together with detailed palaeontological data provided by Haq et al., enhance the ability to recognise genetically related sediment packages in outcrop sections and thereby provide an independent avenue to integrate and augment seismic and diverse surface data. In this approach, sea level rises and falls are manifested by specific physical surfaces that can be used to identify sequences in land-based and offshore marine sections.

LONG-TERM CYCLES IN THE GLOBAL SEA LEVEL CHANGES

Vail et al. (1977) have inferred three megacycles of varying orders of magnitude: super cycle (200–300 m.y.); cycles (10–20 m.y.) and paracycles (1–10 m.y.). Fischer and Arthurs (1977), Ager (1981) and Domon (1971) have suggested qualitatively a mean period of 32 m.y. for Cenozoic, 30 m.y. for Mesozoic and 36 and 11 m.y. for Phanerozoic periods respectively. Several researchers have also suggested very long-term cycles of changes connected to the geo-tectonism and/or changes in oceanic ridges volume etc. (Hallam, 1977). The possible internal sources include tectono-eustacy, orogeny, sea floor spreading, glacio-eustacy, etc. However, earth is recognised as an open system in its galactic environment and has extra terrestrial interactions. These interactions also occur on time-scales of hundreds of millions of years as discussed in chapter 3. They can be deciphered in long-term earth observations and possibility of their linkage could also be explored directly or indirectly. Various spectral studies have been performed on the sea-level changes data from different time scales to explore the evidence of cyclicity.

SPECTRUM OF LONG-TERM SEA LEVEL CHANGES

Record of Vail et al. (1977) shows a typical sinusoid superimposed on which there are some smaller oscillations/harmonics of saw-tooth nature (Tiwari, 1985). These mixed harmonic responses cannot be identified visually as have been done earlier. Hence, to search for cyclic patterns, time series analysis techniques were used. Spectral analyses (Fourier and Walsh) have shown evidence of periodicities centring around (380 – 450 m.y.), 142 m.y., 69 m.y., 43 m.y. and 32 m.y. (Tiwari, 1985). It is interesting to note that some of these periodicities are also similar to those cosmic periodicities discussed in chapter 3 and also found in terrestrial record of geomagnetic reversals as suggested by Negi and Tiwari (1983) and in various other geological records (Rampino and Stothers, 1984).

Negi et al. (1990) performed clean spectral analysis to a more detailed and precise sea level record of Haq et al. (1988) for the last 200 m.y. (Fig. 5.1) to infer periodic evidence, if any. Some researchers however, questioned the tiny 'wiggles' present in the sea level change data of Haq et al. (1988); Gradstein et al. (1988); Mathew (1988). Haq et al. (1988) have maintained that these smaller boundary sequences are accurately resolved and the published sea level record is the best record as yet available for analysing long-term harmonic components. However, the record displayed in Fig. 5.1 (upper curve) clearly shows a visible half sinusoidal trend. To minimise this effect in the spectral estimates it is prudent to remove the sinusoidal trend by appropriately fitting different orders of polynomial curve to the original data. Trying different order of polynomials for better fitting of the data will remove doubt of any dubious spectral peaks. The original sea level record and the detrended curve for the third order polynomial are shown in Fig. 5.1 (bottom curve). The detrended data was used for clean analysis. The results of Fourier and clean spectral analysis, as discussed in chapter 2, are presented in Fig. 5.2(a and b), (Negi et al., 1990). In Fig. 5.2(a) the raw Fourier spectrum of the sea level variation

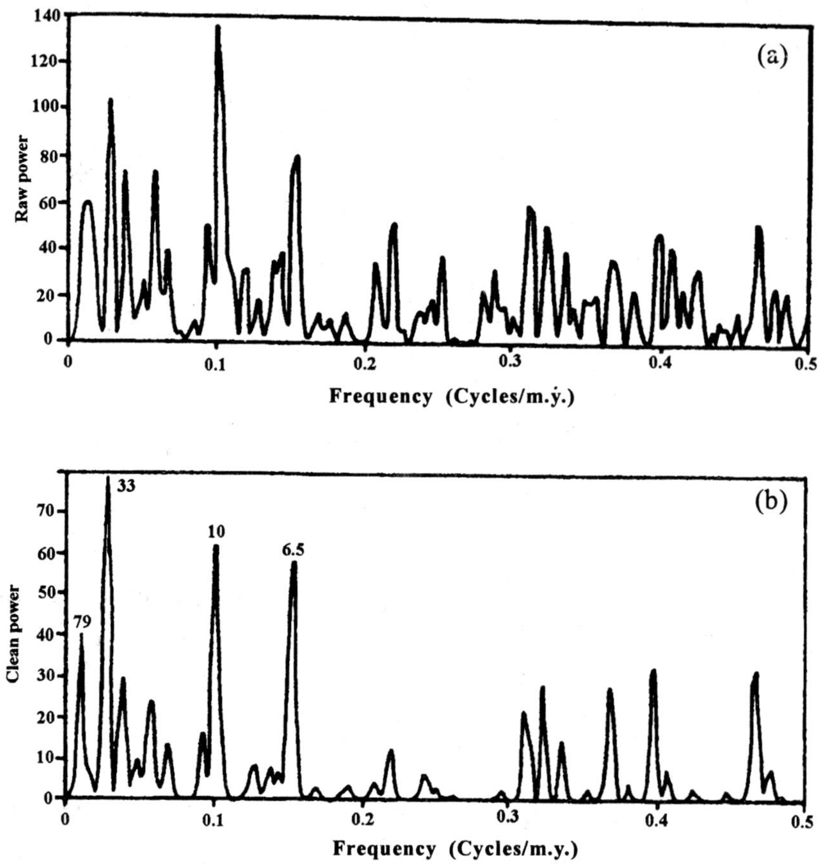

Fig. 5.2 Spectral Analysis of Sea Level Changes Record Shown in Fig. 5.1. (a) Raw Fourier Power Spectrum. (b) Clean Spectra.

(*Source:* Negi et al., 1990, Peaks Shown in Clean Spectra are in m.y.)

record exhibits a noisy spectrum with spectral power at a number of frequencies with considerably high amplitudes at 30, 16 and 10 m.y. The corresponding clean spectrum reveals a dominant spectral peak at 33 m.y. The physical significance of 30 m.y. sea level changes cycle in conjunction with similar other cycles in terrestrial and extraterrestrial processes has significant implication on climate and envirnomental changes.

Further, in order to have some idea of high frequency variability, the sea-level changes data of Haq et al. for the last 30 m.y. was also analysed. The clean spectral analysis of the sea level changes records for the past 30 m.y. separately reveal apparent periodicities in the range of approximately 2 m.y. to 400 k.y. (Fig. 5.3). The cycles are similar to those found in astronomical cycles of orbital eccentricity variations (chapter 3) and also reported in several other terrestrial records of climate variations. High frequency sea level cycles seem to originate from the glacial-interglacial fluctuations induced by orbital varieties. This will be discussed in more detail in chapter 7.

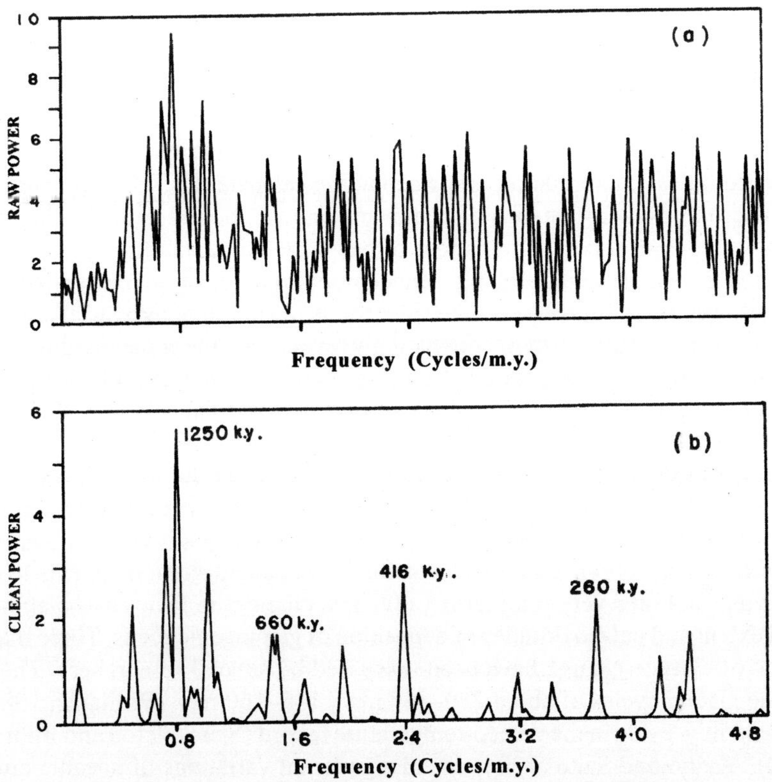

Fig. 5.3 Spectral Analysis to Examine High Frequency Component of Sea Level Fluctuations Using the Data for Last 30 m.y. (Data from Haq et al., 1988). (a) Raw Spectrum (b) Clean Spectra

(*Source*: Negi et al., 1997)

LONG-TERM CLIMATE CYCLES

Evidence of a climatic change decoded from the past remains preserved in the earth formations suggest that earth's climate has changed over time scales ranging from a few thousand years to millions of years. The history of global climate reveals that the climatic conditions that prevailed over last few hundred million years differed markedly from that of the present. The geological history has witnessed most dramatic climatic changes. Evidence indicates that ancient climates of 20 million years BP have been generally warmer than today. The glaciation and interglacials in the Northern hemisphere have fluctuated considerably since then. The past two million years Pleistocene period have seen a periodic sequence of glacial interglacial epochs (Hays et al., 1976). The coming of a new interglacial epoch during recent time was quite sudden. The Holocene (10,000 years B.P.) was accompanied in many parts of the world by fluctuating temperature and increase in precipitation, especially in the sub-tropical deserts of Northern globe. Here we discuss only long-term cycle of order of 10 m.y. and greater. Shorter term climate cycles will be discussed in chapters 7 and 8.

Data Source and Climate Cycles

The proxy record of climatic history of past few million years can be extracted from ecological interpretations of palaeobiological materials, which are mainly the macro-and-micro remains of plants and animals buried in sediments and rocks in geological past. The pattern of past climatic variability may also be obtained from radio-carbon dating, dendochronology, tree rings, oxygen-isotope ratio of oceanic sediment-cores. Besides these, variation of terrestrial climate during geological time is most convincingly revealed by the apparent long-term episodicity of major glaciation. Long-term climate change and associated external and internal physical mechanisms are discussed in detail by Williams, 1981 in his book, *Mega Cycle*. Franlkes (1979) has suggested that at least seven major episodes of terrestrial glaciation occurred during the past 2300 m.y. and perhaps six in the past 1000 m.y. These glaciations seem best regarded as discrete episodes separated by long intervals marked by substantially warmer climates and the absence of ice sheets. Such episodic glaciation appears to be superimposed on a very long-term ~ 600 m.y. changes in facies associations, areal extent and palaeoaltitudes of deposition of glaciogenic rocks. Three main cycles of climate change have been discussed by several researchers. These are the glacial cycles of about 250–300 m.y., 145–160 m.y. (Williams, 1981) and a 30 m.y. cycle in the palaeo-temperature record (Shackeleton and Imbrie, 1990). Kaiho and Saito (1994) found significant variations of oceanic crust production corresponding with variation in the $\delta^{18}O$ of deep water benthic foraminifera, which is proxy for climatic indicator. They reported that maximum value of oceanic crust production correspond with minimum values of $\delta^{18}O$ and minimum production with maximum values of $\delta^{18}O$. They have suggested that such minima and maxima reoccurred quasi-cyclically at regular interval of 33–38 m.y. for the past 100 m.y. (atleast). They have also suggested that changes in land–sea relationships and atmospheric CO_2 related to major fluctuations in oceanic crust production were the main cause of mid-Cretaceous warming and late-Cretaceous cooling. As can be seen, periods marked by high sea level, climate warming, and low-dissolved oxygen phases (X, Y, Z in Fig. 5.4) can be recognised during the latest Cenomanian-Turonian, latest Palaeocene-early-Eocene, and late Oligocene-early Miocene. These times have a periodicity of 33–38 m.y. The quasi-cycles may approximately correspond to quasi-cycles (~30 m.y.) of major geological events demonstrated by Rampino and Calderia (1993).

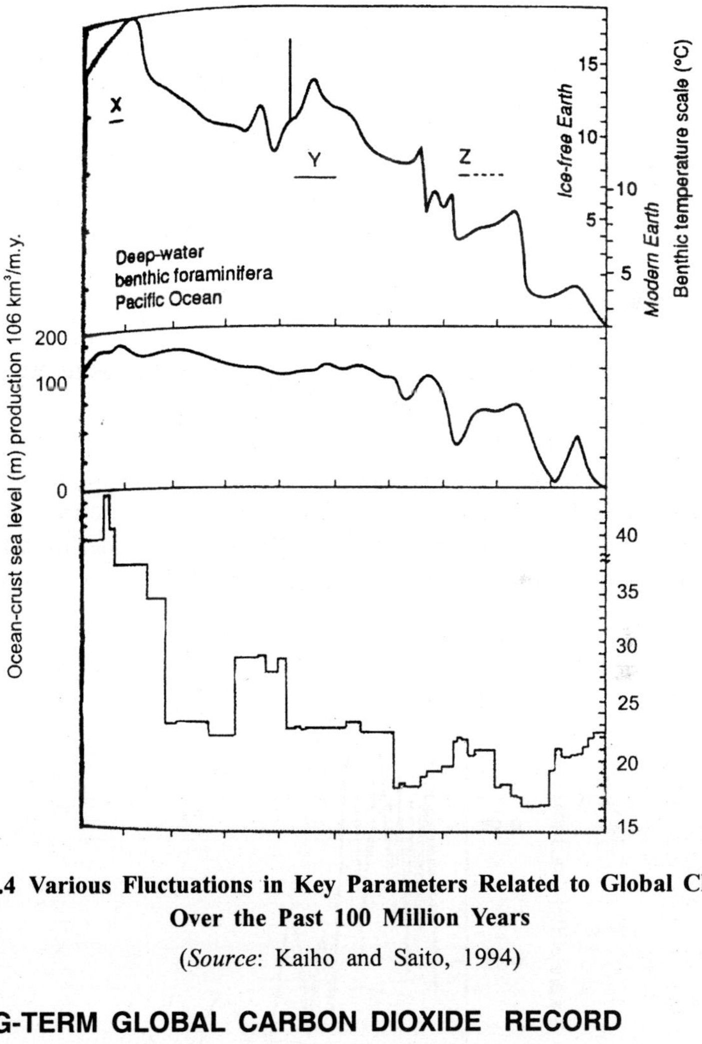

Fig. 5.4 Various Fluctuations in Key Parameters Related to Global Climate Over the Past 100 Million Years

(*Source*: Kaiho and Saito, 1994)

LONG-TERM GLOBAL CARBON DIOXIDE RECORD

The mass of atmospheric CO_2 in the geological past has been consumed in sediments through the processes of weathering and sea level changes. The relative changes in the volume of carbonate rocks in sediments should, therefore, reflect changes in the atmospheric CO_2 (Budyko and Ronov, 1979). A complete inventory of existing sediments on the platforms, geosynclines and rise (pelagic sediments) has been presented by Hay (1985). Some of the carbonates occur in other sedimentary rocks as fossils, cements and minor interbeds. The mass of CO_2 present as carbonate in other sedimentary rocks and carbon contents in pelagic sediments other than carbonate oozes is also included (Hay, 1985). The global and continental CO_2 accumulation rates provide information about the nature of climate evolution over a range of geological intervals. The available CO_2 record entails enough precision with respect to homogeneity and completeness, atleast, for the younger part of the Phanerozoic where much information from the continental margin and open ocean are available. The existing record of Hay (1985) has also been updated based on other recent publications, including Compton and Mallinson (1996).

Visual inspection reveals rhythmic fluctuations with high CO_2 accumulation rates during the Pliocene (2–5 m.y.), Eocene (37–58 m.y.), upper to mid-Cretaceous (66–98 m.y.), lower Permian (258–286 m.y.), Missisippian (320–360 m.y.), upper Cambrian (505–523 m.y.) and lower Cambrian (540–570 m.y.). By considering the timing of initiation of high CO_2 accumulation rates, it appears that the intervals are separated by more or less constant time intervals with low accumulation rates. All these anomalous events are seen to be unbiased and have been recognised as time of high CO_2 turnover and manifested in the connected geological phenomena.

The result of time series analysis using the non-parametric method (Stothers, 1979) of episodic CO_2 data variation is shown in Fig. 5.5. Analysis indicates an apparent 30 ± 3 m.y. cycle (Tiwari and Rao, 1998) rendering a further evidence of long-term common cycle connected with various geo–ocean–atmospheric processes.

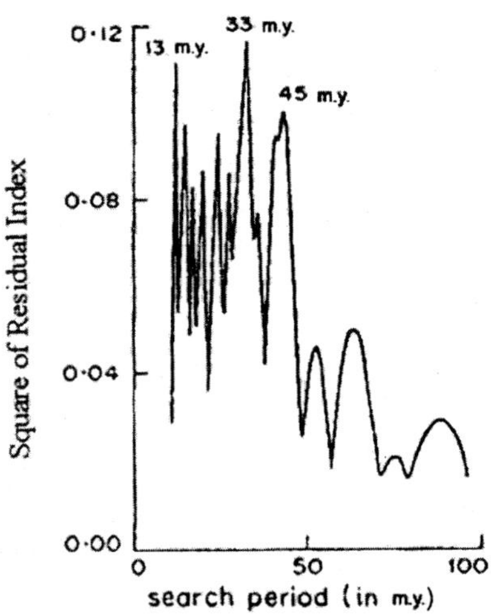

Fig. 5.5 Non Parametric Fit to the Episodic CO_2 Events Given in Table. Best Fit to the Data was Obtained for 33 m.y. Periodicity

(*Source*: Tiwari and Rao, 1998)

Spectral Analysis of Major Sedimentary Episodes

Rampino and Caldeira (1993) analysed some major episodes of geologic changes for the past 260 m.y. to find periodic component and correlation among them, if any. For that, they have compiled the combined data of the major episodes of extinction events, sequence boundaries, flood basalts, orogenic events, sea-floor spreading, oceanic anoxic events and evaporite deposits (Fig. 5.6a and Table 5.1). For the time series analysis they initially used a 10-m.y. moving-window, centred every 0.5 m.y. to the combined age data of the various phenomena and computed the number of dated occurrence that fell within the moving window at 1 m.y. interval. In order to test the effects of window size on the location of these peaks, they also used varying moving windows from 2 to 15 m.y. The peaks remained stable in position despite variations in the size of the moving window. In addition to the above method, they used a second independent method to treat each data in the record as a delta function

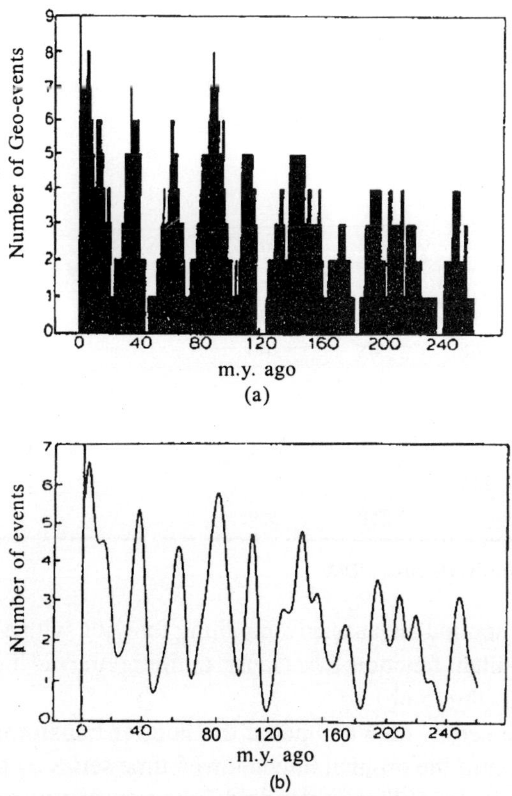

Fig. 5.6 (a) Number of Geologic Events Listed in Table 1, Within a 10 m.y. Moving Window. (b) Number of Geologic Events, Aggregated Geo-events, Weighted by a Gaussian Function with a Scale Length of 5 m.y.

(*Source*: Rampino and Caldeira, 1993)

Table 5.1 Dates of Major Geological Events in the Earth's History

(Age in million years)

Date	*Mass Extinc-tions*	*Anoxic Events*	*Evaporite Deposit*	*Flood Basalts*	*Sea-floor Spreading*	*Sequence Boundaries*	*Orogenic Events*
00–09	1.6		5		2		0.6,2.5,4,5
10–19	11.2			17	10,17	16	12.5
20–29							25
30–39	36.6		36.6	35		30,36.6	
40–49					40		40
50–59					53	52	
60–69	66			62,66	63	60.6	65
70–79					77		
80–89		84				86,88	80,87
90–99	91	91		92	94	97	
100–109							100
110–119	113	113		110	112	113	
120–129							
130–139				130,135		138	
140–149	144		144		148	144	145
150–159		156				154	155
160–169		163					
170–179	176			170		173	
180–189							
190–199	193	193		190		196	
200–209		208	208	200		208	
210–219	216						
220–229						220	220
230–239						230	
240–249	245	245					
250–259			258	250			250

Source: Rampino and Caldeira, 1993

(spike), and then applied a Gaussian smoothing function with a scale-length of 5 m.y. to the resultant function. The Gaussian filter removed high frequency noise components (Fig. 5.6b).

Using this time series, they computed the Fourier transform of the auto covariance function of the original unwindowed time series data, rounded to the nearest million years, utilising a standard Tukey window with a bandwidth of 4.5 m.y. The highest peak in the Fourier power spectrum occurs at a period of 26.6 m.y. (Fig. 5.7). The Fourier transform of the time series data suggests a phase with a most recent maximum of the 26.6 m.y. cycle occurring about 8.7 m.y. ago. They note, however, that the detection of a possible 26.6 m.y.

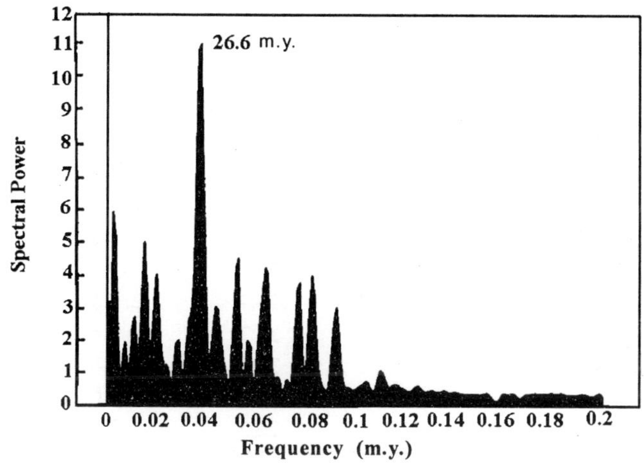

Fig. 5.7 Power Spectrum of Combined Geo-event Data of Fig. 5.6b. Highest Peaks Occur at 26.6 m.y.

(*Source*: Rampino and Caldeira, 1993)

period does not rule out the presence of longer or shorter periods in the geologic record. Sequence boundaries data shown in table reveal a spectrum with strong peaks at 28.3 and approximately 60 m.y., in addition to peaks at 11.2 and 12 m.y. in the high frequency portion of the spectrum (Fig. 5.8). These results are summarised is Table 5.2 with other detailed information of calculations. It is interesting to note that a periodicity of order of 26–28 m.y. is dominating in all geological and sedimentary records.

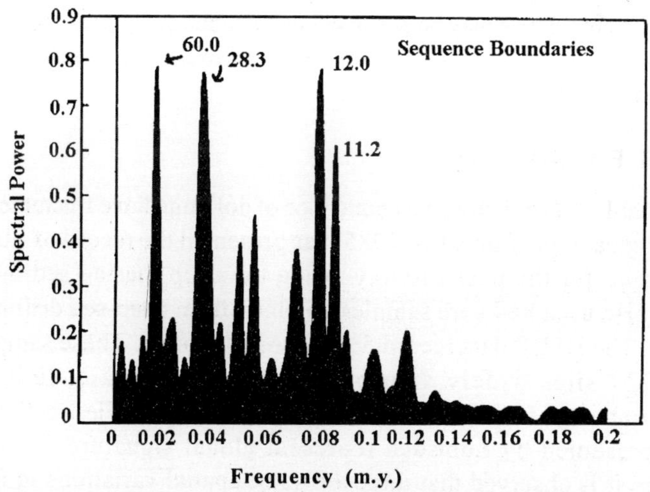

Fig. 5.8 Power Spectrum of Stratigraphic Sequence Boundaries. Highest Peaks Occur at 60 m.y., 28.3 m.y., and 12.0 m.y.

(*Source*: Rampino and Caldeira, 1993)

Table 5.2 Geologic Events with Dominant Periods Phase and Spectral Power at 26.6 m.y.

Geologic Events	*Dominant Period of the Events (m.y.)*	*Dominant Period of Aggregate Events*	*Phase at 26.6 m.y.*	*Power at 26.6 m.y.*
Extinctions Sequence	26.0	26.9	9.0	0.94
Boundaries	28.2	26.5	9.3	0.73
Flood basalts	23.1, 15.4, 26.2	26.7	9.5	0.53
Orogenic events	30.6	26.2	8.4	0.24
Sea-floor spreading	18.4, 26.7	26.6	9.8	0.07^2
Ocean anoxic events	39.7, 26.1	26.9	3.6	0.37
Evaporite deposits	28.2	26.5	11.7	0.04^2

Source: Rampino and Caldeira (1993)

RECORD OF FLUCTUATIONS OF DOLOMITE ABUNDANCE IN DEEP MARINE SEDIMENTS (DADMS)

Dolomite and Process of Dolomatisation

Dolomite is one kind of carbonate rock. The process of dolomitisation can be explained as conversion of limestone ($CaCO_3$) to dolomite by replacing calcium with magnesium. This involves diffusion of Mg inward and Ca outward through extremely slow direct precipitation of calcium. The possible chemical and physical causes of fluctuations of dolomite content in carbonate rocks are worth debating. However, it is speculated that the observed fluctuations in dolomite abundance may be due to either changing sedimentary environments of deposition or geo-chemical variations or due to combined influence of both (Lumsden, 1985).

DADMS Record

The available data indicate that abundance of dolomite have fluctuated through the geological time. Lumsden (1985) has presented the record of fluctuations of dolomite for the past 140 m.y. using the deep marine sediments core samples. He used 844 core samples obtained from deep-sea drilling project (DSDP). The DSDP data record 5% or more dolomite. These samples come from 1227 sites widely distributed over the Atlantic, Indo-Pacific, Gulf-Caribbean, Mediterranean, Red Sea and Black Sea. Hence, the DADMS record presented by Lumsden represent global signature of variability. However, it is observed that there are some spatial variations in individual records. To reduce these spatial variations, he presented a grand total of DADMS record and the sum (combined) data from Atlantic and Indo-Pacific regions (Fig. 5.9). Dot marks show the actual data points. Data indicate the

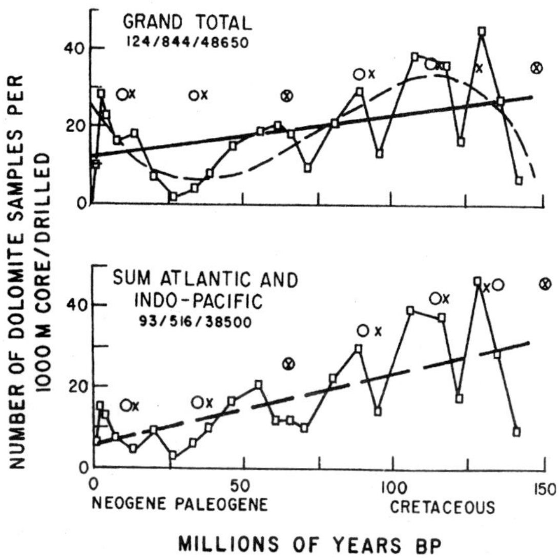

Fig. 5.9 Time Series Representing Global Fluctuations in Dolomite Abundances for the past 150 m.y. (top) Grand Total Record of All Oceans (bottom) of Atlantic and Indo-Pacific Oceans

(*Source*: Lumsden, 1985)

number of dolomite samples counted from sedimentary thicknesses. Thicknesses less than 100 m are excluded (Lumsden, 1985). According to him, this is the most accurate date of dolomite abundance events for the last 150 m.y. The original records are displayed in Fig. 5.9. These two time series are quite reliable for spectral analysis.

Visual inspection of the secular variations in DADMS for the past 140 m.y. (Fig. 5.9) apparently indicates a long-term trend with superimposed quasi-periodic oscillations. This trend may be related to long-term tectonic processes while high frequency oscillations may possibly be related to some quasi-periodic physical processes in ocean. Quantitative analysis of such variations may throw some light on the possible causes of long-term variations in dolomite abundance (Negi et al., 1996).

Spectrum of DADMS Record

The Fourier spectral analyses of DADMS time series indicate long period 'red spectra'. As discussed in chapter 2, in order to compensate for the red spectral effects, it is necessary that the data be suitably detrended. Accordingly, the trend from the data was removed by fitting polynomial. To avoid the possibility that single fit may probably degrade the data characteristics leading to deceptive picture, a different order of polynomial was tried to have better and unbiased spectral results. With this in view, second, third and fourth order of polynomials were fitted to the original DADMS record to account for the long-term trend. After detrending, the Fourier spectra show considerable power variance near 25.5 m.y. terms. The Fisher's test statistics suggest that the 25.5 m.y. peak is significant at more than 95% confidence level (Shimshoni, 1971).

Statistical significance of 25.5 m.y. periodicity in DADMS record using the method of Fourier analysis requires a more detailed examination to test its stability and authenticity using the alternative perspective of spectral analyses. This is due to the fact that the unevenly spaced data set decoded from a limited stratigraphic resolution are somewhat crude and requires an efficient non-linear spectral technique to delineate its periodic behaviour. Clean spectral technique, discussed earlier, in chapter 2 is especially suitable for such data analysis. Clean spectral analysis of unequally spaced data points has advantage of removing the artifacts introduced by low resolution (missing) data in geological records (Robert et al., 1987). In DADMS record, the time interval between successive data points is quite variable. The average value of the spacing is about 6.33 m.y. For clean spectral analysis, we have sampled the data at 3 m.y. interval with Nyquist frequency of 6 m.y. This sampling is appropriate to avoid alliasing and to identify/recover average oscillation present in the data. As discussed earlier, prior to clean spectral analysis, long-term trend was removed from the original record. Figure 5.10 depicts the

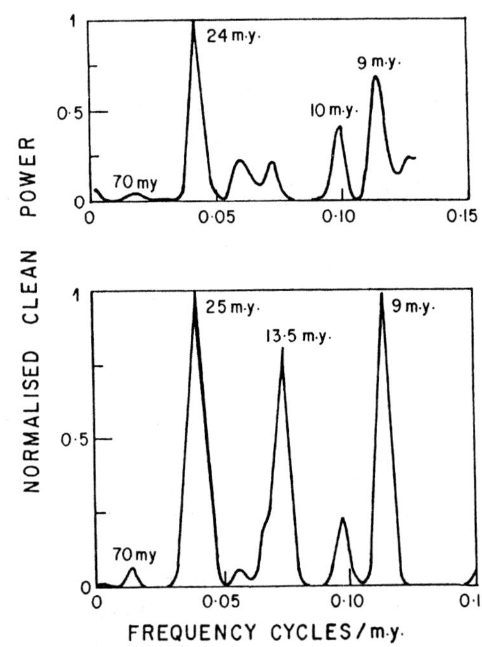

Fig. 5.10 Normalized Clean Spectra of (Top) Global Dolomite Abundance Fluctuations, (bottom) Atlantic and Pacific Ocean

(*Source*: Negi et al., 1996)

results of clean spectral analysis which clearly identifies a strong cycle of order of 25 ± 1 m.y., 12 m.y. and 8 m.y. The result indicates that the resulting peaks are true and stable. The consistency of 25 m.y. periodicity in the spectrum invite for physical explanation in terms of the long term geochemical evolution of ocean processes. The other two spectral peaks at around 12 m.y. and 8 m.y. could be related to the second and third harmonics, respectively, of the main peak. The presence of higher order harmonics in the data, infact, demonstrates the resolving power of clean spectral algorithm. Because of the possibility of high frequency harmonic components in clean spectra, it is difficult to put an exact error bound to these periodicities. However, an approximate error bound may be assigned by fitting a different order polynomial. Accordingly, the above spectral results computed for different orders of polynomials provide definite support in favour of 26 ± 1 m.y. periodicity in oceanic records. This is interesting in view of a similar periodicity in marine mass extinction data and several other oceanic records as discussed above.

A GEOLOGICAL RECORD OF PHOSPHOROUS BURIAL RATE (PBR)

Temporal fluctuations of marine phosphorus burial rate (PBR) is one of the indicators of coupling of climate, continental weathering, ocean circulation and primary productivity under the 'greenhouse and icehouse' conditions and may have significant impact on biotic evolution (Follmi, 1996). The excellent quality of global temporal variations record of PBR with saw-tooth oscillatory pattern for the past 160 m.y. is displayed in Fig. 5.11.

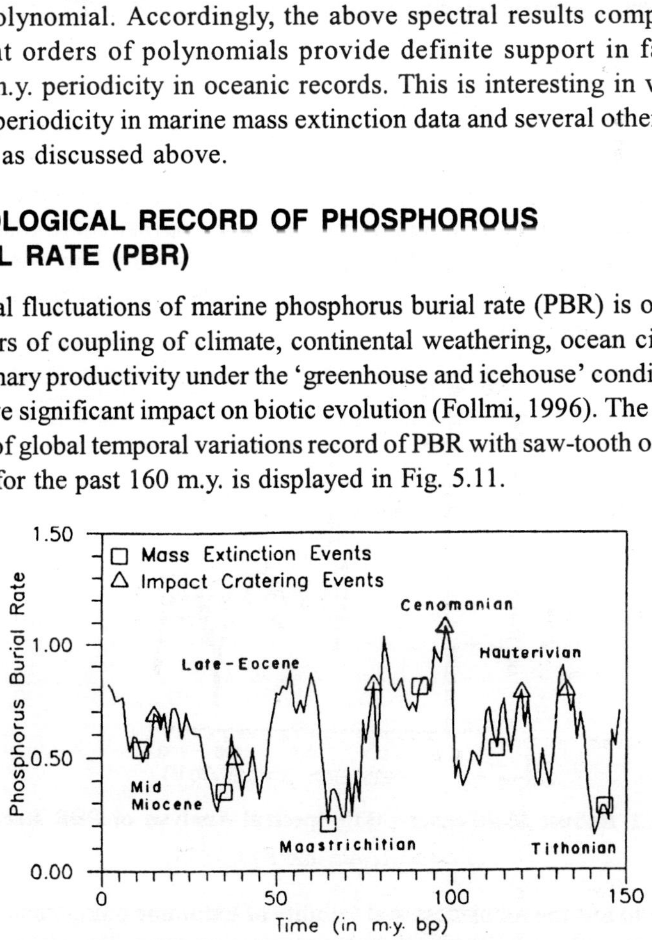

Fig. 5.11 Time Series Representing the Global Phosphorus Burial Rate (PBR).

(*Source*: Follmi (1996), Superimposed on this Curve are Major Mass Extinction and Impact Cratering Events)

Follmi (1996) has discussed in detail the nature and precision of the data. His data compilation is a result of published Deep Sea Drilling Project (DSDP) and Ocean Drilling Program (ODP) phosphate observations totalling about 5648 measurements. Chronological framework is based on numerical ages which were attributed to each measurement using recent biostratigraphic distribution chart and the geological timetable of Harland et al. (1990). The resulting time series indicates significant change in the phosphorus burial rate over the past 160 m.y. and a short-term oscillation superimposed on the long-term secular change. The short-term oscillations might be related to some external or internal triggering processes.

In order to explore the possibility of any periodic component in these geochemical oscillation of the PBR record, the multiple powerful spectral analysis discussed in chapter 2 was employed. The results of various spectral analysis of PBR record reveal a consistent periodicity of order 32 + 2 m.y.

In particular, the most important and powerful method of spectral analysis, which judges the statistical significance of the spectrum as discussed in chapter 2, is the MTM spectal analysis. In Fig. 5.12, high resolution MTM adaptive spectra showing: (i) median smoothed background spectra (ii) analytical robust background noise spectra curve (iii) 95 % confidence level (iv) 99% confidence level are displayed (Tiwari and Rao, 1999). The result of MTM as shown in Fig. 5.12, supports the concept of cyclic geochemical processes.

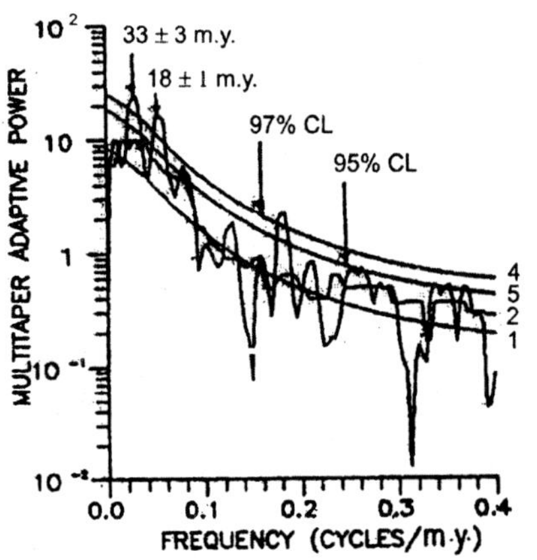

Fig. 5.12 Robust Multi-taper (MT) Spectral Analysis of PBR Record.
(*Source*: Tiwari and Rao, 1999)

In order to test the robustness and stability of harmonic component in PBR record, additional methods of spectral analysis were tried. The clean spectral also clearly reveal a periodicity of about 29 m.y. Maximum entropy spectral analysis (MESA) has been used as one of the most powerful techniques for resolving periodicity from truncated geophysical/geological records. To confirm our clean analysis, MESA has been carried out using the PBR time series. In view of possible implications of length of prediction error filter (PEF) employed in MESA, we used varying lengths of PEF (30–50%) of the total record length to obtain stable results. MESA reveals strong and stable periodicity of the order of 30 ± 1.5 m.y. (Tiwari and Rao, 2003).

BIO-GEOCHEMICAL (CTBF) RECORD

Geo-marine records are one of the best faithful recorders of the earth's evolutionary history. Visual inspection of various geo-marine records such as Calcareous Trochospiral Benthic Foraminifera (CTBF) test size variation, ocean productivity, dissolved oxygen and global climate exhibits a remarkable synchronism. However, identifying the long-term cyclic pattern is essential for understanding the wider ramification of significant physical linkages of the various terrestrial and extra-terrestrial processes. It is known that the amount of deep-water dissolved oxygen and the global climate have undergone major changes during the past 120 m.y. and have influenced significantly temporal variation of test size of CTBF. The study of long-term variation in the test size of CTBF, therefore, may have indirect implication for global climate variation, ocean productivity and organic carbon flux and global deep ocean ventilation (Kaiho, 1998).

The record of deep sea CTBF test size variation is based on the study of 396 samples collected from 31 Deep Sea Drilling Project (DSDP) and Ocean Drilling Program (ODP) sites covering the North and South Atlantic ocean and Indian and Pacific oceans (Kaiho, 1998). Hence, time series displayed in Fig. 5.13 is based on widely distributed samples and well-preserved assemblage showing CTBF test size variation during the past 120 m.y. Kaiho (1998) has given a detailed account of the data characteristics. However, a brief summary of the data is presented here. These data represent palaeodepths of 1500 to 5200 m except for four Aptian samples representing palaeodepth of ~ 800 m. The CTBF test size variation is more or less an unpatched/continuous record except some minor gaps and non-uniformity in sedimentation rate during the Cenozoic period (Kaiho, 1994 and references therein). Stratigraphic distribution of the palaeodepth of the sample and the number of trochospiral specimens studied in each sample were nearly uniform within 2 m.y. average values (Kaiho, 1998). Maximum diameter values vary stratigraphically and geographically because of dissolved oxygen and nutrient level changes with direction of deep-water circulation.

The chronological framework is standardised to the Cretaceous time scale of Gradstein et al. (1994) and the Cenozoic time scale of Berggren et al. (1995). The CTBF data indicate maxima in test size variation during the Pleistocene to Holocene (2–0 m.y.), late-early-Miocene to early-middle-Miocene (17–14 m.y.), early Oligocene (34–30 m.y.), late Maastrichtian (68–66 m.y.) and late Aptian to early Cenomanian (114–98 m.y.), and corresponding minima occurred at the late Miocene to early Pliocene (12–4 m.y.), the late Oligocene to early Miocene (29–20 m.y.), the latest Palaeocene to early Eocene (55–54 m.y.), latest Cenomanian to early Turonian (94–92 m.y.) and the early Aptian (120–118 m.y.). A visual inspection of the raw time series indicates quasi-periodic oscillations superimposed on a long-term exponential decay trend starting from the recent (Fig. 5.13). The secular trend in geo-marine record might be a natural variation of mega geo-biochemical processes (Fischer and Arthur, 1977). However, the apparently superimposed truncated oscillations seem to be related to either some internal ocean-climate dynamics or externally induced-physical processes such as bolide impact events of cosmic origin.

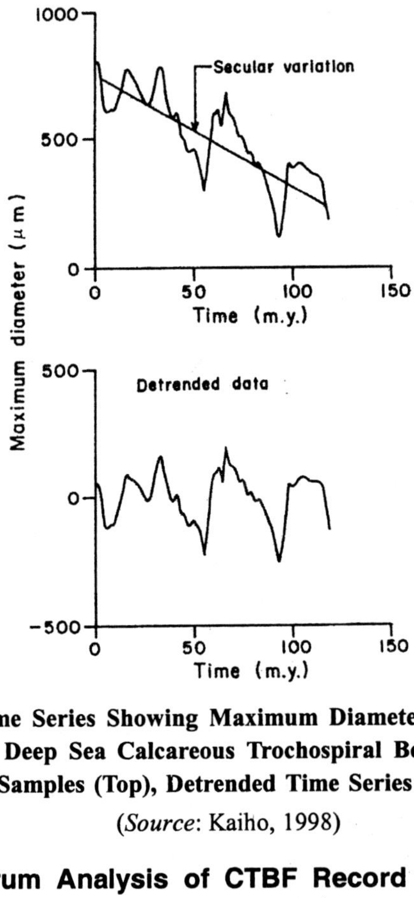

Fig. 5.13 A Time Series Showing Maximum Diameter of the Largest Specimens among Deep Sea Calcareous Trochospiral Benthic Foraminifera (CTBF) Samples (Top), Detrended Time Series (bottom))

(*Source*: Kaiho, 1998)

Multiple Spectrum Analysis of CTBF Record

To obtain the dominant characteristic time scale of CTBF test size variation, multiple spectral techniques were used and following procedures were followed (Tiwari and Rao, 2003):

(i) linear secular decay trend was removed by appropriately detrending the time series and, then

(ii) analysed the resulting time series for statistically narrow band harmonic signal using the Multi-Taper Method (MTM) of spectral analysis (Thomson, 1982; Mann and Lees, 1996)

(iii) invoked the typical climate null hypothesis of 'red noise' by generating robust statistical model (Mann and Lees, 1996).

(iv) In addition to the MTM analysis, maximum entropy method (MEM) and wavelet spectral analysis were also employed to examine the stability and stationarity character of the spectral components.

The MTM spectral analysis of test size variations of CTBF time series reveal a statistically significant and stable harmonic component of 35 ± 5 m.y. (Fig. 5.14) against a robust 'red background noise'. The MEM spectra for

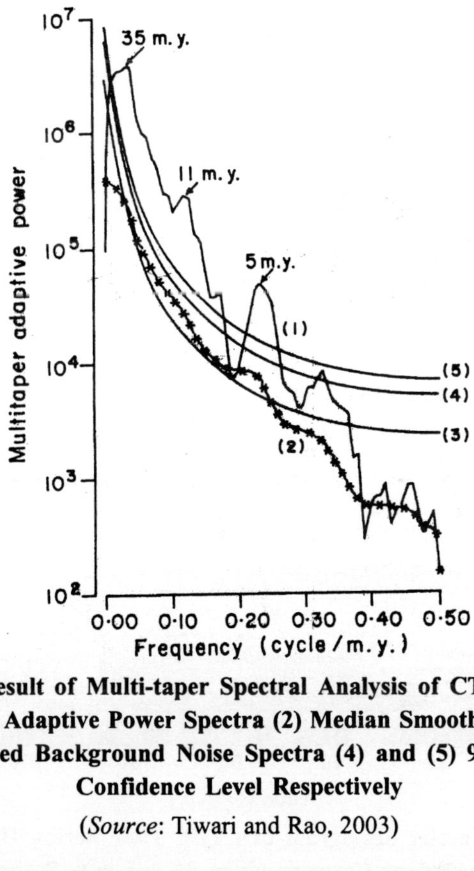

Fig. 5.14 Result of Multi-taper Spectral Analysis of CTBF Record. (1) Multi-taper Adaptive Power Spectra (2) Median Smoothed Background Spectra (3) Red Background Noise Spectra (4) and (5) 95% and 99% Confidence Level Respectively

(*Source*: Tiwari and Rao, 2003)

variable filter length (30–60%), as discussed in chapter 2, are shown in Fig. 5.15. It is clearly seen that those spectral peaks which appeared in MTM spectra are also resolved in MEM spectra and they are stable too. Further, wavelet spectral analysis of this record exhibits a strong stationary power in the above frequency band suggesting temporal persistence of spectral peak through the entire evolutionary period (Fig. 5.16). Slight mismatching in detected harmonics possibly arises due to the different techniques adopted in this study.

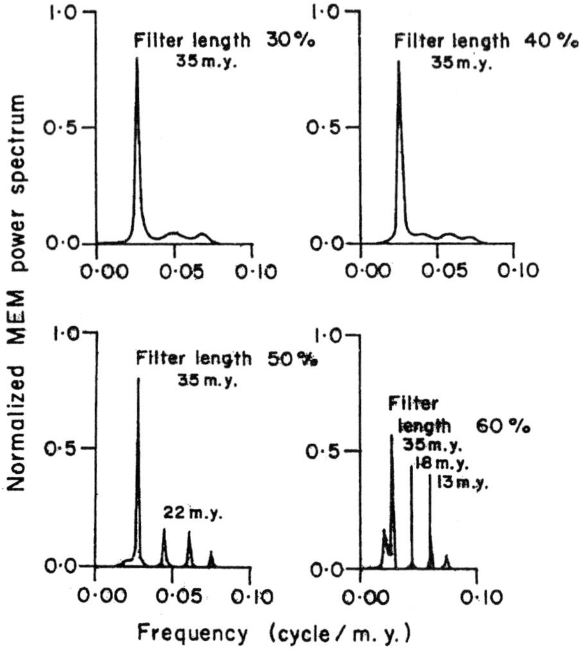

Fig. 5.15 Result of MEM Spectra of CTBF Time Series for Variable Filter Length (30–60%) of the Record Length

(*Source*: Tiwari and Rao, 2003)

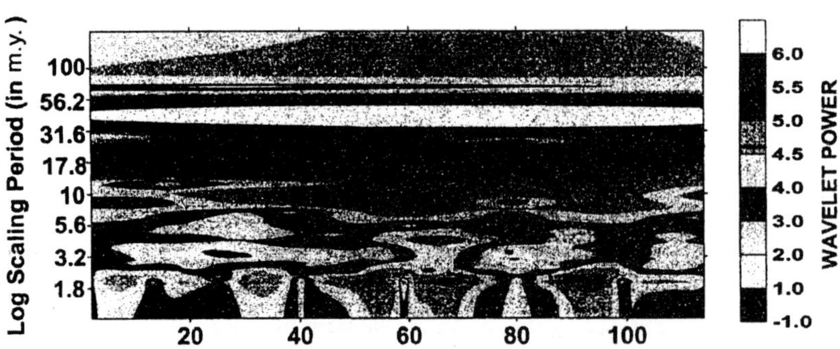

Fig. 5.16 Wavelet Scalogram of CTBF Time Series. Highest Wavelet Coefficient Corresponds to 35 ± 5 m.y. Periodicity

(*Source*: Tiwari and Rao, 2003)

The criteria for discussion of physical significance of a statistically significant spectral peak is based on the following three points (Tiwari and Rao, 2003):

(i) Statistical significance against 'red background noise';

(ii) stability of the spectral peaks in three spectra, and

(iii) stationarity of spectral peak.

Here we observe that only a basic periodicity of 35 m.y. with highest power in all three spectra meet with the above criteria within an error limit of ±5 m.y. Smaller peaks, although significant, in the MTM spectra, do not show up strongly either in the MEM or wavelet spectra and hence seem to be unstable. We, therefore, dwell here only on a 35 ± 5 m.y periodicity which is consistent in all the three spectra of an oceanic record of bio-geoclimate change.

Physical Significance of Bio-Geo-Climate Cycle

There have been several speculations for periodic cometary impacts leading to possible influence on terrestrial processes (Alvarez and Muller, 1984; Rampino and Stothers, 1984a, b; Tiwari and Rao, 2003) and sedimentary/environmental upheaval/stress (Rampino and Caldeira, 1993). Despite arguments against periodic cometary impacts (Weissman, 1985), modulation of the Oort cloud resulting from the vertical oscillation of the solar system across the galactic plane with recent evidence of periodicity of 30–44 m.y. (Mantanari et al., 1998) and 37 ± 4 m.y. (Matese et al., 1995), which is also remarkably close to the periodicity detected here in variation of CTBF, apparently remains a viable mechanism for generating cyclic comet showers and resultant terrestrial processes. However, a detailed mechanism has to be worked out as to how intrinsically coupled climate-ocean atmospheric system have responded directly or indirectly to critical external forcing (Tiwari and Rao, 2003).

It is natural to ask then why do the fluctuations in the test size of CTBF show a periodicity? In a complex oscillatory and globally interacting biogeochemical system, it is somewhat difficult to invoke a specific physical driver and link it with a single physical process, particularly, in view of negative and positive feedbacks (Tiwari and Rao, 2003). However, there is a general tendency of correlation of the recognised minima of CTBF test size variation with climatic warming, high CO_2 level and low productivity and maxima with climate cooling, low level of CO_2 and high productivity (Kaiho, 1998). Interestingly, there is evidence for a 30 m.y. periodicity in oxygen isotope records of climate for the past 180 m.y. (Shackleton and Imbrie, 1990) and CO_2 record of the past 260 m.y. (Tiwari and Rao, 1998). Further, a 30 m.y. cycle is also proposed in the history of biosphere (Fischer, 1981) related to the fluctuation in atmospheric CO_2, possibly caused by changes in the pattern and vigour of mantle convection (Fischer, 1982) with a characteristic time scale of 30 m.y. (Loper et al., 1988). We speculate that the underlying cycle in the CTBF test size variation is a possible consequence of environmental stress coupled with changes in climate, ocean productivity brought about by impact events (Zachos et al., 1989) and/or impact generated volcanism (Rampino, 1987). Climatic oscillations seem to have been operated for most of this geological period with large cyclic shift associated with impact events (Muller and Morris, 1986; Burek and Wanke, 1988). We note here, however, that even if the cause for 35 ± 5 m.y. remains unresolved at this stage, its dominant spectral power and regularity in the three spectra obviously support the cyclic nature of geo-marine processes. Most recently, the spectral analysis of the biological extinction record has shown evidence for a 30 m.y. periodicity (Prokoph et al., 2000). Hence, any link between CTBF and mass extinction cycle will provide useful constraints for understanding bio-geo-chemical cycles. Further, the long-term periodicity in test size variation of CTBF provides insight to the ongoing debate whether geomarine processes are cyclic in nature.

This finding renders further credible support to the concept of cyclic geo-marine processes and will provide stimulus for further quantitative research into the coupling of marine geo-bio-chemical cycles.

FORCING MECHANISMS OF LONG-TERM CLIMATE CHANGE

The long-term climate change and glacial records during the Phanerozoic (570 m.y.), based on climate sensitive sedimentary indicators, exhibit pulsating intervals of cooler and warmer climate epiodes called 'ice house' and 'green house' (Shaviv and Viezer, 2003; Frakes, 1979). The possible causes of long-term climate change and glacial pulsations include both endogenic and exogenic forcing. The endogenetic control of long-term glaciation (Donn and Shaw, 1977; Crowell, 1978) include the movement of continents into high latitudes and arrangements of potential glaciation. Palaeomagnetic data (e.g. Smith, 1973; Smith and Briden, 1977) demonstrate, however, that continental landmasses have, at certain times during the Phanerozoic, occupied very high latitudes without incurring continental glaciation, thus suggesting that endogenetic processes may not be the first-order controls of global climate.

Alternatively several researchers believe that the possible causes of long-term episodicity of glaciation have exogenetic control (Steiner, 1967; Opik, 1965, 1976). The possibility of a connection between the presumed rhythm of glaciation and the sun's galactic period, each taken to be ~200–250 m.y. also known as 'cosmic year' is suggested by Steiner and Grilmair (1973). Several researchers envisaged a strongly eccentric solar orbit around the galactic center and proposed that during 'cosmic winters' when the sun was in the outer, more rarified regions of the galaxy, overall planetary cooling and terrestrial glaciation ensued. Some others hence correlated times of glaciation with the periodic passage of the solar system through the galactic plane (Steiner and Grillmair, 1973; McCrea, 1975; Williams, 1981). In 1967, Steiner presented a wide ranging hypothesis from evidences from the stratigraphic, tectonic, biological and geomagnetic records. They were ascribed to a systematic variation of the gravitational parameter 'G' resulting from the slight eccentricity of the solar galactic orbit. Steiner also estimated a duration of ~ 280 m.y. for the cosmic year. In 1973, Steiner and Grilmair applied this model to episodic and periodic glaciation, which they attributed to successive passages of the Sun through episodes of an eccentric solar orbit. Steiner and Grillmair (1973) has expanded his galactic model to encompass other episodic events in the geological record. Tamrazyan (1967) subsequently presented an elaborate scheme which proposed periodic changes in terrestrial tectonism, volcanism, palaeoclimates, metallogenesis and organic evolution. He ascribed them to a varying gravitational potetnial along an eccentric solar orbit with an anomalistic period of 176 m.y.

In 1975, Williams further reviewed and updated the known age limits of major glaciations during the past billion years. He summarised the present view of galactic structure and dynamics and, based on an estimated length of the cosmic year of ~ 300 m.y., he postulated that major glaciation occurs during passage of the sun (with a period of 145–155 m.y.) through diametrically opposite, little warped regions of the flexed galaxy. It was also advanced in 1975 by the noted astronomer McCrea and advocated by Gribbin (1978). They attribute glaciation to fluctuations in solar radiation consequent upon passage of the sun through dust lanes broadening galactic spiral arms. They considered that a double-arm spiral structure for the galaxy would lead to two solar induced glaciations per cosmic year. According to Williams (1975), available geochronological data indicate that the mean or modal times of major glaciations have recurred about 145–155 m.y., half of the cosmic year term, during the past billion years. Same researchers believe that the estimated duration of the cosmic year, has been increased to a little more than 300 m.y. The former idea of one terrestrial glaciation for each cosmic year, therefore, has been replaced by the concept that two glaciations usually occur on an average every cosmic year. More recently, however, considering the non-circular motion of the solar system around the galaxy, Shaviv and Veizer (2003) suggested an average spiral arm passage period of 137 ± 4 m.y. They argue that this is consistent with meteoritic data showing a periodicity of 143 ± 10 m.y. and the palaeoclimate and palaeotemperature data with a recurrence at 135 ± 9 m.y. The empirical model presented by Shaviv and Veizer (2003) demonstrates that galactic cosmic ray flux (CRF) is closely linked to atmospheric level of CO_2, the main driver of global climates. Detailed calculations reveal that atleast 66% of variance in the palaeotemperature trend could be attributed to CRF variations likes due to passages of solar system through spiral arm of the galaxy. Despite these refinements, recent hypotheses concerning possible galactic causes of glaciation are not without their problems.

Bolide Impact and Climate

Modern bolide impact theory provides another dimension to explain multiple impacts and climate trauma consequent on prolonged stratospheric dusting (Napier, 1998). It is likely that the periodic impactor hit a marine and continental environment to increase in atmospheric pressure which might raise the main global surface air temperature significantly by considerable degree. The spreading dust cloud would subsequently interrupt a substantial fraction of solar radiation. This would tend to produce cooling over an extensive area. It would also reduce the sunlight available for photosynthesis. Periodic impact caused drastic environmental changes and slowly returned to pre-impact conditions (Alvarez, 1987). Several environmental consequences by bolide impacts have been suggested by several researchers also. In a recent study, Kent et al., (2003a) presented a strong case for cometary impact trigger of Palaeocene-Eocene thermal maxima and carbon isotope excursion. Although this particular case was criticized (Dickens and Francis, 2003), Kent et al. (2003b) maintained by providing appropriate evidence that the case for impacts generated environmental deterioration is strong.

CAUSES AND CONSEQUENCES OF COUPLED BIO-GEO-CHEMICAL CYCLES

Atmospheric level of CO_2 are generally assumed to be a principal driver of global climate (Shaviv and Veizer, 2003). A possible evidence of the 32 m.y. periodicity in the global and continental CO_2 variations for the past 260 m.y. (Tiwari and Rao, 1998) may suggest that various earth's processes (volcanism, sea level and climate changes) and impact cratering form a complex interactive loop with a characteristic time scale of 30 m.y. However, while the various time series analyses of geological rhythms exhibit almost a common periodicity of 31 ± 5 m.y., its one-to-one correlation in time domain is somewhat difficult, probably due to different relaxation times of different geological processes.

The available data of CO_2 variation of past 570 m.y. suggest that there have been significant long-term fluctuation in the global CO_2 through the earth's history (Budyko and Ronov, 1979; Hay, 1985). A possible cometary impact generated tectonic (Price, 2003) control on atmospheric CO_2 variation since its concentration is highly sensitive to changes in land area and in the sea-floor spreading rate. Sea-floor spreading controls the rate of outgassing of mantle derived CO_2 and land area governed by sea level changes (Arthur et al., 1985; Kaiho, 1988). They control the rate of continental weathering and hence the carbonate silicate cycle. Increased outgassing of mantle CO_2 during the sea-floor spreading and high tectonic activity may profoundly contribute in the cycling of CO_2 through the geological period. It may be suggested that increased CO_2 production from the greater volcanic activity may have increased the CO_2 concentration of the atmosphere, which, in turn, may have contributed to climate instability and mass extinction (Loper et al., 1988). Campsie et al. (1984) have presented a correlation of biotic crises and tectonic activity.

The high CO_2 level associated with high rates of sea-floor spreading may be directly related to the mantle convection, and hence to the rates of subduction and subduction related-volcanism (Berner et al., 1983 and Arthur et al., 1985). Loper and McCartney suggest a 30 m.y. periodicity in plume activity due to periodic instability of D-layer. During the processes, the hot material from D-layer rise via mantle plumes to the surface where it causes increased volcanic activity, which may affect climate biota (Loper et al., 1988).

Rampino and Stothers (1988) have argued for the periodicity of about 33 m.y. in global volcanism and related it to the impact and other related terrestrial cycles. In reprojection of mass extinctions and volcanic episodes and major dates of CO_2, Tiwari and Rao (1998) found an interesting correlation which shows that the distribution of significant volcanic episodes and high CO_2 accumulation rates match nicely with impacts by age and cycle number for the past 250 m.y. Certain coincidences are rather compelling, e.g., associated volcanism (Arthur, Dean and Schlager, 1985; Vogt, 1972 and Whyte, 1977) correlate remarkably well with the high atmospheric CO_2. During these periods, worldwide sea-floor spreading, rifting of separation and extensive plate reorientation started. Coincidently, Rampino and Stothers (1984) in sea-floor spreading record have suggested a similar periodicity of order of 30 m.y. Tiwari and Rao (2003) have discussed the possible link among the CO_2, seafloor spreading, mass extinction and the impact cratering as they reveal a common cycle of the order of 30 m.y.

Impacts and Sea Level Cycles, Mantle Degassing and Climate

The pattern of abrupt sea level falls and gradual rise is systematic. However, the amounts of the rapid sea level falls are often dramatic. There are 28 rapid sea level falls in the Cenozoic, including the one coinciding with the impact horizons near the K/T boundary (Burke and Wanke, 1988). Oxygen isotopes and other faunal studies show that sea level low-stands represent climatically cold conditions and sea level high-stands represent climatically warm conditions (Vail et al., 1977). In the Cenozoic, there are 28 sea level falls at: ± 66.7, ± 60, ± 59, ±56, ±53.5, ±52, ±49.5, ±45.5, ± 40, ±37, ±29, ±26, ±24, ±22.5, ± 22, ±19, ±16.5, ±15.5, ±13, ± 9.8, ± 6.6, ±6.0, ± 5.2, ±4.2, ± 3.8, ± 2.8, ± 1.7, and ± 0.7 m.y. (Vail et al., 1979). According to Burke and Wanke (1988) and references therein, these could be approximate times of major impacts on earth. Accordingly, an impact establishes freezing conditions, i.e., polar glaciation at the time of the K/T boundary and those subsequent impacts maintained these glaciations (Clube and Napier, 1982; Burke and Wanke, 1988).

Further, palaeo-climatic oscillations are often confirmed as cause for the sea level cycles. Major climate changes during the last 100 m.y. are characterised by gradual cooling associated with three periods of warming during 100–89 m.y., 60–54 m.y., 27–16 m.y. (weak warming) and rapid global cooling events at the Eocene/Oligocene boundary (~34 m.y.) and during the early middle Miocene time (15–13 m.y.). Collectively, these climatic fluctuations suggest a quasi-periodicity of 33–38 m.y. (Kaiho and Saito, 1994 and references therein). Accordingly, the changes in rates of mantle outgassing caused by variations of oceanic crust production, and land–sea distribution may have been the primary causes of mid-Cretaceous warm climate and of the decline in global temperatures over the past 90 m.y. (e.g. Hays and Pitman, 1973; Berner et al., 1983; Barron, 1985; Larson, 1991). The long-term warming during the late Palaeocene-early-Eocene may have been due to a long-term greenhouse effect caused by increases in crustal production (Owen and Rea, 1985; Eldholm, 1990) and hydrothermal activity (Rea et al., 1990). These arguments provide inter link and coupling of sea-level change, climate and CO_2 with earth's interior process.

Further, it is suggested that impact generated increases in the mantle degassing at mid-oceanic ridges (and associated with ocean-plateau eruptions), related hydro-thermal activity and metamorphic de-carbonation of sediments at subduction zones leading to increases in atmospheric CO_2 produced greenhouse effects and associated increases in global temperature (Sheridian, 1983, 1987; Owen and Rea, 1985; Lyle et al., 1987; Rea et al., 1990; Arthur et al., 1991; Rampino and Caldeira, 1993). This adds another dimension to our understanding and emphasises the role of impact in shaping various earth's system processess periodically.

CONCLUSIONS

Quantification of cyclic pattern in long-term ocean-atmospheric records may provide a predictive tool and form a basis to go backward in time to compare available data. Evidence of periodicities and their linkages with various other processes may help understand how the earth worked even when the oceanic record has been completely destroyed in the subduction such as in the Palaeozoic (Sheridan, 1987). The analyses of oceanic records which are supposed to be more reliable provide strong evidence and further strengthens the case for cyclic behaviour of earth's processes. Hence, the general conclusion drawn here based on gross systems analysis of various geological and tectonic records is that these results may provide a working model to understand the evolutionary history of coupled earth-ocean and atmospheric processes considering a long period cycle of 30–40 my. This unifying theory may explain the possible link between the various processes of earth's interior oceanic and atmospheric processes.

6

Mass Extinction Cycle, Correlations and Forcing Mechanisms

INTRODUCTION

One of the most debated topics of research in the earth sciences has been the pattern and processes of biological extinction and diversity (Hoffman, 1989). Geological evidences indicate that several times in the history of the earth there have been catastrophic episodes, which have abruptly disrupted the biological life and ecosystems. Besides many, Permian and Cretaceous extinction events, in particular, have been the main focus for research. It has been widely discussed that at the end of Cretaceous, about 65 m.y. ago, a large meteorite or comet struck the earth. Coupled, perhaps with other long-term geophysical/ geological processes such as sea level and climate change, geotectonic processes such as mantle plume and volcanism etc. led to the catastrophic mass extinction—with the disappearance of about 40% of fossilisable genera and perhaps 75% species (Marshall, 1998). During the past two decades, the subject of extinction has been the main focus of intensive research activity. Raup (1986) discusses various fundamental issues in his review. Accordingly, one of the questions that have bothered scientists is regarding the secondary causes that translated the bolide impact into mass extinction. Another central issue has been posed by Raup and Sepkoski (1986) that whether the Phanerozoic mass extinction record shows a consistent repeatable pattern or is it confused amalgam of signals produced by many independent evolutionary histories (Lewin, 1988).

Raup (1986) mentioned that strong conventional wisdom exists in the field of palaeontology and geology since the days of Lyell and Darwin. The basic tenets are as follows:

(i) Extinction is a complete gradual process driven by the multiple interactions among and between species and their physical environment.

(ii) Major extinction events are randomly distributed in time.

(iii) All major extinctions are independent and different.

Surprisingly, however, several recent works have shown that the extinction process is quasi-poriodic phenomena (Raup and Sepkoski, 1984; Rampino and Stothers, 1984) and extinction pattern is uniform atleast among the limited groups of submarine organisms (Lewin, 1988). The order of extinction seems to reflect something about the true physical process underlying the evolutionary history of groups of organisms. Further, the results suggest that extinction is a process, driven physically rather than biologically, deviating from the conventional wisdom of Lyell and Darwin Raup (1986). Underlying physical processes could either be internal or external and is at present a matter of great debate.

Several studies indicate that few billion species of plants and animals have lived in the last 600 million years. However, according to current evidence, many of these are extinct and there are only a few million species living today. The available data suggest that large number of species suddenly disappeared and/or became extinct (Sepkoski, 1989). The current information is based on the database of extinction analysis, which reveals that the history of evolution of life on earth can be characterised by a succession of catastrophic extinctions. The estimated mean residence time of various species is short, relative to geologic time. This has tempted the geologists to assume that the process of species extinction throughout the geological history has been abrupt and/or catastrophic in nature. The issues of catastrophe and evolution have been hotly debated during the Mason meeting of the Royal Astronomical Society at Oxford in 1988 (Davies, 1989). The end of the Cretaceous extinction was profound and shows relatively sudden and rapid turnovers at very short time scales. Abruptness of extinction pattern is also debated. The Cretaceous-Tertiary (K-T) mass extinction assumes great interest due to the fact that the best-documented extinction events are available for this period which especially include the demise of the dinosaurs. Episodes of mass extinction have been associated with terrestrial events such as sea-level fluctuations, enhanced volcanism, climate change, ocean anoxia and changes in productivity level, orogenesis and ocean salinity variations, global warming etc. The terrestrial causes have been challenged with more modern view of external forcing mechanism such as bolide impact (Alvarez et al., 1980) and radiation (Kitz, 1996).

In addition to the causes and effects of biological extinction processes, the time series analysis of the detailed available marine extinction data for the past 250 m.y. have shown evidence of periodicity, which is also one of the hotly debated issues in the earth science (Raup and Sepkoski, 1984). There have been counter arguments also, regarding evidence of periodicity in extinction record. In this chapter, we discuss evidence of periodicity of extinction processes and its possible causes. Understanding the causes of this mass extinction is critical to explaining the history of life.

TEMPORAL VARIATION IN INTENSITY OF MASS EXTINCTION

The palaeontologists and geologists have defined the intensity of extinction as measure of total taxa that have become extinct in a selected unit of time (a geologic stage). For appropriate definition, they considered the following three points (Raup, 1986):

(a) extinctions in a time unit as a percentage of standing diversity,
(b) number of extinctions per million years, and
(c) percentage of extinctions per million years.

The percentage of extinctions per million years are considered the most appropriate due to the fact that it normalises both for standing diversity (taxa available for extinction) and for absolute time (Raup, 1986). It is worth mentioning, however, that all are subject to sampling problems and persistent biases. This is due to the fact that the time dimension is an estimated duration of a stage or other conventional time unit, and stage durations are generally found to be uncertain, as they are based on interpolations between widely separated calibration points (Raup, 1986). Additionally, the fully normalised metric carries the assumption that extinction is a continuous process, which, in fact is not true. Further, the extinction process is dominated by widely spaced point events and also there is probable errors in stage durations.

MASS EXTINCTION RECORD

There are atleast twelve major extinction episodes, in addition to several minor events, that have been recorded in the history of the earth. Some geologists, however, suggested that atleast five extinction episodes out of twelve are considered as the largest and hence conventionally labelled mass extinctions (Raup, 1986). These are terminal Ordovician (~440 m.y. ago), late Devonian (~365 m.y. ago), terminal or late Permian (~250 m.y. ago), terminal Triassic (~215 m.y. ago) and terminal Cretaceous (65 m.y. ago). The Permian extinction event among these emerges as the largest one (Erwin et al., 2002). Recent studies, however, recognise the twelve extinction events as real (Table 6.1) and have invariably used these data for the time series analysis. The 'background' extinction, seem to appear as the normal or spontaneous rate of replacement of one species by another. A plot of varying intensity of extinction for the 79 generally recognised stages of the Phanerozoic suggests the simple binary classification of extinction which may possibly indicate that extinctions are continuous and qualitatively different from each other (Raup and Sepkoski, 1986). The Fig. 6.1 shows percentage of extinction for marine animal families and genera in stratigraphic interval between the mid-Permian and the recent.

Table 6.1 Ages of Extinction Events with Dates from Three Time Scales

System and Stage	*End of Interval (x 10^6 years ago)*			*Event*	
	Harland et al. (17)	*DNAG (19)*	*Odin (18)*	*Families*	*Genera*
Tertiary, Pliocene	2.0	1.6	1.9		Possible
Tertiary, middle, Miocene	11.3	11.2	11	Significant	Present
Tertiary, late Eocene	38	36.6	34	Significant	Present
Cretaceous, Masetrichtian	65	66.4	65	Significant	Present
Cretaceous, Cenomanian	91	91	91	Significant	Present
Cretaceous, Aptian	113	113	107		Possible
Cretaceous, Hauterivian	125	124	114	Doubtful	
Jurassic, Tithonian	144	144	130	Significant	Present
Jurassic, Callovian	163	163	150	Doubtful	
Jurassic, Bajocian	175	176	170	Doubtful	
Jurassic, Pliensbachian	194	193	189	Significant	Present
Triassic					
"Rhaetian"	213	208	204		Present
Norian	219	216	209	Significant	
Triassic, Carnian	225	225	220	Doubtful	
Triassic, Olenekian	243	240	239	Doubtful	
Permian					
Dzulfian	248	245	245	Significant	
Guadalupian	253	253	252		Present

Source: Raup and Sepkoski, 1986

Selectivity of Extinction Events

The selectivity of extinction is another important issue in palaeontology (Raup, 1986). There is considerable evidence that extinctions can be selective processes depending upon the variety of ecological, geographical and physiological variables (Lewin, 1988). Such selectivity could be expected to effect different groups in different ways yielding a random non-uniform response to extinction. Surprisingly, major extinction events are selective, i.e., the victim and survivors are not random sample of the pre-extinctions biodata. A full understanding of selectivity is the most promising tool to discovering the environmental stresses that cause extinctions (Raup, 1986).

Gradual versus Catastrophic Extinctions

One of the fundamental issues in geological science is 'whether the extinction process is continuous or episodic?' One view suggests that the disappearance of late Permian (Erwin, 1993) and the Cretaceous-tertiary boundary (Sloan et al., 1996) was gradual. The other view is that species extinctions were catastrophic at these boundaries. For instance, Sheehan et al. (1991) performed statistical analysis on family diversity of dinosaurs in the Hell Creek formation and concluded that there was no evidence of gradual decline of dinosaurs at

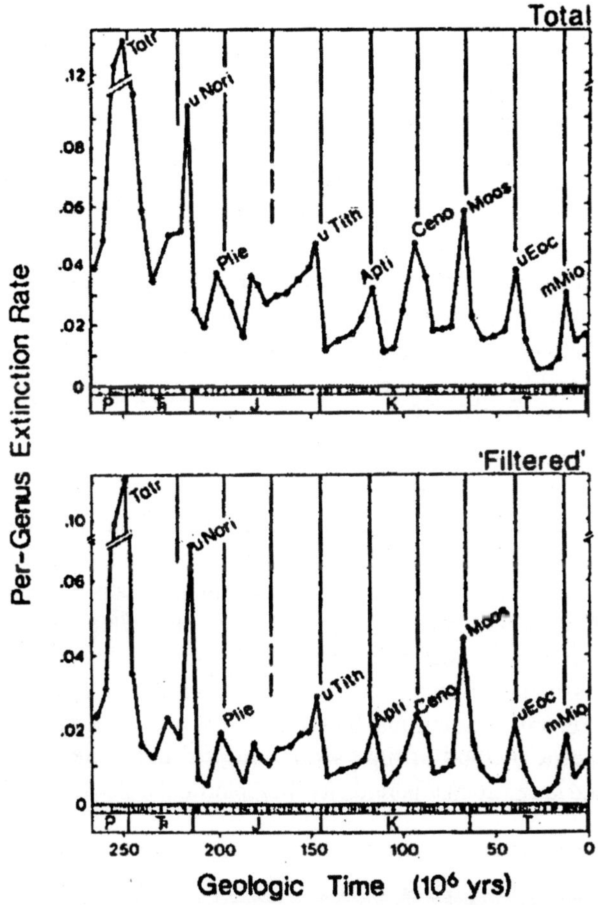

Fig. 6.1 Shows per Genus Extinction Rate (in Units of Extinctions/Genus/ Ma.) for 49 Sampling Intervals from the Mid-Permian to Recent. The Upper Time Series is for the Entire Data Set of 17,500 Genera. The Lower Time Series is a Filtered Version of a Subset of 11,00 Genera

(*Source*: J. John. Sepkoski, Jr., 1989)

the end of Cretaceous and that their findings are in good agreement with an abrupt extinction event such as the one caused by an asteroid impact. Recent study has suggested that there is indication of very abrupt end of Permian event among organisms in marine and terrestrial environment (Retallack, 1995). Signor and Lipps (1982) have demonstrated that a random distribution of errors at the end points of biostratigraphic ranges can produce range truncation and gradual decline preceding a sudden extinction boundary. This truncation causes a smearing of the record of an extinction event backward in time and is called the 'Signor-Lipps' effect (Raup, 1986). Rampino and Adler (1998) used published biostratigraphic data on ranges of fusulinids and other foraminifera from Permian-Triassic boundary sections in the Southern Alps, (Italy) to test the effects of sampling and species abundance on the record of the latest

Permian extinction. Their results are found in agreement with simulations for abrupt extinctions, as opposed to gradual or step-like extinction scenarios. On the basis of estimated sedimentation rates for the Italian sections, they have shown that the end Permian extinction could have been very sudden (< 30 k.y.) and the disappearance of most latest Permian foraminifera was most likely coincident with a worldwide ecological stress event, identified by global negative carbon-13 (δ^{13}). Although the ongoing debates on gradual versus catastrophic extinctions have not been demonstrated in the general case, forgoing discussions imply that extinction events are catastrophic in nature. However, in view of the poor quality of dinosaur's fossil record, the question remains important and bears its effect on other questions of mechanism.

Periodicity of Extinction

The evidence for periodicity has come from compilations of the stratigraphic ranges of marine taxa. A number of analyses have suggested that major extinction events are regularly placed in time. Perhaps, the first claim of marine mass extinction periodicity was made by Fischer and Arthur (1977) who proposed qualitatively a 32 m.y. cycle in diversity minima in pelagic ecosystems on the basis of global data for various species, including ammonite genera, and large, open ocean predators. Raup and Sepkoski (1984) performed time series analysis of fossil marine animal record and made a claim for a 26 m.y. periodicity in extinction record. Rampino and Stothers, (1984a) analysed the same data using a non-parametric method of time series analysis and suggested a 30±2 m.y. periodicity and later a 26 m.y. periodicity (Rampino and Caldeira, 1993) (Fig. 6.2).

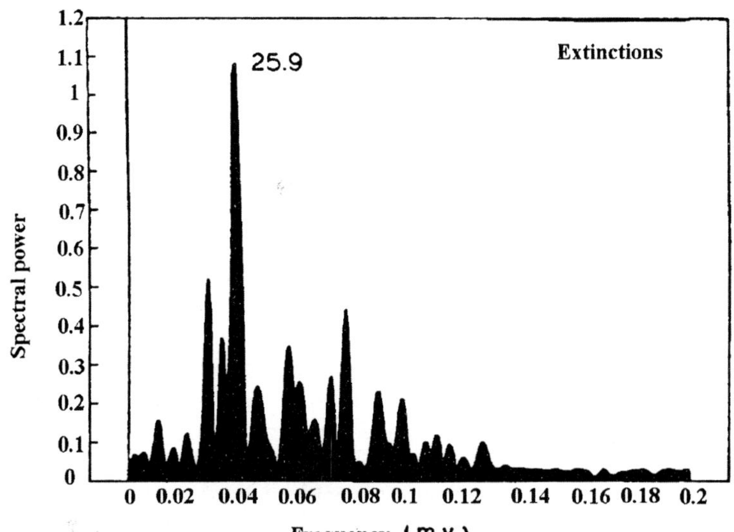

Fig. 6.2 Result of Power Spectral Analysis of Eleven Extinction Events. The Highest Peak is at Period of 25.9 m.y.

(*Source:* Rampino and Caldeira, 1993)

The data set mainly consisted of 567 extinct families in 39 stages of late Permian to Miocene age. Extinction intensity was measured as the percentage of families becoming extinct in each stage. The resulting time series exhibited 12 local maxima, or 'peak', of extinction, many of which corresponded to well-documented mass extinctions.

The claim of a periodicity in marine mass extinction produced considerable controversy (Hoffman, 1988; Stigler and Wagner, 1987). They posed several questions pertaining to its statistical significance and physical reality. Mainly, they have challenged evidence of periodicity in the extinction record on the ground that the available extinction data are incomplete and poorly dated. Indeed, Stigler and Wagner (1987 and 1988) analysed the extinction data and rejected the statistical significance of the underlying periodicity. They have argued that 26 m.y. periodicity embedded in the geological time scale is such that any random assignment of extinction events using that time scale has a non-trivial likelihood of showing a 26 m.y. periodicity. Secondly, in their analysis, they have demonstrated that tendency for random series to exhibit the 26 m.y. periodicity is enhanced by applying a certain kind of moving average filter. Evaluating the statistical significance, they discovered that the Harland time scale used by Raup and Sepkoski in their original paper exerts a peculiar bias towards a best fitting period of 26 m.y. and the test was as sensitive to measurement error of a type known to be present in the data as it was truly periodic signal, given the noise levels expected with these data. Several researchers (Raup and Sepkoski, 1984, 1988; Rampino and Stothers, 1984, 1988; Stothers and Rampino, 1990; Rampino and Caldeira, 1993) suggested that there existed as many as 25 extinctions that have been detected in the geological record during the last 570 m.y. (Phanerozoic). Out of these 25, there are about 11 events, which may be considered to be relatively major extinctions during the past 260 m.y. According to the latest compilation, these events are centred around 1.6 m.y., 11.2 m.y., 36.6 m.y., 66 m.y., 91 m.y., 113 m.y., 144 m.y., 176 m.y., 193 m.y., 216 m.y. and 245 m.y. These extinction peak dates (in the per genera extinction rate) are based on 17,500 genera (Sepkoski, 1989). Rigorous time series analysis of available updated data reveals a periodicity of 26–30 m.y. (Raup and Sepkoski, 1984, 1986; Rampino and Stothers, 1984, 1988). They maintained that major extinction events are regularly spaced in geologic time. Evidence of periodicity in extinction data has been further strengthened by analysing more updated extinction data using the method of wavelet analysis as discussed below (Prokoph et al., 2000).

Periodicity and Chaos in Mass Extinction Processes

Prokoph et al. (2000) analysed a high-resolution fossil record using multiple spectral and non-linear time series analysis. Their fossil record of 622 planktic foraminifera (PF) contain data from 200 stratigraphic stages spanning over past 127 m.y. The results of wavelet spectral analysis (Prokoph et al., 2000) are reproduced in Fig. 6.3. From Fig. 6.3A and F, it is clearly evident that the wavelet scalograms represent high wavelet coefficients for stationary 30 m.y. periodicity of PF diversity (Fig. 6.3B) and relative diversity changes/m.y. (Fig. 6.3F). Wavelet analysis also reveals additional pronounced peak at around 4.7

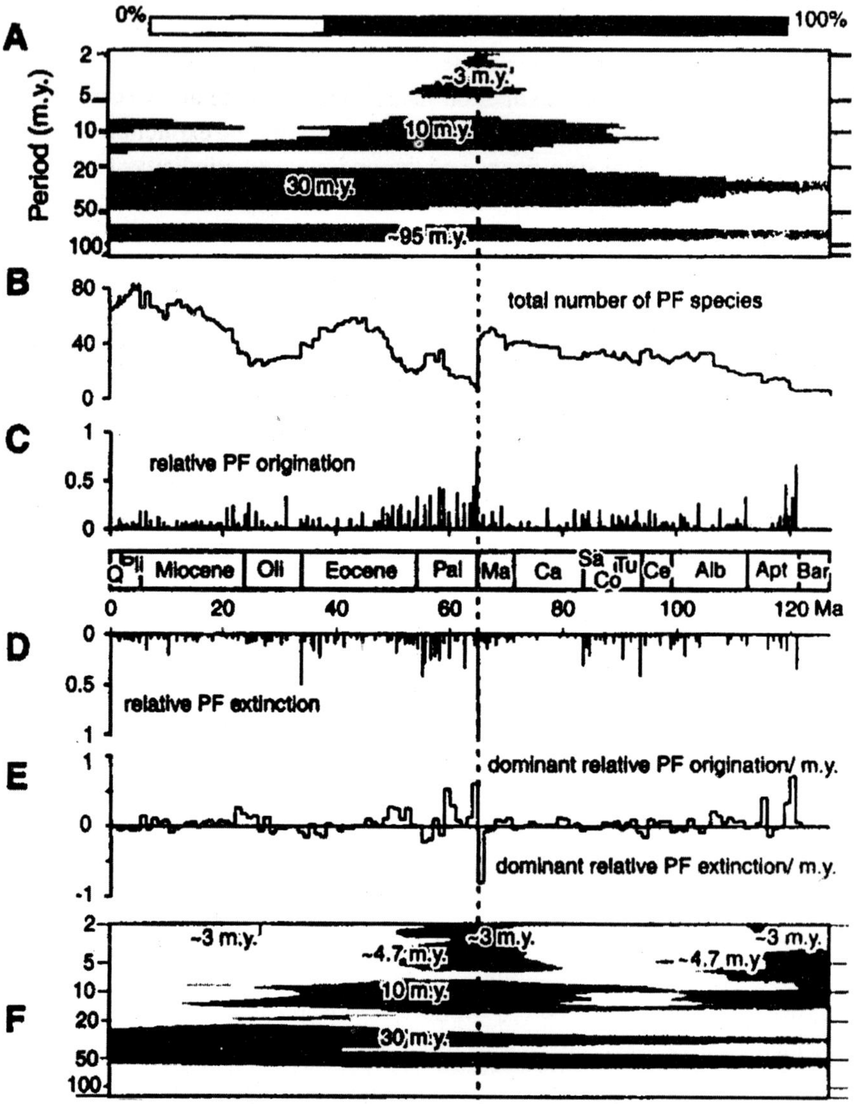

Fig. 6.3 (A) Wavelet Analysis of Total Number of Planktonic Foraminiferal (PF) Species in the Past 127 m.y. Vertical Axis is Logarithmic Scaled Periodicity, Horizontal Axis in Time

(*Source*: A. Prokoph et al., 2000)

m.y. and 10 m.y. which are apparent from the Aptian–Albian and Campanian to Eocene respectively. It is significant to note that a wavelength of around 3 m.y. in the scalograms of FS (6.3A and F) also corresponds to the sharp reduction in PF diversity at KT boundary and strong diversity increase at early Aptian and Miocene. Fourier analysis of relative PF origination data and extinction data reveal only the power of 31 m.y. They have also performed strict statistical test to examine the significance of these various peaks and stability in extinction record.

Further, in order to distinguish between random and deterministic components in the fossil record, they used correlation function analysis based on the concept of non-linear dynamical theory (Fig. 6.4). Their analysis suggested that there occurred a complex deterministic low dimensional pattern with dimension between 5–6 in the long-term planktonic evolution, in particular, in the extinction record. They have suggested that the occurrence of intense diversity fluctuations with 3–10 m.y. periodicity evident in wavelent spectra after major extinction events may be attributed to non-linear, self-organised evolutionary response to the availability of new escape. These coupled non-linear-periodic scenario may explain the recurrence of similar morphtypes in ~30 m.y. intervals (Prokoph et al., 2000).

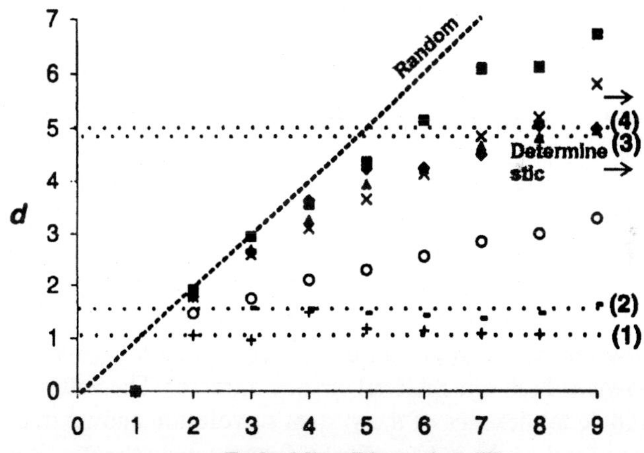

Fig. 6.4 Correlation Function Analysis: Correlation Dimension Vs Embedding Dimension Plot of Modeled and Planktonic Foraminifera Time Series. Dash Lines Mark Theoretical Approach of Purely Random Processes; Dotted Horizontal Lines (1–4) Show Approached Stabilisation Level of Dimensionality

(*Source*: Prokoph et al., 2000)

Modern View on Chaotic Mass Extinction Processes

Chaos is a notion of complexity and disorder in a dynamical system. The non-linearity is the chief cause of chaos. However, in many cases low dimensional chaos is believed to be predictable/determinstic to some extent.

The chaos defines the physical situation somewhere between the periodic and stochastic random processes. In most of the cases, chaotic processes are completely unpredictable whereas in some other cases quasi-determinism and/ or a quasi-periodicity can not be ruled out. This, however, depends on the degree of chaos present in the underlying physical system. The presence of chaos, however, certainly rule out the presence of randomness. It is now widely recognised that most of the geophysical and geological processes have some degree of chaos. And, hence, along with the search of cyclicity, it is imperative to look for chaotic behaviour of these processes also.

Patterson and Fowler (1996) have shown a possible evidence of self-organisation in plankton foraminiferal evolution records. They used three different types of data sets: number of species in time, percentage of extinction in time and percentage of speciation in time. They analysed these data using the techniques from the non-linear dynamical system theory. The analyses of a set of 196 extinction events and the same number of speciation data sets reveal the power law behaviour suggesting that these processes are self-organised in their evolution. 'Self organisation in physical system represents non-random processes dominated with high dimensional chaos.' More recently Dimri and Raviprakash (2001) examined the scaling behaviour of power spectra of extinction events in fossil record. The results appear to be quite useful for understanding the fractal pattern in the fossil record.

Further, the comparison of return maps of extinction and speciation data with random data show clear distinction between these processes (Fig. 6.5 A,B,C). For instance, the foraminifera data exhibit an affinity towards low value. This clustering can be indicative of determinism (Patterson and Fowler, 1996). In contrast to this, random data scatter uniformly across the return map and as expected does not show any direction or orderly behaviour (bottom). As pointed out earlier, the deterministic systems tend to evolve to an 'attractor' that can be point line, limit cycle or torus. This type of attractor is called 'low dimensional attractor'. There is also a highly complicated attractor known as high dimensional chaotic attractor. The nature of attractor depends on the complexities of the system's evolution and its dimension.

In order to visualise the nature of high dimensional attractor, it is essential to perform more rigorous non-linear analyses. Thus, using a predictor technique to search for correlations in higher dimension they found that the extinction data are consistent with low dimensional attractor. Further, their non-linear predictive analyses reveal that comparatively the extinction data are more highly deterministic than speciation data (Patterson and Fowler, 1996). As expected, the random data show zero predictability (Fig. 6.5D). Based on these results, they tested five ongoing hypothesis (Patterson and Fowler, 1996 and references therein):

(1) Periodically recurring logistic style events wherein the interdependencies caused by competition necessitate large drop-offs at population maxima; these population avalanches are followed by sharp rises of new successors;

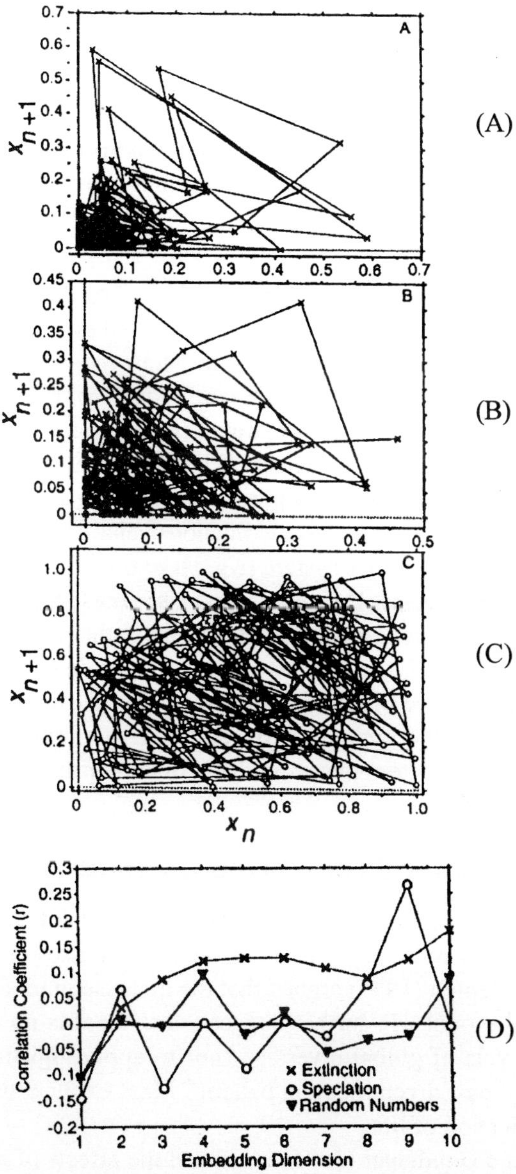

Fig. 6.5 Return Maps [Xn] Vs [Xn+1] for (A) Planktic Foraminiferal Extinctions (B) Planktic Foraminiferal Speciations and (C) Uniform Random Data. (D) Wimplex Time Series-predictor Analysis for Planktic Foraminiferal Extinctions, Planktic Foraminiferal Speciations and Uniform Random Data. Ordinate is Linear Correlation Coefficients for Predicted Vs Actual Values of Time Series as Function of Embedding Dimension (Abscissa)

(*Source*: R. T. Patterson and A. D. Flower, 1996)

(2) Periodic large extinction events driven by external events as predetermined by another system (e.g., large body impact.
(3) Completely random extinction driven by either internal or external stochastic events.
(4) Internally driven extinction processes where spontaneously derived interdependencies cascade through the ecosystem (Plotnick and McKinnney,1993); and
(5) Some combination of hypothesis 2 and 4.

Patterson and Fowler results rejected hypothesis 1 on the basis that speciation as well as extinction events in such a well-regulated system would have to be deterministic, or at least behave similarly. They also rejected hypothesis 2 because the data are not strictly harmonic. However, it is conceivable that pure results may be explained by extraterrestrial forcing at all scales, from micro-tectonics to large body impacts. They rejected hypothesis 3 because their extinction data did show organisation. They argued that their analyses did not preclude quasi-periodic large extinction events driven by external forces as pre-determined by another system (e.g. large bolide impact), or internally driven extinction processes. Their data support is in the form of a model whereby the internal organisation of an ecosystem regulates the response to changes in a deterministic manner, the relative scales of disturbances and extinctions depending on the degree of interdependency within the system. In total, these researchers rejected the hypothesis of randomness of extinction events and support for the quasi-determinisms in the extinction processes (Patterson and Flower, 1996).

EFFECTS OF MASS EXTINCTIONS ON BIODIVERSITY

One of the basic questions that has cropped up is whether the mass extinctions processes have any role to play on biodiversity processes or not. Based on a new database, Benton (1995) argued that the increase in biodiversity follows an exponential model. He further suggests that there is no need to assume equilibrium levels of global diversity, 'nor to apply logistic models to the investigation of past diversification pattern'. Mass extinctions played no role in the diversification of life.

Courtillot and Gaudemer (1996) examined the effects of mass extinctions on biodiversity using the database of Sepkoski. Contrary to the above arguments put forward by Benton (1995), these researchers find that diversification can be predicted by a simple model in which long period of growth obeying a logistic equation are sporadically interrupted by brief events leading to mass extinction (Fig. 6.6). They have suggested that the initial value of diversity immediately after a mass extinction, initial growth rate and subsequent equilibrium level, determine each segment of growth curve. The larger mass extinction that come after long periods without one seem to have been followed by a larger equilibrium level of diversity.

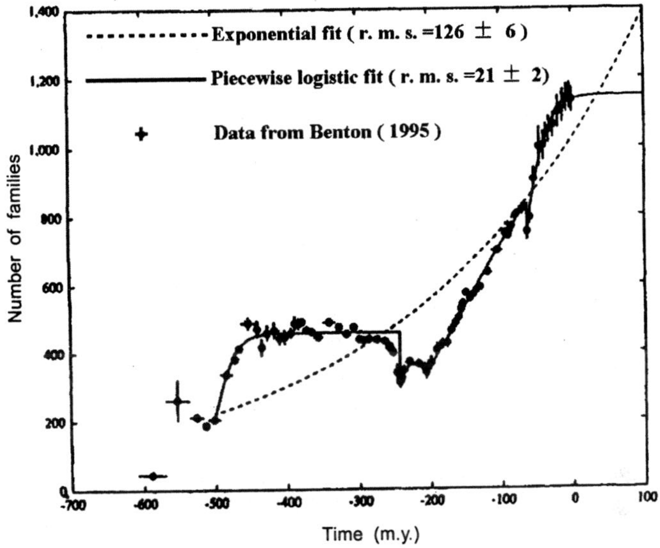

Fig. 6.6 Diversification of Families of Marine Organism through Time: Data and Models. Dotted Exponential Fit; Piecewise Logistic Fit

(*Source*: Courtillot and Gaudemer, 1996)

POSSIBLE CAUSES OF MASS EXTINCTION

There have been major controversies on the possible cause of biological extinction (Koeberl et al., 2003), Alvarez (1987) has discussed at length the possible causes of biological mass extinction and has argued that bolides impact was the main cause of biological extinction. He has provided several arguments and evidences against internal causes. Besides this study, however, there are several others who strongly believe that the main cause of extinction was not external but of internal origin (Officer et al., 1987). These arguments are presented here.

Internal

The possible cause of mass extinction has been a matter of great debate. Most of these debates are mainly based on the coincident correlation of extinction episodes with that of other geological/geophysical phenomena. Apart from this, there are several views which suggest that there are multiple causes of biotic crises. The cause and effect scenario is not very clear at present. The commonly cited terrestrial causes for the mass extinctions include climatic deterioration (Stanley, 1984; 1988), sea-level changes (Hallam, 1984) and global volcanism (Officer and Drake, 1983; 1985) and changes in the long-term CO_2 level, geomagnetic reversals etc., (Muller and Morris, 1996). Figures 6.7 and 6.8 show some qualitative correlations of these various terrestrial phenomena

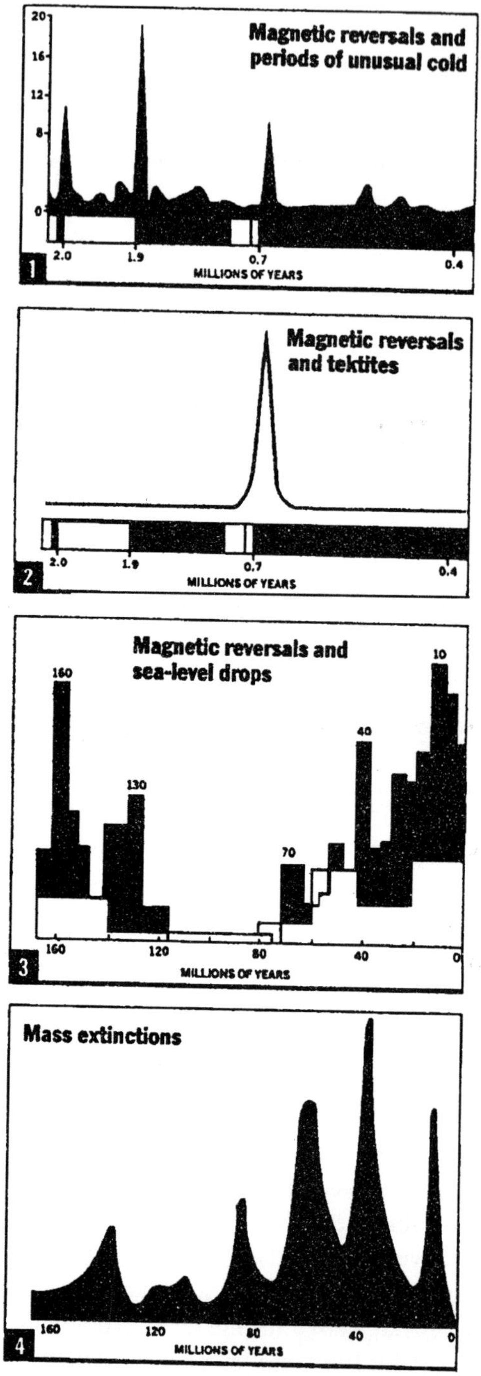

Fig. 6.7 Temporal Correlation of Magnetic Reversals, Sea Level Changes and Mass Extinctions

(*Source*: Fischer, 1988)

and mass extinction. Figure 6.7 gives some impression that extinction events correlate with other terrestrial events like sea level change reversals and CO_2, but this does not prove as cause and effect.

Climate and Sea Level Changes and Mass Extinction

There is evidence to suggest that sea-level and climatic changes could be connected with biological mass extinctions. Strong evidence exists for an association of extinctions with the time of lowering of sea level (Hallam, 1989). Loss of habitable conditions on the marine continental shelves have been suggested as a primary cause of extinctions. Climate deterioration has been claimed as the proximal cause in a number of cases, most notable in a recent study by Stanley (1984), of a regional extension event in the Pliocene.

Selective biological mass extinctions may be due to either sea-level fall or subsequent rise. Some of the well-known extinction events happened at the time of sea-level lowering and glaciation. However, these lows are also related to climatic deterioration and it is therefore somewhat difficult to resolve, whether cause of extinction was climatic deterioration, or sea level lowering or both. A strong correlation exists between the extinction of marine organisms throughout the Phanerozoic and major regressions and anoxic episodes associated with the subsequent sea-level rise (Hallam, 1984). The main regressive event took place not at, but shortly before the boundary (very close to the boundary) and estimated sea level fall and rise is to have been about 80–130 metres (Haq et al., 1987). There are many documented time of sea level lowering, some apparently caused by glaciations and others not, and many indications of long climate change independent of sea level. There are also many extinction events. Just to establish cause and effect is difficult and requires a rigorous assessment of probability in any complex time series. Considering an individual case of cretaceous Teritiary boundary (K-T), it is suggested that the K-T extinction event was preceded by several million years of global cooling and substantial sea level lowering. However, due to poor database for dinosaurs, it is not yet clear as to how could a regression cause their demise.

Alternatively, however, the frequent recurrences of regression–transgression couplets due to mantle disturbances and/or anoxic event may provide an important clue to global mass extinction events (Hallam and Wignall, 1999). They noted that the impact of super plumes on the base of the continental lithosphere may generate broad regional regressions due to the widespread doming effect prior to eruption and dome collapse. Conversely, plume generated doming of the oceanic lithosphere should generate transgression on continental shelves. Accordingly, the latter event is associated with the regression and superplume eruption within the Pangean supercontinent (in eastern North America and West Africa) and the former with the transgressive phase and possibly the development of super plume within the Panthalassa Ocean (Hallam and Wingall, 1999). Even if there are some such coincidences, the association of physical and biological event by itself does not prove cause and effect (Raup, 1986).

In connection with marine mass extinctions, researchers firmly believe that the calcareous plankton, foraminifera and coccolithophorides underwent a catastrophic extinction (Erwin, 1994). Furthermore, evidence indicates that the extinction of phyto-and zoo-plankton and disappearance of the carbon isotope gradient is abrupt and occurs over a period of less than 10^4 years (Alvarez, 1987 and references therein). This supports the case for catastrophic extinction and if so, is unlikely that a gradual environmental change occurring over a longer period of time could have resulted in a rapid and extreme collapse of marine ecosystem, unless some unknown environmental threshold was exceeded. The instantaneous environmental stress brought about by a bolides impact provides a possible explanation for the collapse of primary productivity and its coincidence with an abundance of spherules and platinum group metal at K-T boundary (Zachos et al., 1989; Marshall, 1998). Several researchers now seem to accept the impact hypothesis and are trying to explain how abrupt environmental changes induced by impact events lead to extinctions.

It has been shown, in context of sudden mass extinction, that the climate can abruptly change as a result of slow variations in, for example, the carbon dioxide content of the atmosphere (Crowley and North, 1988). Researchers suggest that such changes happen because the earth's climate may have two or more stable states for a given set of conditions. For example, with a particular concentration of carbon dioxide in the atmosphere, the earth might be stable either with or without polar ice cap. If the carbon dioxide concentration declines, a point may be reached where the ice-free state is no longer stable. The system must then flip into the state with an ice cap. Then, the concentration of carbon dioxide increases, the climate need not change back because the state with ice cap can still be one of the stable options at higher carbon dioxide concentrations (Crowley and North, 1988).

Flood Volcanisms and Mass Extinction

Rampino and Stothers (1988) suggested a correlation between the onset of continental flood basalt volcanism and mass extinctions in the last 250 m.y. with a possible periodicity of 32 m.y. for both phenomena. They suggested that the flood basalts are triggered by cometary impacts and subsequently the environmental effects of volcanism cause mass extinctions. Atmospheric and marine disturbances are produced directly by the impacts and/or indirectly via the induced volcanism. Burek and Wanke (1988) proposed specific model involving impact-induced phase change in the upper mantle leading to uplift and volcanism.

Although the theory of impact generated volcanism is contradicted recently (Ivanov and Melosh, 2003), Rampino (1987) has suggested that volcanic activity triggered by bolide impact can cause mass extinction (Stothers and Rampino, 1990). However, volcanism can be triggered internally also, without the intervention of impacts (Courtillot and Ciswoski, 1987; Loper et al., 1988). Flood basalts have identifiable plume tracks caused by the motion of the surface plates relative to a persistent source of magma (Richards et al., 1989; White and McKenzie, 1989). These researchers argue that an impact cannot promptly produce chemically distinct, persistent magmatism as seen in the flood basalts. Such magmatism must be produced internally, independent of surface disturbances. An internal mechanism with the capability of triggering the initiation of plumes and having a period close to 30 m.y. similar to extinction periodicity has been suggested by Loper and McCartney (1986), based on the analysis of the stability of the D-layer by Loper and Eltayeb (1985). Recently Morgan et al. (2004) suggested that large explosive carbon rich gases (eg CO_2, SO_2) released from cratonic lithosphere provide keeling mechanism of mass extinctions. Such deleterious gases are released by mantle plume incubation beneath cratonic lithosphere and typically found to be associated with the onset of continental rifting. They also stress that such explosive deep lithospheric blast could create shock waves leading to biological upheaval.

However, as noted by several researchers (Alvarez, 1987 and references therein) the above various hypothesis are in strong disagreement with three observations—the worldwide distribution of spherules, the greatly enhanced iridium concentration in the K-T boundary layer and the widely observed presence of shocked quartz in the layer. There is a possibility that bolide impact triggered volcanic eruptions would have registered about 13 on the Ritcher scale but such eruption alone cannot account for the extinctions. In a recent study, Becker et al. (2004) discovered a Permian-Triassic (P-T) boundary impact crater in part of the southern hemisphere (offshore of Northern Western Australia) coincident with well-known P-T extinction. They suggest that bolide impact may induce a volcanic outburst if the bolide strikes at a pre-existing hotspot. This might imply that, although an impact may not be a direct cause of large volume of flood basalt eruption or mass extinction, it may still act as a 'trigger' of event.

Geo-Chemical Events and Mass Extinction

It has been suggested that productivity crisis could have been one of the possible causes of biological crisis during geological times. A plot of major extinctions and impact events (triangle and square) and phosphorus burial rate (PBR) data according to cycle number depicts a correlation of the major mass extinctions with low PBR, with repeated cycle of 30 m.y. during the past 160 m.y. (Tiwari and Rao, 1999) (Fig. 6.8). A common periodicity and striking coincidence of low PBR and mass extinction dates seem to suggest a direct

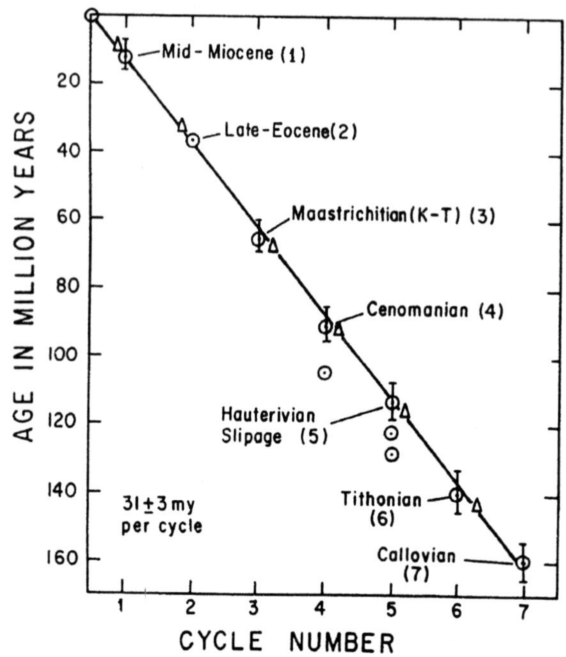

Fig. 6.8 Shows the Plot of Cycle Number with Age for CO_2, Phosphorus Burial Rate and Mass Extinctions

(*Source*: Tiwari and Rao, 1998)

or indirect link between them. However, how this PBR cycle links with mass extinction, needs to be explored. Modern views favour involvement of multiple geological/geophysical processes as causative factors for regulating cyclic biological diversification and extinction pattern. Most discussions have also concentrated on impact cratering coupled with episodic geochemical events as one of the possible pace-makers of biological extinction (MacDougall, 1988). He suggested that large bolide impact at the end of Cretaceous would have produced significant amount of nitrogen oxides by shock heating of the atmosphere. He examined two possibilities, (i) that the elevated $87_{Sr}/86_{Sr}$ rates was due to dissolution of bolide in sea water, and (ii) that the impact ejecta (vapour and solid) were the major source of the structure. He concluded that the resulting acid precipitation would have increased continental weathering greatly and could be an explanation for the observed high ratio of $87_{Sr}/86_{Sr}$ in sea water at K-T barely as he observed in his data. In another study Bhandari et al. (1994) have studied chemical profiles in K-T boundary section of Meghalaya (India) to explore the possible cause of enhanced geochemical anamolies. They observed that extinctions are confined to a short span but are probably not as sharp as anticipated by the asteroidal impact hypothesis. It appears that the Indian researchers are somewhat confused, perhaps due to truncation or 'smearing' of their record, popularly known as 'Signore-Lipps' effect as discussed in previous section. It is, therefore, somewhat difficult at this stage to single out any specific cause of biological mass extinction.

Whatever may be the driving mechanism, however, the available data from various sources indicate that a quasi-periodic geochemical cycle (PBR), possibly induced by impact events, may have played some role in driving biotic crisis (Tiwari and Rao, 1999). The regulation of ocean productivity is important in the global carbon cycle, which supports the contention that 'biological clock' is largely controlled by ocean fertility changes of the marine carbon reservoir. Possible link between fluctuation in PBR, biological productivity and vertical circulation of oceans have been suggested (Follmi, 1996). Accordingly, ocean productivity influences carbon supply to sediments and its burial efficiency, and organic carbon at a number of geological boundaries associated with a drop in productivity and biological mass extinctions.There is a dearth of rocks bearing fossil urchins just after the K-T boundary. These data indicate that there was a sustained drop in primary productivity after the K-T boundary, a conclusion also reached by others. The result indicates that sea urchins extinction was driven mainly by a nutrient crisis, that the adult life stage was more severely affected than the larval stages (Marshall, 1998).

Recent debate and discussion on the role of enhanced CO_2 at the end of Triassic and major mass extinction episodes have further highlighted the significant role of CO_2/global warming as cause of mass extinctions (Caldeira and Rampino, 1991; Beerling, 2002; Retallack, 2002). Impact generated volcanism is also probably relevant to the episode of global warming associated with the end-Palaeocene mass extinction event. Eldholm and Thomas (1993) have suggested that CO_2 output of North Atlantic volcanic province basalts could have triggered rapid environmental change through increased high latitude surface temperatures, in turn, changing deep-sea circulation and productivity. Hallam and Wignall (1999) have speculated that eruption at the time of Siberian Traps, the largest continental flood basalt province, had a primary role in generating a significant long-term increase in atmospheric CO_2.

External

However, the inferred link between extinctions and various earth's processes became quite controversial, and a large number of criticisms and counter arguments have been put forward against a terrestrial cause of biological mass extinction. Consequently, during the past decades, traditional views have changed and researchers have found compelling evidence that suggest that external forcing mechanisms might have played a critical role in modulating the extinction process. There are two mechanisms related to external forcing of biological mass extinctions. One is quasi-periodic bolide impact theory which has recently gained widespread popularity in explaining major mass extinctions (Rampino and others, 1984; Alvarez and Muller, 1984; Hut et al., 1984; Alvarez,

1987 and references therein) at different geological boundaries and the other is related to radiation hypothesis (Kitz, 1996).

Bolide Impact and Mass Extinction

Alvarez et al. in 1980 reported the first hand evidence of the iridium anomaly of exogenic origin at the K-T boundary. The main arguments in favour of Alvarez et al., hypothesis is embodied in interpretations of iridium enrichment, osmium isotope ratio, shocked quartz and spherules interpreted to be microtectites. They have discussed the possible cause of extinction in connection with K-T event. It has been noted that mass extinctions of the magnitude of the K-T events are relatively rare in Phenerozoic time. Similarly, collisions with ten km object are also rare, perhaps ten such events occured during last 600 m.y.

Hsü (1989) noted that the existence of large impact craters prove that collision with bodies releasing much larger impact energy have taken place recurrently during the geological history. Figure 6.9 shows the frequency and energy of impact follow the same log-log/inverse law for naturally occurring event (Wetherill and Shoemakar, 1982; Grieve, 1982). There is also general agreement that impact with earth of 6–24 bodies equal or larger than 10 km

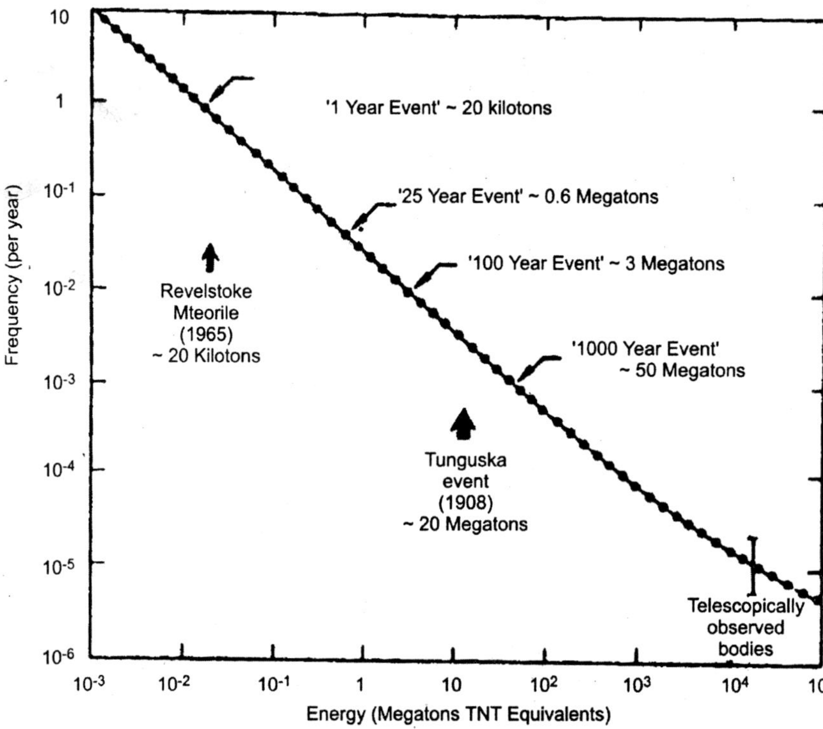

Fig. 6.9 The Log-log Relationship of Frequency vs Energy Release of Meteorite Impact Based Upon Data on Meteorites and Impact Craters

(*Source*: Shoemaker, 1983 and Hsü, 1989)

across should have occurred during the phenerozoic (Shoemaker, 1984). Evidence indicates that a number of major extinction events in the geological history have been found to be associated with stratigraphy of large body impact events (shocked minerals, tektites/micro-tektites and iridium anomalies) (Alvarez et al., 1980). Most of the researchers have suggested that perturbations of comets in the Oort clouds raise the probability of periodic comet impact shower on the earth (Rampino and Stothers, 1984; Davis et al., 1984; Alvarez and Muller, 1984). Accordingly, the collision of earth and a large piece of solar system debris such as meteoroid, asteroid or comet caused the great extinctions at different geological boundaries. A wealth of compelling evidence provided by Alvarez and his colleagues suggest that 65 m.y. ago a huge-sized object hit earth and caused the extinction of most of the then existing species bringing a close to the Cretaceous period of geological history and opening the Tertiary period. According to these researchers, a bolide 10 km in diameter struck the earth and enveloped it in an opaque blanket of dust leading to a number of consequences of the collision. For instance, they note that the fallen dust cut out the sunlight globally, thereby stopping photosynthesis. As a result, it was extremely cold during the period of darkness. Most of the animal life on earth died of starvation. After an impact, the temperature would go down but would increase as a result of 'greenhouse' effect and the heating was a major contributor to the killing.

Alvarez (1987) suggest another important killing mechanism, which involves shock heating of the atmosphere by expanding fireball. Accordingly, the high energy released by the impact would give rise to the production of huge amounts of nitrogen oxides in the atmosphere, leading to high acidic rain and surface ocean waters with pH low enough to dissolve calcium carbonate shells of marine invertebrates (Alvarez et al., 1980). The association of the hypothesis of a periodic extraterrestrial impact with the proposals of 26 m.y. periodicity of extinctions in the geological past (Raup and Sepkoski, 1984) and of a quantitative as well as qualitative difference between background and mass extinctions (Jablonski, 1986a, b respectively) has led to the concept of the biosphere being decisively shaped by mass extinctions triggered by more or less regular collisions of the earth with large celestial bodies.

Based on the impact theory, it has been predicted that an iridium layer would be seen worldwide, as reviewed by Alvarez, 1987. It is predicted that the impact of a bolide, 10 km in diameter, is very probable on a million year time scale. Accordingly, it is interesting to mention that iridium enhancements have now been seen at about 75 locations throughout the world at the K-T, and that all were laid down simultaneously in the same magnetic reversals zone. The iridium abundance of 0.5 parts per million found in certain primitive meteorites as representative of solar system debris testify the size of bolide about 10 km in diameter (Alvarez, 1987). The other evidences that support the impact theory are the presence of high temperature and shocked quartz in the K-T boundary layer. In addition to this, the K-T boundary clay includes an amount of soot that could only result from the burning of an appreciable fraction of all the biomass present on earth 65 m.y. ago. Alternatively, however, there is some counter argument which suggests that fire were set not by infrared radiation from the fireball over the impact crater. However, inspite of these arguments, there are some interesting findings based on chemical analysis, which provide convincing evidence to support the impact theory. For instance, researchers have reported that there are identical 'chemical fingerprint' of the boundary layer clay globally, when it was deposited (Ganapathy, 1982).

Radiations and Mass Extinction

Kitz (1996) has presented convincing argument that radiation-induced genetic loading of species, which leads to a high proportion of unsuccessful mutations, is a factor of mass extinction. According to him, the severity of these is dependent on the galactic component of cosmic rays, which is a variable element. He provides geological evidence that indicates that 'Killer cycle' is almost 74 m.y. He further suggests that the 'Killer cycle' is asymmetric, and crossing of the galactic plane occur alternatively at 44 m.y. and 38 m.y. intervals with its visible impact on biological extinctions. Mass extinction also occurs at these intervals because the level of radiation at the galactic plane is estimated to be higher than anywhere else in the galaxy. During the last 75 m.y., the earth, as a part of the solar system, traversed zones of the galaxy where the level of radiation was much greater than it is currently. This evidence is crucial because the time period almost coincides with the most famous Cretaceous mass extinction, author argues. Therefore, it would appear that mass extinction takes place at times of heightened cosmic ray activity. Same author also argues that during the geological time the level of cosmic rays impinging on the earth may have changed significantly on many occasions at the time of weak magnetic field or reversals, (Simpson, 1966) (seems to have been fluctuating periodically) causing severe increases in the genetic loading that affects all living things on the planet. The upper curve in Fig. 6.10 shows the sun's oscillation about the galactic plane, assumed to be periodic at 74 m.y. In this figure, vertical lines indicate times when the sun crossed the galactic plane and deduced for the periodicity of mass extinctions. Each occurred at the end of the period or epoch. Lower curve shows the extinction events. Crossing of the galactic plane by the sun takes hundreds of thousands of years. During this time, the level of cosmic radiation on the earth becomes abnormally high and has a profound effect on the biota. Some part of living species become extinct for environmental reasons while a great majority come to an end for genetic causes. It has been shown that the major factor in changes was the galactic component of the background radiation due to high-speed sub-atomic particles bombardment from space.

The hypothesis of periodicity of marine mass extinction is concerned intimately with catastrophic models of mass extinction. This is support for viable astronomical mechanism for extra terrestrial impact (Alvarez and Muller, 1984; Rampino and Stothers, 1984). However, this time structure of extinction also matches with that of other geological/geophysical phenomena suggesting possible coupling and multiple causes of biotic crises. Possible terrestrial

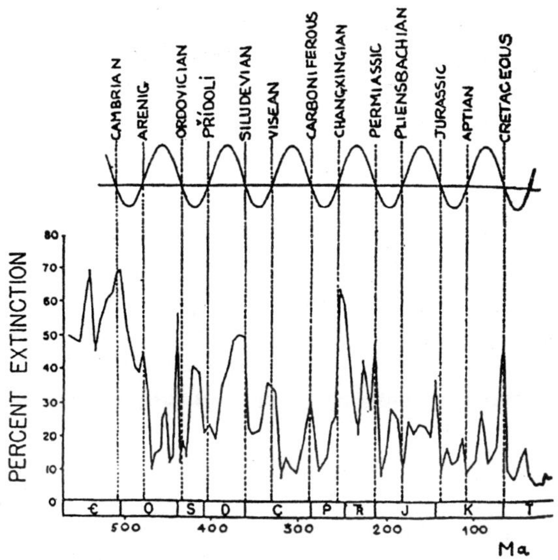

Fig. 6.10 Relation of Mass-extinction Events to the Sun's Asymmetric Oscillation about the Galactic Plane

(*Source*: Kitz, 1996)

causes, possibly impact generated, include flood basalts, sea-level changes; drop in ocean productivity and global warming (Rampino et al., 1987; Tiwari and Rao, 1999). Geological evidence also supports that the impact induced extinction evidenced by the iridium anomalies, shocked quartz and micro-tektites found at these temporal boundaries (Rampino et al., 1987; Rampino and Stothers, 1984; Stothers and Rampino, 1990; Alvarez and Muller, 1984).

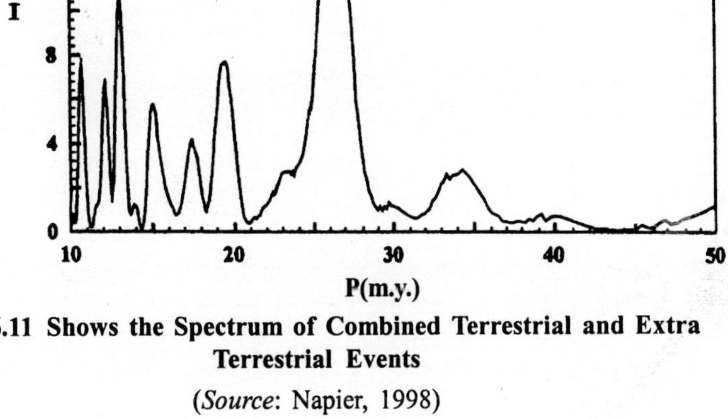

Fig. 6.11 Shows the Spectrum of Combined Terrestrial and Extra Terrestrial Events

(*Source*: Napier, 1998)

Time series analysis of combined records of terrestrial and extraterrestrial (impact events) reveal approximately 28 m.y. periodicity (Fig. 6.11) and may suggest that on this time scale all terrestrial processes are coupled and governed by a common extra-terrestrial process (Napier, 1998).

CONCLUSIONS

Several studies based on the empirical evidence support the case for periodicity in the marine mass extinction record while some others provide arguments against it. Better data set is required to strengthen the scientific facts. However, the present data available and empirical evidence strengthen the case for periodicity or atleast deterministic pattern with some degree of chaos, and clearly reject the Darwinian view and theory of randomness of extinction record. The new available data and their analyses using the modern techniques demonstrate the deterministic pattern in the evolution of biological processes. The possible causes of extinction cycle are also debated. Several researchers have provided explanations for the possible link of mass extinction in terms of external and internal forcing mechanism. It appears that galactic theory/ impact generated earth processes and their consequences on mass extinction seem to be most viable at present. The available extinction record and scientific arguments support the view that biological extinction are physically driven process.

7

Orbital Cycles and the Earth's Processes

INTRODUCTION

The influence of astronomical cycles as the earth's evolutionary processes are not confined to the mega-scale 10^7–10^8 years cyclic variations that we have discussed in the previous chapters, but also cover a much wider spectrum. Orbital forcing, for instance, is widely recognised as one of the main controlling factors for inducing cyclic patterns not only in climate records but also in several other geophysical processes. Periodic variations in the earth's orbit on the time scale of 10^4–10^5 years play crucial role in modulating terrestrial climate and other global earth's processes. The knowledge of the climate variability on this time scale is crucial to the understanding of how earth's environment and external forcing are coupled to transfer the heat input from the sun into climatic variations. This in turn, will help predict future global climate. Understanding the nature of climate variability is also important particularly because knowledge of natural climate variability on this time scale may provide some idea about the recent anomalous anthropogenic contributions and global change. This field of research is quite exciting and also presents one of the most challenging topics of inter-disciplinary researches in coupled earth–ocean–atmospheric system. The subject naturally focuses our attention on the complex, interacting components of the various dynamical earth's processes including the coupled global bio-geochemical cycles, solid earth, ocean and atmospheric processes, changes in chemical composition of ocean and atmosphere, and variations and the interplay of solar driven processes. In this chapter, we critically review the role of cyclic orbital variations on various earth processes, in particular on climate change, based on the high quality of published results.

Figure 7.1 shows the schematic view of the cyclic variations in the earth's orbital parameters that cause significant changes in insolation on the surface of the earth and atmosphere. This, in turn, causes significant changes in atmospheric pattern, ocean circulation and climate etc.

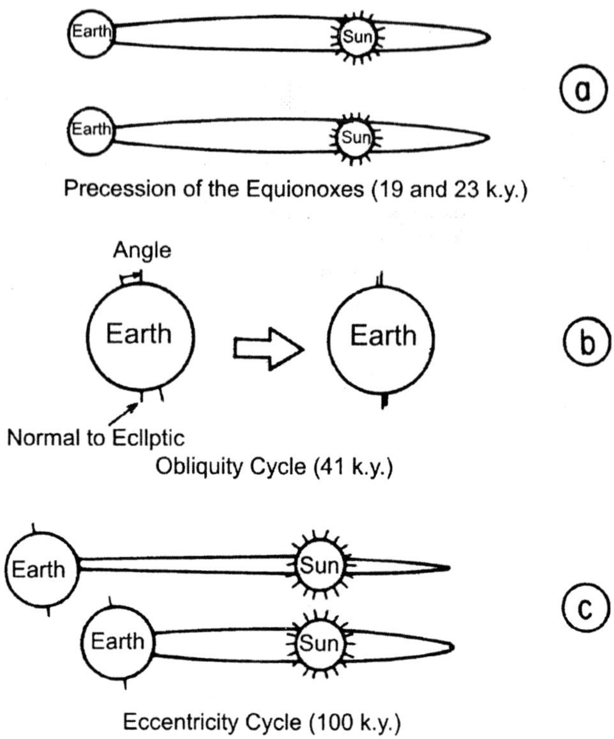

Fig. 7.1 A Schematic Diagram of the Cyclic Variation in the Earth's Orbital Parameters. (a) Precession (b) Obliquity and (c) Eccentricity

(*Source*: http://deschutes_gso.uri.edu/rutherfo/milankovitch.html)

The three main orbital cycles—eccentricity, obliquity and precession are discussed in chapter 3 and their beat frequencies arising due to non-linearity are the basis of the astronomical theory of climate change. Signatures of orbital cycles are recognised in the palaeoclimate variations and also in several other geophysical processes (e.g., geomagnetic secular variation, sedimentary sequences, sea-level changes and in the variation of CO_2. The physical basis for understanding the climate changes in terms of different astronomical parameters are:

(i) climate modelling in accordance with the catastrophic abrupt and periodic nature of climate variability (Alley et al., 2003),
(ii) consideration of contributions of various coupling and feedback processes and inherent non-linearity in the climate dynamics,
(iii) understanding the role of stochastic noises (white, red) (Alley et al., 2000),
(iv) separating the contribution of carbon dioxide and sea level changes and orbital variation etc.

MILANKOVITCH THEORY AND CLIMATE CHANGE: A HISTORICAL PERSPECTIVE

Although the development of the orbital theory for climate change was started during the nineteenth century by Croll (1864), later Gilbert (1895) observed a possible connection between sedimentary cycles and astronomical cycles. Bradley (1929) provided evidence of 21,000 years precession cycle in sedimentary column of the Eocene green major formation in the central and western United States. The theory was only recognised by the work of Milankovitch, and hence now widely known as Milankovitch theory, which was perhaps the first to present a full detail of astronomical theory of Pleistocene ice age. Yugoslav Vakian scientist, Milankovitch (1941) published new calculations of the orbital variations and the changes in amount of solar radiation, which is the ultimate driving force of climate change. He demonstrated that real changes in insolation affect the climate by computing the orbital elements. It becomes one of the major themes and a new paradigm in earth science.

However, important revolution took place more than 25 years ago in the field of climate research when, Hays et al. (1976) presented convincing support for astronomical cycles (Milankovitch cycles) of glacial rhythms. Berger (1977) presented full spectrum of astronomical data and astronomical cycles. Imbrie and Imbrie (1980); Imbrie, (1982) and Imbrie et al. (1989) provided a detailed account of orbital-climate theory and model to explain the Pleistocene ice ages. According to the Milankovitch theory, the ice ages are the consequence of secular variations in insolation caused by changes in the earth's orbital parameters. These orbital parameters are eccentricity (shape of the earth's ellipse), obliquity (tilt of the earth's axis), and precession (axial precession and precession of the equinoxes). Variations in these orbital parameters of the earth would produce variations in the seasonal and geographical allotment of insolation reaching the upper atmosphere tending to enhance the waxing and waning of large ice sheets and substantial alterations in terrestrial oceanic and atmospheric systems (Elkibbi and Rial, 2001). These studies have demonstrated that the astronomically induced changes in incoming solar radiations may be the most direct reason behind the growth and retreat of the quaternary ice-sheets. Berger et al. (1988) have shown that long-term insolation patterns are generally found to be quite consistent with climate variations. Figure 7.2 shows temporal variations of three main orbital parameters (e.g. eccentricity, obliquity and precession) over the past 800,000 years from numerical simulations. ETP in Fig. 7.2 (bottom) represents a normalised and summed combination of the eccentricity, tilt and precession. Shown on the right side of these curves are also the power spectra of each curve with dominant periods in thousands of years (Imbrie et al., 1984). The main orbital cycles are: 100,000 years (eccentricity); 43,000 years (obliquity); and 23,000 and 19,000 years (precession). According to Milankovitch theory, variations in terrestial climate depend on the insolation distribution at the different latitudes on the top of the

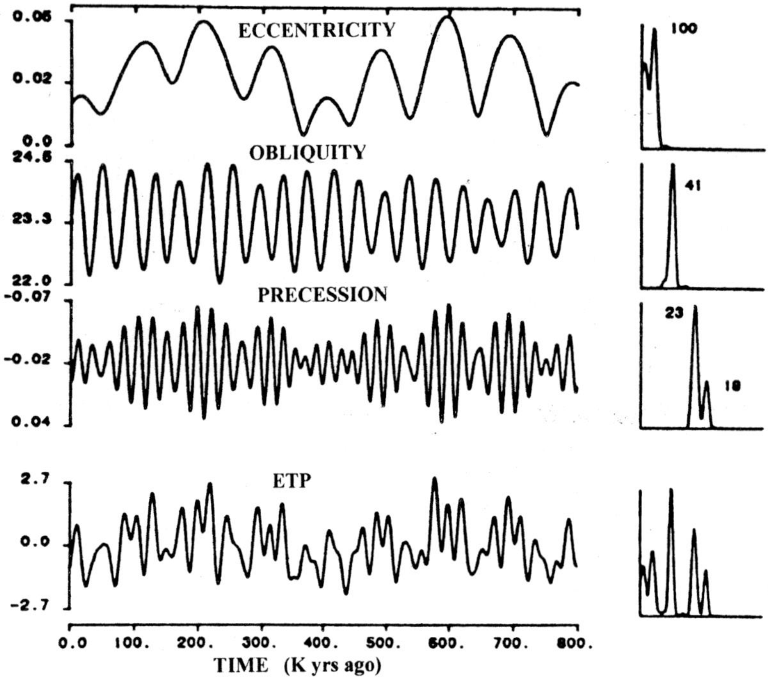

Fig. 7.2 Long-term Variations of Eccentricity, Precession and Tilt Obliquity Over the Past 800000 Years. Shown at Right Side of the Curve are Power Spectra of each Curve

(*Sources*: Imbrie et al., 1984)

atmosphere, which is influenced/forced by the changing position of earth's orbits around the sun. During the past years, several studies have established that the ice age and major climate rhythms are linked with such variations.

Climate Time Series

Temporal climate variation records are obtained by decoding the continuous record of plankton embedded in the ocean sediments, as the death rate of planktons is sensitive to climate conditions. It is posible to date and compare the magnetism locked up in the deep-sea sediments at the time of their formation with the magnetism of continental rocks that have already been dated by other means. The earth's magnetic field varies considerably over geological time. The pattern of changes of geomagnetic field, especially the intermittent reversals, provides a distinctive fingerprint. A stretch of sediments drilled from the seabed shows a particular magnetic fingerprint that matches well with the dated pattern of magnetism found on the continents. This method provides more or less precise dates for sediments on the sea floor (Gribbin, 1989).

It is widely recognised that sediment core act as a fossil thermometer. Researchers have shown that biochemistry of foraminifera is sensitive to changing global temperatures and hence can be used for constructing the climate history (Jouzel, 1999). According to Gribbin (1989), types of $\delta^{18}O$ isotopes and $\delta^{16}O$ are present in air and seawater (H_2O). Stable isotope ratios of $\delta^{18}O$ isotopes and $\delta^{16}O$ extracted from foraminiferal shells of marine organisms constitute a powerful tool for simulating former ice sheet volume (Bates and Jackson, 1980). $\delta^{18}O$ is heavier than molecules of seawater. Heavier molecules of water freeze more readily than lighter molecules. As a result, during an ice age proportionately more of the heavy molecules are locked away in ice. During an ice age, sea creatures take up lighter oxygen from the surrounding water to help to build their shells. By measuring the way the proportions of the two oxygen isotopes vary in different layers of sediments, geologists can infer the exact temperature at the time the sediments were deposited and also the extent of global ice cover (Gribbin, 1989). Thus sediments provide a direct record of the pulse of ice age-history.

It may, however, be noted that these ratios are not only simply proportional to past ice volume but they are also a function of seawater temperature and salinity. This is due to the fact that the foraminiferal shell becomes less enriched in $\delta^{18}O$ with increasing seawater temperature. Shackeleton and Opdyke (1993) suggested that in order to minimise the seawater temperature and salinity variability effects and to isolate the glacial volume control over the oxygen isotope ratios, it is appropriate to analyse foraminiferal samples from the tropical Pacific ocean or from deep sea environment showing rather minor temperature and salinity variations (Cronin, 1999).

MILANKOVITCH CYCLES AND SPECTRA OF CLIMATE RECORDS

Geological evidences indicate that there are a regular rhythm of ice ages and abrupt change in temperature with a roughly 100 k.y. years long interval. The Milankovitch cycles in climatic record due to variable solar radiation is well established for the Pleistocene age. Hays et al. (1976) reported statistically significant periodicities of 100 k.y., 41 k.y., 23 k.y. and 19 k.y. This evidence provided enough confidence to understand and accept the Milankovitch theory of climate change. Spectral analyses of climate records provided substantial evidence that a considerable fraction of the climate variance is driven in some way by insolation changes particularly accompanying changes in the earth's obliquity and precession. (Imbrie and Imbrie, 1980; Berger, 1978a,b, 1980). Researchers have compared the prominent peaks of the spectra of climate data with those of the astronomical cycles and have successfully explained much of the variance, atleast, at the periods corresponding to obliquity cycle (43 k.y.) and precession (21 k.y.). Several studies (Hay et al., 1976, Ruddiman and McIntyre, 1981; Imbrie, 1982), however, noted that the observed spectral peaks near the obliquity and precession depend upon the location of the core and the nature of the climate parameter analysed. These spectra depend upon the way the climate system react to the insolation forcing and upon the type of insolation to which it is sensitive (Berger, 1988). According to the Milankovitch theory, the high polar latitudes must record the obliquity signal whereas low latitudes record only the precessional one. Berger and Pestiaux (1984), however, note that the latitudinal dependence of the insolation parameters is more complex. According to these researchers mid-month high-latitude insolation in summer displays a stronger signal in the precession band than in the obliquity band.

However, the astronomical theory faces serious difficulties to account for the longer period variations in the eccentricity bands, especially 100 k.y. and 413 k.y. cycles. Although the climate spectra exhibits a dominant spectral power corresponding to the eccentricity cycle, the 100 k.y. variations in insolation forcing due to eccentricity change are too small (the eccentricity variations alter the total sunlight falling on the earth by only 0.2% at the most: (Berger, 1988) to be the direct cause of the great ice ages (Berger and Loutre, 1990, Imbrie et al., 1993). Researchers have further noted the absence of significant spectral power corresponding to 413 k.y. cycle in the oxygen isotope data in the last two million years (Imbrie et al., 1993, Mix et al., 1995a, Clemens and Tiedemann, 1997).

Figure 7.3 shows spectrum of climate variation over the past 0.5 m.y. (Berger, 1988). The vertical scale shows the relative amplitude of different climate cycles in the isotopic record of two Indian Ocean cores. The figure shows three main orbital cycles of 100 k.y. (eccentricity), 43 k.y. (obliquity) and 24 k.y. and 19 k.y. (precession). Figure. 7.4 shows coherency and variance spectra calculated for records of climate and orbital variations spanning the past 780 k.y.

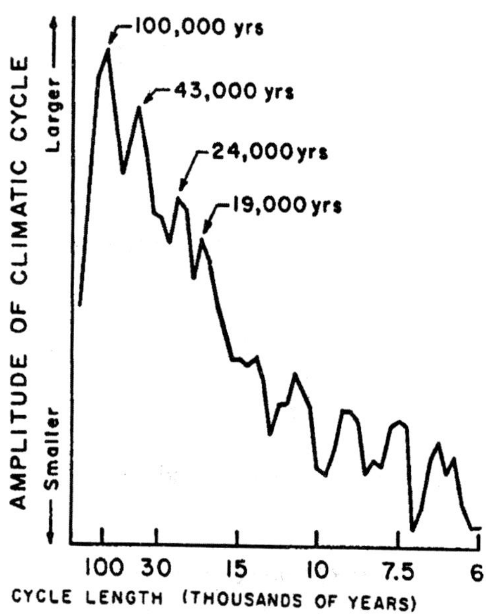

Fig. 7.3 Spectrum of Climate Variation Over the Past 0.5 m.y. The Vertical Scale Shows the Relative Importance for Different Climate Cycles in the Isotopic Record of two Indian Ocean Cores

(*Source*: Imbrie and Imbrie, 1979; Berger, 1988)

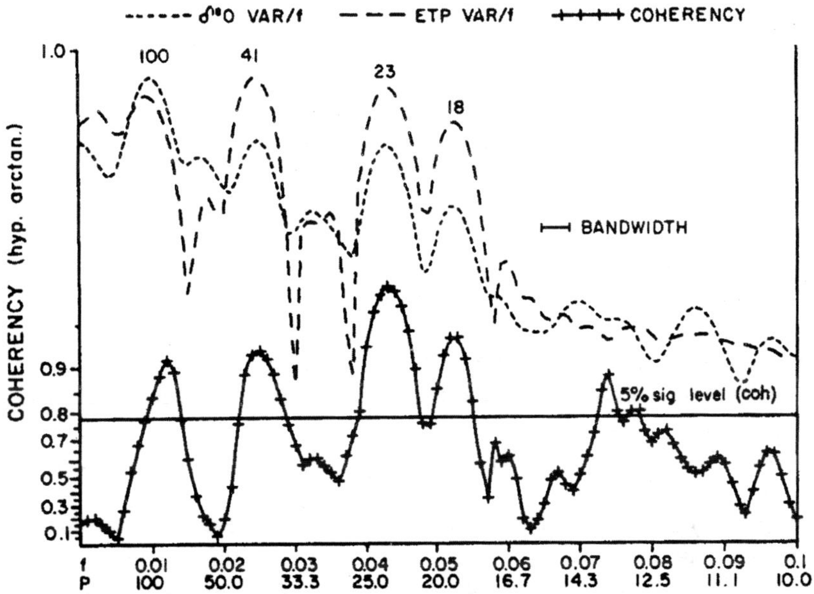

Fig. 7.4 Shows Coherency and Variance Spectra Calculated for Records of Climate and Orbital Variations Spanning the Past 780 k.y.

(*Source*: Imbrie et al., 1984; Berger, 1988)

It is prudent to mention here that the most dominant variance components in almost all the climate records centred near eccentricity cycle (100 k.y.) is found to be in phase with eccentricity cycle. However, large amplitude/power near this frequency needs additional non-linear explanation for its amplification mechanisms: by the deep-sea circulation, the glacial ice, the carbon dioxide and related mechanism such as ice albedo feedback, the elastic lithosphere, the viscous mantle and the ocean ice interactions etc. (Berger, 1988).

The observations also indicate that ice age is succeeded by a slightly warmer interval, called an interglacial, which lasts between 10 k.y. to 20 k.y. years. This longer Milankovitch cycle has repeated approximately 20 times in a little more than the past two million years. According to astronomical theory, we are living near the end of a natural warm spell, an interglacial that began a little over 18 k.y. ago (Berger et al., 1988).

Milankovitch Cycles through Geological Time

Milankovitch cycles have invariably been detected through almost all the geological periods. An important and thorough review to this account has been presented by Berger and his co-workers (1989a,b). The stratigraphic records of Milankovitch cycles from different parts of the earth's history have provided important observations, which suggest that different orbital cycles have dominated the global climate. In the late Pleistocene, for instance, the 100 k.y.

cycles are dominant. The early Pleistocene and Pliocene deep-water oxygen-isotope records are dominated by obliquity cycles (Hays et al., 1976; Imbrie et al., 1984; Hilgen, 1991a,b and 1989). Late-Miocene record is dominated with strong precessional signals (Tiedemann and Franz, 1997; Shackleton and Crowhurst, 1997; Hilgen et al., 1999) whereas early and mid-Miocene oxygen isotope records are dominated by obliquity signals. There is also evidence of higher order eccentricity cycles during this period (Tiwari, 1987a), which, as discussed above, might represent some non-linear climate response or some combination tones of obliquity or precessional cycles (Luc-Beaufort, 1994, 2001). Several studies indicate that deep-sea stratigraphic record of Oligocene period have been dominated with obliquity signals even at very low latitude. This, however, has been interpreted as the possible influence of Antarctica ice sheets on bottom water production and circulation (Weedon et al., 1997; Zachos et al., 2001). Further studies indicate that the late Eocene was dominated by precessional signals with a switchover to obliquity cycles at around 36 m.y. (Palike et al., 2001; Sloan and Huber, 2001). A large number of facies and every period of the Phanerozoic have revealed a remarkable evidence of Milankovitch cyclicity backed up by time series analysis (Olsen, 1986; Weedon et al., 1999; Melnik and Smith, 1989 and Park et al. 1993). Several studies have also noted that Mesozoic and Jurassic climate have also been dominated by different orbital cycles (Herbert, 1986; Hinov and Park, 1998; Olsen and Kent, 1999).

Deceptive Noisy Signals in the Climate Spectra

Presence of different kinds of noise in the climate time series, particularly in the unevenly spaced data series, poses one of the major problems in the climate data interpretation. Figure 7.5A shows two sinusoidal waves with periodicities of 100 and 41 k.y. embedded in white noise and Fig. 7.5B shows spectral analysis of unevenly spaced signal in Fig. 7.5A. Periodic components are clearly identified and peaks are statistically significant at 97% confidence level. (M. Schulz and K. Statteger, 1997).

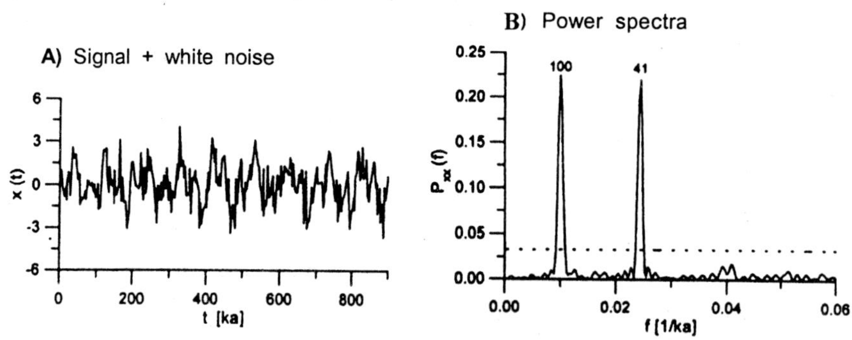

Fig. 7.5 (A) Two Sinusoidal Waves with Periodicities of 100 and 41 k.y. Embedded in white Noise. (B) Spectral Analysis of Unevenly Spaced Signal in (A). Periodic Components are Clearly Identified and Peaks are Statistically Significant at 97% Confidence Level.

(*Source*: M. Schulz and K. Statteger, 1997)

Figure 7.6A shows two sinusoidal waves with periodicities of 100 and 41 k.y. embedded in red noise. Figure 7.6B shows spectral analysis of unevenly spaced samples in Fig. 7.6A. It is clearly seen that, in addition to eccentricity and obliquity periods ≅ at 97 k.y. and 41 k.y. respectively, additional spurious peak of about 74 k.y. appears which is also significant (M. Schulz and K. Statteger, 1997). One should therefore be careful while interpreting climate spectra otherwise misleading information will be added up.

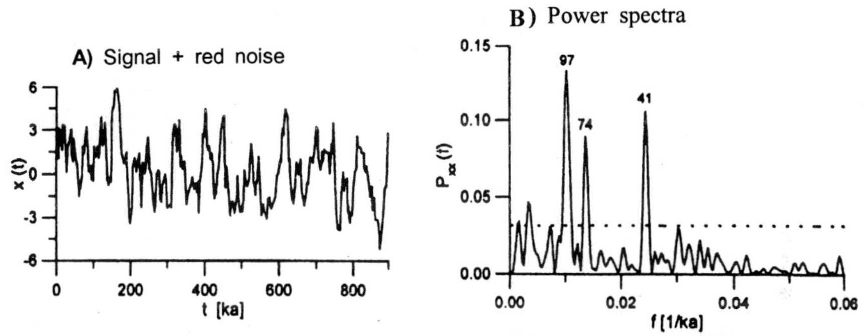

Fig. 7.6 (A) Two Sinusoidal Waves with Periodicities of 100 and 41 k.y. Embedded in Red Noise. (B) Spectral Analysis of Unevenly Spaced Samples in (A). It is Clearly Seen that, in Addition to Eccentricity and Obliquity Periods ≅ at 97 k.y. and 41 k.y. and Additional Spurious Peak About 74 k.y. Appear which is also Significant.

(*Source*: M Schulz and K Statteger, 1997)

Combination Tones, Beat Cycles and Non-linearity in the Climate System

The coupled atmosphere, ocean and cryosphere display the complex climate system variability. In addition to the main orbital cycles (e.g. eccentricity, obliquity and precession), spectral analysis of climate indicator records reveal the presence of significant variance at frequencies other than the main cycles. These are higher order eccentricity and beat cycles and combination tones of main orbital cycles, which probably arise due to non-linear interactions of various orbital terms. (Yiou et al., 1991; Ghil, 1994). The presence of these additional peaks possibly indicate that the problem of non-linearity and hence 'chaos' is inevitable in climate phenomena. Although quasi-cyclic patterns do exist in climate records, the long-term extrapolation of the present system of motion back into the distant past may, therefore, not be possible due to the presence of chaos (Lasker, 1989). From the theoretical point of view, a non-linear response of an oscillator is characterised by the creation of new frequencies and coupling among the frequency bands, which seems consistent with the observation (Elkibbi and Rial, 2001). Although several attempts have been made to explain such additional appearance of spectral features in the climate spectra, it is as yet unknown how and what non-linear mechanisms produce the combination tones of the orbital forcing. Researchers have carried out several theoretical experiments to understand whether the climate system might actually generate those frequencies internally with minimal external influence or such non-linearity has totally responded to external forcing.

Spectral, studies also indicate that besides three main orbital frequencies $10^{-5} - 10^{-4}$ cycles per year (cpy), higher frequencies in the range of 10^{-4} cpy to 10^{-3} cpy have also played a significant role in coupled climate variability. Since there are several other explanations for this range of frequencies, this places major limitations on our understanding of the complex climate dynamics in terms of astronomical cycles. The most obvious effect of non-linear system is splitting of spectral peaks and harmonic components. The underlying non-linear dynamical system produces the frequencies at sum and difference, when the system is excited at two or more frequencies eg., non-linearity and internal forcing mechanisms.

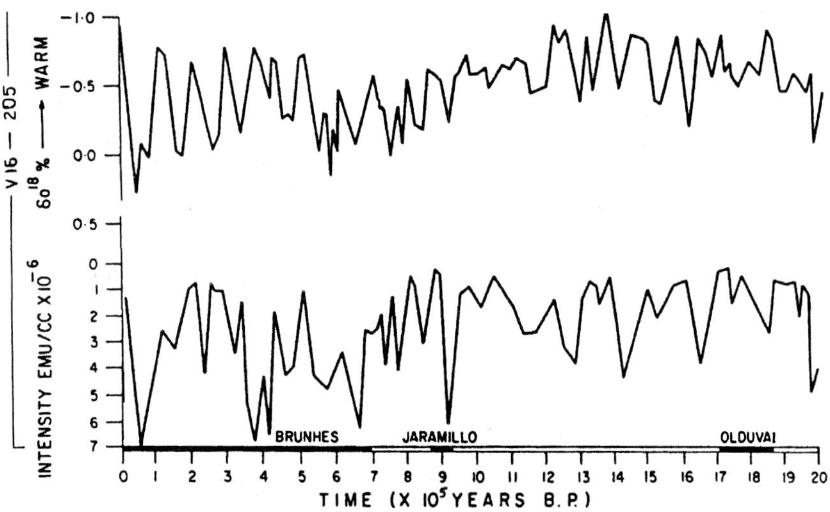

Fig. 7.7 Time Series of Ice Volume Changes as Reflected in Oxygen Isotope $\delta^{18}O$ Varieties (a) Climate Change Indicator; (b) Geomagnetic Intensity Variations During the Past Two Million Years

(*Source*: Wollin et al., 1978)

Figure 7.7 shows time series of climate variability (oxygen isotope and palaeomagnetic intensity variations) for the past 2 m.y. Figure 7.8 shows corresponding Walsh power spectra which, interestingly reveal an integrated picture of orbital periodicities ranging from 413 k.y. to 8 k.y. The above figure provides examples of main orbital cycles, splitting and combination tones etc. (Negi et al., 1993). A complex system composed of interconnected subsystems with different characteristic times can produce a response with quasi-periodic behaviour at linear combinations of the forcing frequencies. The choice between the possible combinations depends on the type of non-linear ties and on the values of the internal parameters (Elkibbi and Rial, 2001). The amplitude

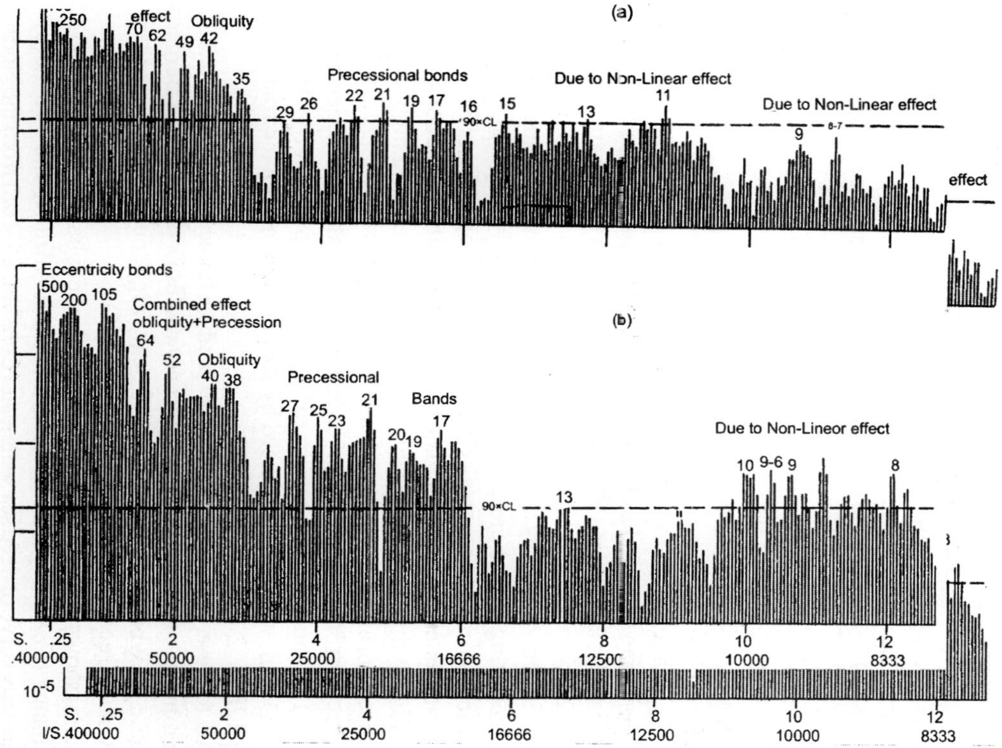

Fig. 7.8 Walsh Power Spectra of (a) Climate Change and (b) Paleo-magnetic Intensity Variations. These Spectra Show Clear Evidence of Integrated Picture of Orbital Cycles, which Ranges from 400 k.y. to 8 k.y.

(*Source*: Negi et al., 1993)

of the higher harmonics will depend on the non-linearity of the system but will, in general, decrease rapidly with increasing frequency. Wigley (1976) and Ghil and Letreut (1981) demonstrated through the mathematical models the effect of non-linearity in climate system. Following Wigley (1976), let us consider the effect of a non-linear response mechanism of a simple input signal with spectral peaks of frequency ω_1 and ω_2. It is expected that the interactions of these frequency will produce an output signal with frequency at $\omega_1 + \omega_2$, ω and $\omega_1 - \omega_2$. Wigley (1976) has mathematically demonstrated the case of spectral splitting of precessional cycle and eccentricity cycle in climate signals. It was witnessed that the orbital inputs spectra show a split peak near the precession period of 21,000 years and periods of 23,100 and 18,800 years. Wigley (1976) showed that non-linear interaction between these two frequencies could be expected to give an output signal with period 101,000 (ω_1–ω_2) term. The ω_1+ω_2 frequency corresponding to the 23,100 and 18,800 years precession peaks give a period of 10,400 years. These periods are also invariably present in climate spectra.

Pestiaux et al. (1988) have investigated the palaeoclimatic variability using the core with a high resolution sedimentation rate. Using a set of four deep-sea cores from the Indian Ocean, they have defined three frequency bands of high palaeoclimatic variability (Kerr, 1996). These are centred on the spectral maxima located at 10.2 k.y., 4.6 k.y. and 2.3 k.y. The frequency localisation of the spectral peaks is then refined by high-resolution spectral analysis. Interestingly, some of the resulting cycles are close to those previously detected in other palaeoclimatic records. Other peaks are also in good agreement with the response of a non-linear climatic oscillator forced by insolation variations. This link might indicate that the climatic system is characterised by a highly non-linear response to orbital forcing and that the precessional contribution of the forcing interacts strongly with the precipitation-temperature feedback used in the model. This kind of quasi-non-linear interaction is particularly obvious in the Indian Ocean, which is influenced by the periodic monsoon circulation (Kutzbach, 1981).

In addition to these, non-linear effects also produce higher order frequencies. There are a number of higher order eccentricity cycles as proposed by Berger (1977). Some of which are the result of beat frequency of longer order eccentricity and precession cycle (Stothers, 1987).

Figure 7.9(a) shows the time series of climate variations based on Globoratalia Menardii change (*Core*, Vol. 26-41) from the North Atlantic, and Fig. 7.9(b) Magnetic intensity curve of deep sea sediments (*Core* Vol. 26–108) from the North Pacific (Wollin et al., 1971) for the last 1.2 m.y. Figure 7.10 shows corresponding Walsh power spectra of Figs. 7.9(a) climate curve and 7.9(b) Palaeomagnetic intensity variations curve respectively. The two spectra clearly reveal strong power at 60 k.y. besides a 24 k.y. cycle possibly corresponding to precessional cycle. Here in climate spectra 150 k.y. cycle may be related to beat cycle of precession. It is significant to note that combined effect of precession and obliquity at 60 k.y. is dominant in these records (Negi and Tiwari et al., 1984).

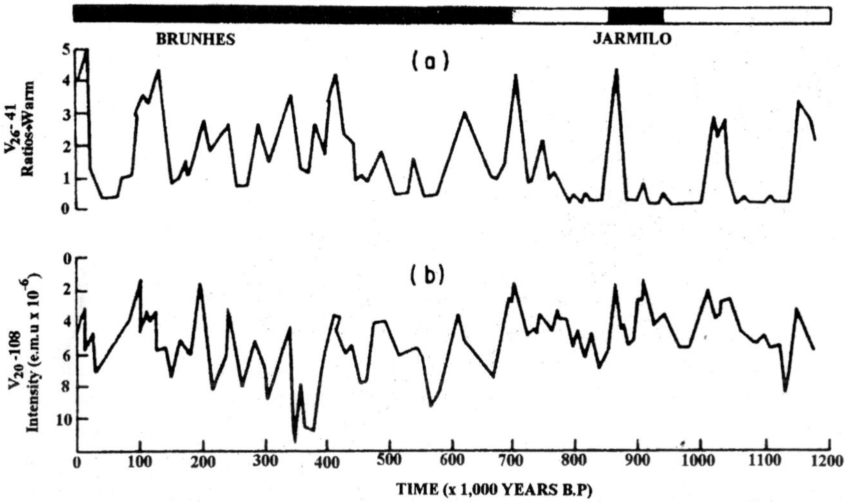

Fig. 7.9(a) The Globoratalia Menardii Climate Change (*Core*, Vol. 26-41) from the North Atlantic and (b) Magnetic Intensity Curve of Deep Sea Sediments (*Core* V26-108) from the North Pacific (Wollin et al., 1971) for the Last 1.2 m.y.

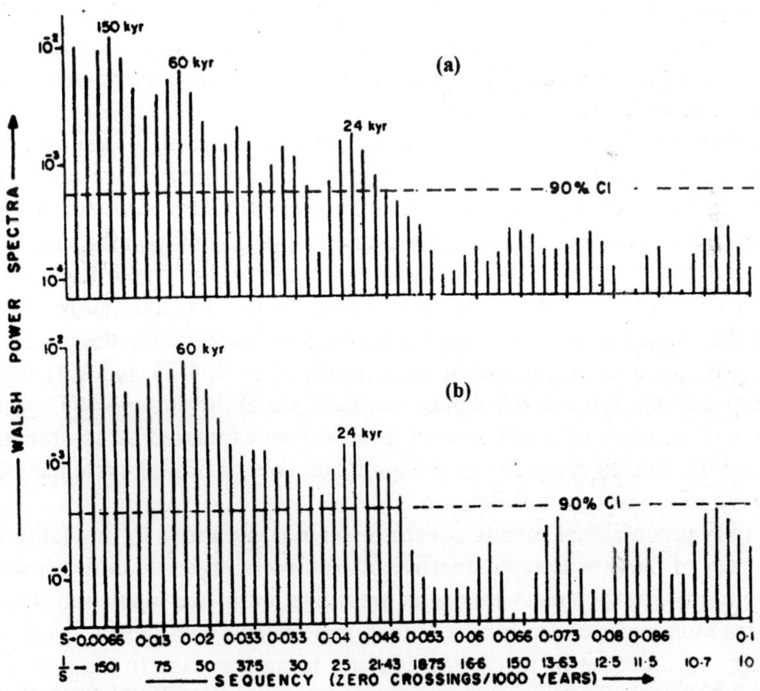

Fig. 7.10 Walsh Power Spectra of (a) Climate Curve and (b) Palaeomagnetic Intensity Variations Curve. It is Significant to Note that Combined Effect of Precession and Obliquity at 60 k.y. is Dominant in these Records

(*Source*: Negi and Tiwari, 1984)

Further, Tiwari (1987b) and Mix et al. (1995) found evidence of a 30 k.y. cycle in oxygen and carbon isotope spectra. Since this period is not the main component of orbital forcing, it may be related to the non-linearity of climate records. Mix et al. (1995) explains this cycle as a non-linear coupling of the 41 k.y. and 100 k.y. signals as the combination tone of 1/41+1/100 k.y. gives a 30 k.y. cycle. Alternatively, Rial (1999) accounts for the presence of these non-orbital cycles by frequency modulation, which creates new frequencies as combination tones of forcing frequencies. In addition to this, Tiwari (1987b) also found evidence of 57 k.y.–60 k.y. cycles in palaeoclimate indicator records. This, he suggested, might arise due to combined effect of precession and obliquity cycles as also proposed by Berger (1977).

A physical interpretation of these high frequency bands of climatic variability may be associated with non-linear response of the climatic system forced by insolation variations at the top of the atmosphere. The first feature of the 10.5 k.y., 5 k.y. and 2.5 k.y. quasi-periodicities is that they are roughly harmonics of themselves and of Milankovitch periods of 41 k.y., 23 k.y. and 19 k.y. found in the insolation variations (Berger and Pestiaux, 1984). However, in the case of a non-linear system, the problem is much more complicated. The response often contains a component at the forcing frequency, but this oscillation is no longer necessarily a sinusoid. It can be distorted, more or less asymmetrically. One can say that non-linear oscillators are characterised on the one hand by harmonics, sub-harmonics and combination tones of the forcing frequencies and on the other hand by a frequency-amplitude dependence of the response (Rial 1999).

Le Treut and Ghil (1983) have demonstrated the sensitivity of a simple non-linear climatic oscillator to orbital forcing developed by Kallen and co-researchers (1980) and by Ghil and Le Treut (1981). This oscillator includes a large part of the components of the climatic system acting on the time scale of glaciation cycles, namely, the global temperature, the global ice volume, and the deformation of the earth's crust under the ice. The internal frequency (δ) of the oscillator has been found to be roughly 6 k.y. Taking the insolation forcing frequencies δ_1, δ_2 and δ_3 respectively at 1/19, 1/23 and 1/41 k.y.$^{-1}$, the response exhibits amplified palaeoclimatic variability around 105 k.y. and 6 k.y. with clusters of peaks around these values expressed as combination tones of the forcing frequencies δ. Significant role of orbital cycle has been observed in monsoon amplification.

Mainly precessional effects govern insolation variations in low latitudes. The order of the precessional coefficients is always higher than the order of the obliquity coefficient. A monsoon amplification of the insolation forcing has been found independently by sensitivity experiments with general circulation models: an increase in insolation leading to an increase in temperature, precipitation and cloudiness in the monsoon countries (Kutzbach, 1981; Royer et al., 1983). This finding confirms the importance of the precipitation-temperature feedback, particularly strong in the Indian Ocean due to the monsoon circulations.

The Problem of Eccentricity Cycles in Palaeoclimate Records

In thorough reviews, Berger, 1988 and recently, Elkibbi and Rial, 2001 have discussed important issues of eccentricity cycle in palaeoclimate records. Accordingly, one of the major unresolved issues in the climate research has been the explanation of 400 k.y. and 100 k.y. eccentricity cycles. Several numerical studies and modelling based on pertubation of ocean chemistry, atmospheric CO_2, solid earth and non-linear physics have been put forward to explain the problem of eccentricity signal in climate record. Accordingly, the mid-Pleistocene transition was accompanied by the disappearance of the ~ 400 k.y. cycle from the data, in spite of its being the most prominent component of the eccentricity forcing in the last 600 k.y. The spectral analyses of the $\delta^{18}O$ records from 1.2 m.y. to the present reveal insignificant power at ~ 400 k.y. cycle but a highly amplified ~ 100 k.y. cycle. But, before 1.2 m.y., the data exhibits 413-, 124- and 95-k.y. periods, which correspond to the three main eccentricity cycles (Elkibbi and Rial, 2001). Researchers have found that the variance in the insolation forcing, mainly that centred in the 23 k.y. band, was shifted into the modulation eccentricity band (Clemens and Tiedemann, 1997). Spectral analysis of the sedimentary record of biogenic silica Lake Baikal over the last 5 m.y. reached to similar conclusions (Williams et al., 1997). Because the ~ 400 k.y. and the ~ 100 k.y cycles are all part of variations in orbital eccentricity, any model that simulates the ~100 k.y. cycles must also account for the absence of climate response at periods near 400 k.y. The main problem, however, is that most climate models are generally unable to simulate the 100 k.y. cycle without generating a proportionally powerful 400 k.y. signal. Imbrie et al. (1993) presented a thorough review of the multitude of hypothesis that have been proposed to explain the large amplitude late-Pleistocene 100 k.y. cycles. It is well known that the 400 k.y. cycle is virtually undetectable during this period and a 100 k.y. eccentricity cycle causes small variations in insolation compared to precession and obliquity.

The frequency modulation hypothesis, however, solves the puzzle of the absence of the ~ 400 k.y. cycle from the climate response (Rial, 1999). He argued that the spectrum of deep-water oxygen isotopes has the combination tones on either side of the main spectral peak near 100 k.y. that are characteristic of frequency modulation. The spectrum of a frequency-modulated signal does not exhibit the modulating frequency as a spectral peak, but rather as the separation between the carrier and its sidebands (Elkibbi and Rial, 2001). Imbrie et al. (1993a) propose that the climate system shows some sort of resonance at ~ 100 k.y. long cycles. To explain, they have generated a non-linear model with a single time constant combined with slow rates of ice sheets growth and fast rates of ice sheets decay. This model generates a 100 k.y. cycle with the saw-toothed variations matching with the characteristic of the late Pleistocene. Ridgwell et al. (1999), however, uphold the view of Imbrie and Imbrie, 1980 and emphasise that the spectrum of deep water oxygen isotopes can be explained by reference to saw-toothed characteristics.

Several other researchers, (Ghil and Childress, 1987b) however, assume that the non-linear response of the climate to the 23- and 19 k.y. precession signals produces combination tones that generate ~100 k.y. glaciation cycles with the right amplitude, while others (Mix et al., 1995) note that the precession bands in δ^{18} O and in δ^{13} C data from deep sea sediments do not evolve like 100 and 41 k.y. bands. Clark et al. (1999) propose that eccentricity may play a role in triggering delegation by modulating the precession cycle. This implies that the effect of precession is decreased in magnitude during periods of low eccentricity (Valdes and Glover, 1999). Imbrie et al. (1993a) have discussed the possible cause of amplification of the 100 k.y. cycle. Accordingly, the climate system seems to respond as a nonlinear amplifier to the ~100 k.y. eccentricity cycle. The non-linear energy response may possibly channelise from other frequencies into the ~100 k.y. band through feedback processes. Some recent evidence, however, suggest that the 100 k.y. cycle may not be the only one influenced by a non-linear response (Mix et al., 1995b). Rial and Anaclerio (2000) examined four time series of deuterium, sodium, CO_2 and methane concentration from Vostoke ice core and found evidence, which indicate that even climate response corresponding to 41,000 years obliquity cycle is actually non-linear. Shackleton (2000), found evidence of 100 k.y. cycle in marine records and claimed that non-linearity seems to be related to the deep-sea temperature rather than ice volume variation, as commonly assumed in various studies. Alternatively, several other researchers believe that the 100 k.y. glaciation cycles may be the result of non-stationary synchronisation between obliquity and eccentricity, with maintained synchrony reducing 100 k.y. periodicities in the climate response over the past million years (Elkibbi and Rial, 2001).

According to Ridgwell et al. (1999), the 100 k.y. cycle could possibly be paced by an integer number of orbital oscillations. Figure 7.11 shows a comparison between: (a) the SPECMAP stacked δ^{18} isotopes composite, and (b) insolation curve for 65° N, June 21. In Fig. 7.11, the arrows indicate quasi-periodic exceptionally weak insolation maxima which occur ahead of glacial-interglacial terminations by one precessional cycle (Ridgwell et al., 1999). Evidently, the most coherent pattern with the δ^{18}O spectra is a match consisting of every fourth or fifth precessional cycle (Fig. 7.11). Keeping in view complexity of climate dynamics, they however, do not rule out other permutations producing an equally good spectral match (Elkibbi and Rial, 2001)

Based on the ensuing discussions, it may be surmised that whatever may be the cause of dominance or amplifications of 100 k.y. cycle in climate spectra, the mechanism should be also to account for their apparently sudden appearance ~ 1 Ma BP at the mid-Pleistocene transition. It is interesting to note that the dominant 41 k.y. ice volume variations shifted to dominant 100 k.y. variations under essentially the same orbital forcing and the obliquity signal remains unchanged during this transition period, whereas the 100 k.y. eccentricity signal actually diminishes (Berger, 1988; Mix et al., 1995; Elkibbi and Rial, 2001)

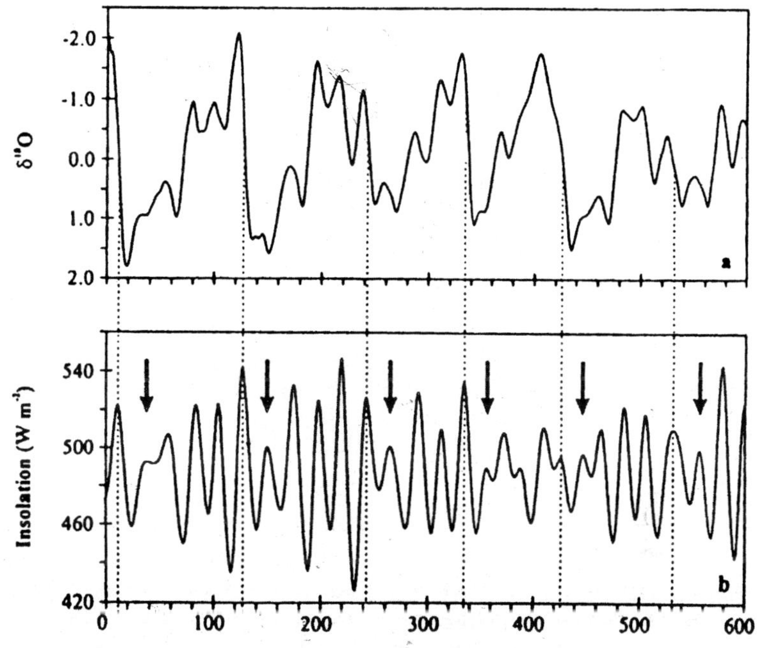

Fig. 7.11 Shows a Comparison Between (a) the SPECMAP Stacked Oxygen Isotope –18 (δ^{18}) Composite and (b) Insolation Curve for 65° N, June 21

(*Source*: Ridgwell et al., 1999)

(Fig. 7.12). Figure 7.12A shows marine oxygen δ^{18} data from Ocean Drilling Programme (ODP) site 849 over the last two million years (Mix et al., 1995b). Upper curves in Fig. 7.12B and C show filtered version of δ^{18} record using 100 k.y. and 41 k.y. filters respectively. The lower curve in B and C represent the filtered eccentricity and obliquity indices respectively (using the same filter as in 7.12A for δ^{18} records) (Berger, 1978 a,b). It is clearly evident that there is large amplitude corresponding to 100 k.y. periodicity in the ice data which started at about 1.2 m.y. while the amplitude of the 100 k.y. eccentricity cycle decreased and that of the obliquity remained unchanged (Mix et al., 1995a; Clark et al., 1999).

Modelling studies carried out by Maasch and Saltzman (1990) show that the ~100 k.y. cycles can be produced even without any external orbital influence. These models show that a non-linear climate can bifurcate in response to the atmospheric concentration of CO_2 crossing a critical threshold. They claim that this is a feature that mimics the mid-Pleistocene climate shift remarkably well. The main drawback of this model however is that one has to look for precise critical threshold level of CO_2, which is not possible back in time.

A coupled model of the northern hemisphere ice volume, which links insolation, atmospheric circulation and CO_2, the ocean mixed layer, sea ice, land-based sheets, and other continental surface conditions, provide a strong

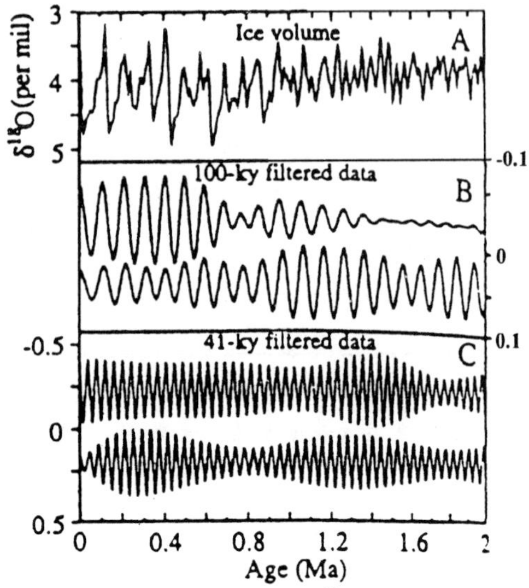

Fig. 7.12 (A) Marine Oxygen Isotope δ^{18} Data is from Ocean Drilling Program (ODP) Site 849 Over the Last Two m.y. (Mix et al., 1995). Filtering this Curve Yields the Upper Curve in (B) (100-k.y. filter) and (C) (41 k.y. time which are in Percentage and Offset from the Scale in (A). The Lower Curve in (B) and (C) Represent the Filtered Eccentricity and Obliquity Indices Respectively. (Berger, 1978 a, b)

(*Source*: Mix et al., 1995; Clark et al., 1999)

100 k.y. climate cycle between 1 m.y. and the present (Berger et al., 1999). The model shows that by decreasing linearly the CO_2 trend instead of quasi-cyclic CO_2 concentrations yields reduced amplitudes in climate oscillations (Gallee et al., 1992). The simulation outcome is comparable to the $\delta^{18}O$ series, in that, it exhibits a clear gradual trend of ice volume increase.

Figure 7.13 shows comparison between oxygen δ^{18} time series from Pacific site 849 plotted to the GSS97 time scale and the insolation curve (July 21 at 65°N; Lasker et al., 1993). Below the broken line on the oxygen δ^{18} data, 100-k.y. ice is detected. In Fig. 7.13, Roman and Arabic number labels indicate termination and selected isotope stages. The vertical broken lines show the mid-point of terminations based on GSS97 time scale. The inclined solid line on the insolation curve denotes the level below which 100 k.y. ice would nucleate and be stable (Elkibbi and Rial, 2001 and shaded areas after Raymo, 1997). Elkibbi and Rial (2001) however, note that there is no clear evidence of a steady decrease in CO_2 concentration over the past 3 m.y. Under low CO_2 conditions, observed after ~ 1 m.y. BP, only the largest insolation maxima, resulting from the combination of high eccentricity, high obliquity and minimum climatic precession, would produce ice sheet melting, paced with dominant ~ 100 k.y. cycle (Berger et al., 1999). There are several other suggestions also such as perturbation in ocean carbon chemistry, abrupt reorganisation of ocean circulation (Deblonde and Peltier, 1991) to explain

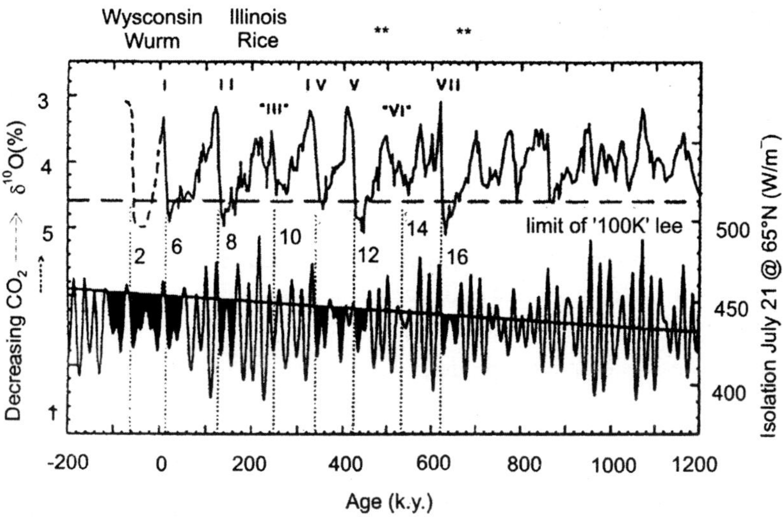

Fig. 7.13 Comparison Between Oxygen Isotope δ^{18} Time Series from Pacific Site 849 Plotted to the GSS97 Time Scale and the Insolation Curve (July 21 at 65°N after Lasker et al., 1993)

(*Source*: Elkibbi and Rial, 2001)

mid-Pleistone transition. Evidence indicates that the flipping was accompanied by a ~ 0.3 % decrease in mean ocean $\delta^{18}C$ between 0.9 and 1 m.y. BP, while the Atlantic-Pacific $\delta^{18}C$ gradient remained unaffected (Raymo et al., 1997). Alternatively, the underlying geologic substratum may provide the clue to understand the connection between the ice-sheet dynamics, the mid-Pleistocene transition, and the 100 k.y. glaciations (Clark et al., 1999).

The Devil's Hole $\delta^{18}O$ records (land record) indicate increase in the duration of the glacial cycles from about 84 to 120 k.y. during the middle-to-late Pleistocene. This has tempted Winograd et al. (1992) to suggest that these variable climate terminations/shifts are aperiodic in nature. According to them, this observation is inconsistent with the astronomical theory. However, they conclude that Pleistocene glaciations originate from internal nonlinear feedbacks within the atmosphere ice sheet ocean system. Rial (1999), however, shows that in the time domain, frequency modulation of a single frequency signal generates an output with periods varying slowly with time. In this case, frequency modulation of the high frequency 100 k.y. eccentricity signals by the lower frequency 413 k.y. cycle affected the observed increase in the duration of glacial cycles from about 80 to 130 k.y. (Elkibbi and Rial, 2001). The glacial-interglacial cycles may also be described as 'quantum' since the time interval between two terminations corresponds to either four or five precessional periods (Raymo, 1997).

Yet another suggestion to explain dominant eccentricity signal in climate spectra is given by Muller and MacDonald (1995). According to the

Milankovitch theory, the 100 k.y. glacial cycle is caused by changes in insolation (solar heating) brought about by variations in the eccentricity of the earth's orbit. However, as discussed above, researchers found it quite difficult to explain the large amplitude of 100 k.y. cycle. This has led them to seek for an alternative hypothesis. According to Muller and MacDonald (1995, 1997a,b) there are serious difficulties with orbital theory and hence they attributed the 100 k.y. eccentricity cycle to the inclination of the earth's orbital plane. They compared the power spectra of orbital models based on three potential drivers of the climate with the spectra of oxygen isotope records ranging from untuned to fully tuned (Fig. 7.14). The three simulations were based on:

(i) a linear eccentricity model,
(ii) non-linear precession based model,
(iii) orbital inclination model.

Their analysis show that the spectrum of orbital inclination model showed a single peak at the period of 100 k.y. which compares favourably well with ice data (Fig. 7.14). Based on this observation they argue that 100 k.y. glaciation cycles are forced by the variation in the inclination of the earth's orbital plane relative to invariant plane of the solar system. The mechanism that could link periodic oscillation of orbital inclination to climate is extra-terrestrial accretion of meteoroids or dust which in turn induce ice ages. Accordingly, if this mechanism is correct, they predict that a 100 k.y. cycle should be seen in ice and sedimentary records of extra-terrestrial accretion also. Several other

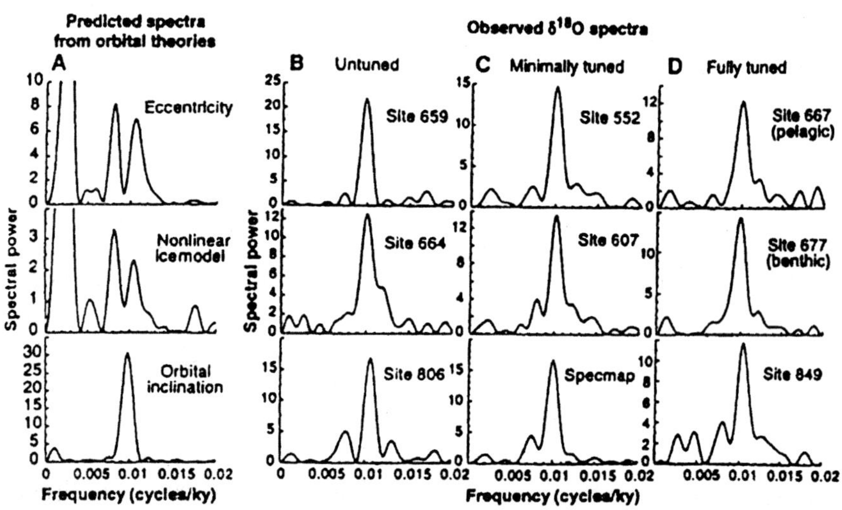

Fig. 7.14(a) Exhibits Power Spectra of three Orbital Models Based on Eccentricity (Linear), Precession (Non-linear) and Orbital Inclination. (b,c,d) shows the Power Spectra of Oxygen Isotope-18 Data Ranging from Untuned to Fully Tuned. The Spectra Corresponds to the Last 600 k.y.

(*Source*: Muller and MacDonald, 1997a)

researchers however, believe that such mechanism is unnecessary to explain the terrestrial climate record (Ridgewell et al., 1999; Rial, 1999; Elkibbi and Rial, 2001; Clark et al., 1999).

Changes in Periodicities Through Geological Time

Due to the gradually increasing earth–moon distance (partly due to tidal dissipation), the average period of the precession and obliquity are not fixed and both these cycles have lengthened at the rate of a few per cent per 100 million years (Berger et al., 1992). Several recent studies have shown that the two main periods of orbital parameters, precession and obliquity have slowed down/changed due to continued evolution of the earth–moon system. For instance, the values of precession parameter from about 16 k.y. and 18.5 k.y. in the early Palaeozoic changed to their present values of 19 k.y. and 23 k.y. The theory that describes the gravitational effects of the sun, the moon and the planets on the earth's orbit suggests that earth–moon distance has changed during geological times and could affect the period of astronomical forcing. The changing earth–moon distance, length of the day and dynamical ellipticity of the earth has influenced precession and obliquity. Table 7.1 (Berger et al., 1992) gives estimates of past value of obliquity and precesion. However, the eccentricity cycles are the product of interplanetary gravitational forces, hence they have probably remained stable through atleast the last 600 m.y. (Walker and Zahnle, 1986). However, several studies have noted that the dynamical ellipticity of the earth, affected by changing ice mass volumes and distributions and mantle convection, also affects the history of orbital change and hence insolation history (Dehant et al., 1990; Rubincam, 1995; Mitrovica et al., 1997; Forte and Mitrovica, 1997).

Table 7.1 Estimated Past Values of the Main Astronomical Periods, Derived Taking into Account the Slow Variations of the Earth–moon and Planetary Systems.

Age (m.y.)	*Periods (in years)*			
0	19,000	23,000	41,000	54,000
50	18,800	22,600	39,900	52,100
100	18,500	22,300	38,800	50,200
150	18,200	21,900	37,700	48,500
200	18,000	21,500	36,600	46,700
250	17,700	21,200	35,600	45,000
300	17,400	20,700	34,200	42,900
350	17,000	20,200	32,900	40,700
400	16,700	19,700	31,600	38,700
450	16,300	19,200	30,300	36,800
500	16,000	18,700	29,000	35,000

(*Source*: Berger et al., 1992)

Higher Order Eccentricity Cycles

Higher order periodicities (2000 k.y., 1200 k.y., 800 k.y., 600 k.y.) have been reported in climate and stratigraphy records e.g., in sea level and climate changes records (Tiwari, 1987a, 1993, 1997), in the range of 790 to 860 k.y. record (Park and Herbert, 1987), and in the range of 600 to 750 k.y. (Tenkate and Sprenger, 1989). Van Tassel (1987) has also reported similar cyclicity in carbonate/insoluble residue ratios of hemipelagic and sedimentary records (Olsen and Kent, 1999). Evidence of higher order periodicities was also suggested in mesoscale Phanerozoic sedimentary rhythms (Algeo and Wilkinson, 1988). Recently, Hein et al. (1992) and Negi et al. (1993) suggested higher order orbital cycles in ferromanganese-crust of the Pacific Ocean and short-term reversals wiggles, respectively (Table 7.2). It is also interesting to

Table 7.2 Correlation of Sea-level Periodicities with Similar Periodicities in Other Geophysical Records

Periodicities in Global Sea-level (k.y.)	*Climate (Tiwari, 1987) (k.y.)*	*Geomagnetic Reversals Beat Negi et al. (1993) (k.y.)*	*Ferro-manganese Crust Hein et al. (1992) (k.y.)*
2000	–	–	2040
1250	1270	1310	1310
880	980	800	800
660	–	–	–
416	416	400	–
260	200	–	–

note that Oslen (1986) analysed one of the best and exquisitely detailed sedimentary record of the orbital variation decoded from Triassic lake sediments. This analysis clearly reveals the cyclicity of the order of 100 k.y. 1.2 m.y. and 2 m.y., which also match with that of Tiwari (1987a). Most recently Kashiwaya et al. (2001) reported dominant orbit-related cycle of 400, 600 and 1000 k.y. in long-term continental record of Lake Baikal. It appears that changes in insolations were closely related to long-term climate globally. The range of all cycles corresponds well to the range of periods of the earth's orbital parameters as predicted by the Milankovitch insolation theory (Berger, 1977). The higher order periodicities found in the climate and sealevel changes records closely match the theoretical estimates of eccentricity beat cycles (Berger, 1977). This provides strong empirical evidence of orbital forcing or their control over the stratigraphic patterns in the past geological time. Although the physical link between climate and sealevel changes may be more intricate and need more explanation, a clear evidence of identical higher order orbital periodicities in the climate record and sea level obviously indicate that both processes are governed by a common cause. Glacial and deglacial cycles substantially modify the sealevel fluctuations on a global scale (Heckel, 1986; Mathew and Poore, 1980; Hodell et al., 1986). If we consider the 400-k.y. cycle corresponds to the principal eccentricity cycles (*Ep*) of the orbital variations, the higher order cycles of the sea-level oscillations may also be linked to the compound and amalgamated periods of lower order eccentricity as well as to higher order eccentricity cycles (e.g., *Ep/2, 3Ep/2, 2Ep, 3Ep, 5 Ep*).

In addition to this, Luc-Beaufort (1994) has related higher order cycles identified in marine climate record of Miocene periods as a possible result of long-term modulations of eccentricity and obliquity parameters. Luc-Beaufort (1994) performed spectral analysis on two palaeo-oceanographic time series for the past 10 m.y. His analysis revealed a number of possible cycles of order of 3.39 m.y, 2.30 m.y., 2.170 m.y., 1.615 m.y., 1.20 m.y., 1.164 m.y., 1.14 m.y., 0.86 m.y. and 0.780 m.y. He interpreted these frequencies in terms of modulation of eccentricity and obliquity cycles. Table 7.3 summarises the

Table 7.3 Correlation of Sea Level Periodicities with Different Versions of Orbital Cycles

Sea-level change (k.y.)	*Eccentricity Berger (1977) (k.y.)*	*Beat period Stothers (1987) (k.y.)*	*Obliquity/ eccentricity beat Luc Beaufort (1994) (k.y.)*
2000	2035	2031	2380
1250	1233	1282	1140
880	–	–	800
660	600	604	–
416	413	413	–
260	260	–	–

different views and explanations of higher order orbital cycles and the correlation with sea level changes periodicities. In view of striking similarities of our periodicities in the sea-level changes and orbital periodicities, Tiwari et al., 1997 suggest a combined role of orbital as well as modulation cycles of orbital variations in these processes. There could be two possible interpretations of the 1250 k.y. cycle that appears in the clean spectra (Tiwari et al., 1997). First, it could be simply the approximate integral multiple of 412 k.y., eccentricity component (412 × 3 k.y.), which also (1333 k.y.) appears in eccentricity expansion series (Berger, 1977). The second possibility could be that several weak cycles in sea level and climate records are the result of non-linear interaction of precessional cycles. Thus, it might also appear as a beat cycle due to non-linear interactions of two precessional cycles of 23.716 k.y. and 23.23 k.y. In addition to this, two other sets of precessional cycles e.g., t = 18.76 k.y. and t = 19.1555 k.y. will produce dominant non-linear periodicity of 2 m.y. which is also equivalent to slow variation in eccentricity cycles (Berger, 1977). The 'beating phenomena' of the two very close precessional frequencies (Wigley, 1976) or non-linear glacial-interglacial response to periodic forcing may possibly produce higher order sea-level periodicity.

Besides the above interpretations, it may also be suggested that dominant periodicities of 416 k.y. and 880 k.y. cycles in the sea-level data and climate, could be related to the basic eccentricity cycles of the earth's orbit. Although the explanation for the role of these eccentricity cycles in climate system and sea level processes are debated (Pollard, 1982; Wigley, 1976; Ghil and Letreut, 1981; Letreut and Ghil, 1983), theoretical climate models (Benzi et al., 1982; Nicolis, 1982) demonstrate the possibility of amplification of the eccentricity signal by the climate system itself. The external forcing due to orbital variations and internal stochastic mechanism could amplify the forcing effect on long-term alterations of climate and sea level changes.

Milankovitch Cycles in Other Geophysical Processes

In CO_2 Records: Orbital cycles have also been globally found in the variation of ocean primary productivity (Luc-Beaufort, 1997); temperature variations (Jouzel, 1987) and CO_2 records (Shackleton and Pisias, 1985).

This might indicate that the amount of carbon dioxide in the atmosphere is in some way linked to the orbital variations. Evidence also comes from the spectral study of CO_2 data decoded from Vostoke ice core samples (Barnola et al., 1987; Gentham et al., 1987) which reveal different orbital cycles. This might suggest that the natural greenhouse effect appears to have an adjustable thermostat, whose temperature setting is partly determined by astronomical influences. The finding is important from the point of view of coupling of sea level, climate and CO_2 variation with external forcing. It shows how changes in the concentration of carbon dioxide in the atmosphere affect the temperature of the globe. The evidence of orbital cycles in CO_2 record is particularly significant since the recent build-up of carbon dioxide caused by human activities is jumping against the natural trend, throwing in the working of the weather machine. The precise orbital cycle might help in separating natural and man-made anthropogenic effect.

***Magnetic Intensity Variations*:** The impact of Milankovitch cycles is not limited to the domain of climate, ocean, atmosphere and sedimentary rhythms. The orbital signatures have been well identified in secular variation record of earth's magnetic field. Kent and Opdyke (1977) spectrally examined the long-term secular time series of geomagnetic field variation obtained from deep-sea core sample and suggested a strong evidence for 43 k.y. periodicity. They suggested that the geomagnetic field intensity may vary with the earth's orbital obliquity due to its effect on precessional forces on the earth's core. It has earlier been proposed that precession plays an important role in the energy budget of the earth's geodynamo. Some other researchers, however, argued that the energy available from precession is atleast an order of magnitude less than that required for driving the geodynamo. Later, some researchers also suggested that there are statistically significant periodicities of the order of 43 k.y. and 25 k.y. in normalised intensity variations. Negi and Tiwari (1990b) suggested 41 k.y. and 25.8 k.y. periodicities in palaeomagnetic intensity variation and associated with the obliquity and precession cycle of the earth's orbit, respectively. Some researchers have also presented result spectral analysis of magnetic susceptibility records in which they found spectral peaks centred at around 42 and 23 k.y. This is an important result because for the first time a proxy data from terrestrial record strongly reflected indication of orbital cycles (Napoleone and Ripepe, 1989). Further, Yamazaki and Oda (2002) have examined variation of magnetic susceptibility time series of climate and suggested possible evidence of Milankovitch frequency and it's bearing on palaeoclimate. Fourier analysis of 2.5 m.y. long susceptibility record revealed strong spectral peaks of quasi-period of 100 k.y. and 40 k.y. corresponding to the eccentricity and obliquity cycle. More recent reports based on spectral analyses of sedimentary records of relative geomagnetic palaeointensity from different locations of North Atlantic sites reveal significant periodicities centred at eccentricity (100,000 years) and obliquity (41,000 years). These researchers also found identical eccentricity signal in the record of bulk magnetic properties, such as susceptibility, and therefore, attributed eccentricity signal to lithological variations controlled by eccentricity- driven climate change. Their analysis, however, did not reveal obliquity signal in bulk magnetic properties and hence they conclude that this cycle seems to be a property of geomagnetic field itself implying that orbital variations corresponding to this cycle play a role in forcing geomagnetic field intensity (Channel et al., 1998). Some other records of quaternary geomagnetic field intensity variations also show non-stationary orbital cycles (Saito and Kobayashi, 1989 and Saito et al., 1998). In another interesting study, Iorio et al. (1998) performed spectral analyses of long magnetic profiles of Cretaceous age in the Phoenix and Hawaiian lineation within the Pacific basin. Their analysis has shown two to five non-harmonic wavelength peaks at more than 90% confidence level within the range of several tens to hundreds of kilometers for each profile. For each profile with four-peak wavelength, their ratios correlate strongly, ('r' more than 0.98) with the ratios for the earth's orbital eccentricity periodicities for this timespan. This study might suggest the earth's changing orbital parameters, probably through their influence on the internal geomagnetic field, influences that magnetic anomaly signal.

The physical implication of these results can be understood as follows: (i) There is possibility that orbital cycles identified in palaeomagnetic intensity variations record could be of climate origin (ii) orbital cycles might influence core magnetic field. Controversy can be resolved only by analysing appropriate case samples. For the second case the following explanations/mechanisms have been suggested. The earth's magnetic field is believed to be generated by velocity differences in the outer liquid core and/or at boundary of core and mantle (Muller and Moris, 1986; Doake (1977). It seems to be logical to believe that orbital parameters must lead to changes in the torques (exerted by the sun, the moon and other planets) and therefore, they may well influence perturbations in the core-mantle dynamics. Consequently, any variations in coupling will be reflected directly in the properties of the earth's magnetic field (Mulkus, 1966; deBoer and Smith, 1994). In particular, Mulkus (1966) has suggested probable influence of precession of equinox on the core processes. Several others, however, have questioned the efficacy of gravitationally powered geomagnetic dynamo. An alternative explanation has been given where it is noted that variations of latitude and annual mean insolation induced by orbital variations would affect the earth's globally averaged temperature and hence, the surface distribution of ice and water. Doake (1977) and Muller and Morris (1986) suggested that change in latitudinal mass loading could alter the earth's moment of inertia (core process). Hence it is clear that orbital frequency may have some direct and indirect role on earth's core processes and hence on its rate of rotation. It could disturb the geodynamo and even lead to reversals.

Explosive Volcanic Activity: Evidence of orbital cycles is not restricted in geomagnetic and climate record but it is also clearly found in volcanic eruption record (Paterne et al., 1990). Power spectral analyses, of marine volcanic record indicate strong power variance at a frequency corresponding to a precessional period of about 23,000 years. However, due to poor statistical quality of the radiometric data, similar spectral analyses have not shown significant peaks in the continental records. According to these researchers, the value of the spectral peaks in marine data is more precise since the number of dated events in the marine environment is significantly higher than that of the continental data. The observed volcanic periodicity is very similar to that of the precession of the earth's, which is also exhibited in various other geophysical phenomena. Paterne et al. (1990) have also provided physical significance of this periodicity in astronomic terms. Accordingly, the solar and lunar attraction on the earth's equatorial bulge leads to changes in the earth's rotational motion and orbital revolution, periodically. It creates solid earth tides, which modify the speed of rotation with shorter time constants. The solar and lunar tides account for quasi-periodic stresses in the earth's crust of less than 10^{-2} bar with a maximum change rate of 10^{-5} bar/s. Stress accumulations ranging from 10^{-7} to 10^{-4} bar/s were observed in some volcanic areas prior to eruptions supporting the hypothesis of a tidal triggering of the eruption (Paterne et al. 1990). A detailed volcanic record extending back to 190 k.y. in a single area has shown some very interesting results. Spectral analysis of the past global ice volume changes as recorded by the $^{18}O/^{16}O$ variations of foraminifera shows well-marked peaks corresponding to periodicities of about 23 k.y. (Precession) and 45 k.y. (Obliquity) (Paterne et al., 1990) which are very similar to explosive volcanic periodicity indicating that both phenomena may be linked to the precession cycle of the earth's orbit. The volcanic change does not show a marked periodicity at 45 k.y. however. Hence the authors cautioned that the results must be viewed with care, especially, because a global phenomenon like orbital variations and climate change have been invoked as an association and explanation. Similar periodicity in explosive volcanic activity must be investigated in other areas. Nevertheless, the data described here appears to provide encouraging consideration of oscillating volcanic activity probably related to a periodical replenishing of shallow magma chambers.

***Some Other Environmental Records*:** Recognition of orbital cycles in various other environmental records has been reviewed by deBoer and Smith (1994). Accordingly there are also some evidences of significant orbital cycles in the temporal records of lake deposits, aeolian deposits, palaeosols, shallow marine carbonate, storm intensity, calciturbidites, terrigerious clastics and deep sea benthic species diversity (Cronin and Raymo, 1997).

CONCLUSIONS

The Milankovitch theory of climate change, particularly its impact and understanding during the last two decades, has been one of the major revolutions in the earth science after the plate-tectonics. The theory has provided several important clues to understand the palaeo-climate variability on this time scale. As has been discussed, evidence for these astronomical cycles are also accumulating in several other geophysical processes. The evidence for orbital cycles has been found not only in Pleistocene but also almost throughout the geological times, thus suggesting a basic forcing mechanism of terrestrial climate and its direct/indirect impact at another geo–bio–ocean–atmospheric phenomena. However, there are several controversies too, which are yet to be resolved. For instance, the case of 100-k.y. cycles is still one of the major controversial problems in climate dynamics. If these issues are resolved, the theory can be used more confidently for tuning the geological records as well as in interpretation and prediction of climate change not only in the Pleistocene but throughout the geological history.

8

Solar and Tidal Cycles in the Earth-System Processes

INTRODUCTION

Understanding the natural variability of climate and related earth's phenomena on a still shorter time-scale is of immediate concern for the society in atleast three ways (Overpeck, 1995):

(i) by rendering information of the patterns, ranges and causes of natural climatic variability.

(ii) providing a base line of observations that is long enough to unambiguously separate natural from anthropogenic environmental change.

(iii) for preparing a framework for testing the ability of predictive model of climate system response in different conditions.

It is now well recognised that earth-system (earth–ocean and atmospheric) processes are coupled dynamical system. Variations in the response of coupling significantly influence the oceanic–atmosphere–biosphere system and hence prediction of shorter-term natural climate variability. Hence, a clear understanding of the coupling between the atmosphere and the ocean is of fundamental importance for predicting the climate variability over seasonal to millennial time scales. This is one of the most difficult tasks facing climatologists. One of the reasons for the poor understanding of the climate behaviour is that the climate that we experience results from complex non-linear system that stem from both ordered and chaotic forcing mechanisms. This complex non-linear behaviour is dominated with characteristics of multiple interactions between many different components of internal and external forcing. This hinders the prime objective of long-range (decadal-scale) climate prediction and requires not only sound data, robust techniques for data analysis but also appropriate interpretations and physical linkages.

Palaeoclimatic information can be obtained from a variety of archives of palaeoinformations like Lacustrine and marine sediments, tree rings, speleothems, ice core and peat bogs besides from instrumental historical records. Historical and proxy records decoded from such sources suggest that there have been significant quasi-periodic changes and variability not only, in climate and temperature but also in various other terrestrial phenomena such as, sea level, secular variation in the earth's magnetic field etc. Various studies conducted during the past years suggest that pattern of decadal and interannual millennial bands of climate variation over the past years are generally compatible directly or indirectly with one or combination of several hypothesised causes (Schimel and Sulzman, 1995; Clark et al., 1999). These are:

(i) random variability in the atmosphere,
(ii) inherent or forced variability in the ocean dynamical system,
(iii) solar variability and tidal forcing,
(iv) variability in volcanic and anthropogenic aerosol loading of the stratosphere and the troposphere. (Overpeck, 1993).

It is well recognised that tidal waves generated by 'the planetary orbital motions are dominated by contribution from Jupiter, Saturn, Uranus and Neptune. These planetary orbital motions transmit momentum by gravitational torques causing changes in velocity and spin rate of successive planets and to the sun itself (Fairbridge, 1984). Several researchers have also suggested that besides ocean-atmospheric influences, tidal and solar cycles (solar irradiance) have visible impact on our terrestrial climate. Hence, during the past few years, the search for a link among the solar, tidal and earth–ocean–atmospheric processes, on a time scale of 10°–10^4 years, has been very encouraging. The noted astronomer, William Herschel, in the early nineteenth century, was perhaps the first to explain weather variation and climate change by extra-terrestrial means as early as in 1956. Heinrich Schwabe (1843) discovered that the number of sunspot varied in a periodic manner. He also provided some credential support to the weather and solar cycles relationship. Sunspot cycles, reflect basic changes in solar behaviour and provide quite satisfactory explanation to various climate and weather phenomena. Burroughs (1992), in his excellent monograph, *Weather Cycle: Real or Imaginary* has reviewed the influence of tidal and solar phenomena on the weather cycles.

The sun and moon raises tides in ocean as well as in solid earth due to their gravitational attraction. The gravitational force depends on distances following the inverse square law. The centre of gravity is very close to inside the earth's centre compared to the sun, the moon tide is three times greater (Fairbridge, 1984). The tidal bulge in the ocean tends to stay aligned with the slow moving moon because the earth is orbiting around the sun and the moon is orbiting around the earth. The gravitational torques/attraction of sun and moon generate, therefore, tidal forcing on the earth resulting from the combined pull of the moon and the Sun (Weedon, 2003). These tidal forcings are quite complicated phenomena and widely affect the movement of the earth–ocean–atmosphere. The changing positions of other planetary bodies also generate tidal effect but play a minor role. Hence, there are longer tidal cycles too widely known as 'planetary conjunction cycles', which may also significantly affect the earth-atmospheric and oceanic processes (Negi and Tiwari, 1987). The tidal effect of planetary conjunction cycles is quite small. The effect arises possibly due to short-term effect on basic energy flow from the earth's moon and sun system due to changes in their orbital angular momenta and disturbances in orbital configurations of all other planets.

In this chapter, we focus on the possible influence of external forcing (role of solar and tidal cycles) in controlling or triggering/modulating the terrestrial climate and related phenomena. Our emphasis here is to discuss the results in frequency range of month to year and decade to several centuries. In some of the phenomena, such as climate and sea level changes, there are some convincing evidences of presence of tidal and solar cycles while in many other cases this evidence is not very clear and may be indirect only. We summarise these results and discuss their physical linkages.

EVIDENCE OF TIDAL AND SOLAR CYCLES IN VARIOUS EARTH–OCEAN AND ATMOSPHERIC PROCESSES

Sea Level Changes

It is widely recognised that sea level changes globally on varying time-scales of a year to a century. There are several internal and external causes of this variability. The internal causes include the changes in ice mass on land due to global warming, the temperature of ocean water and water aquifers or surface reservoirs. The external causes might include the tidal and solar variability. Understanding these various forcing mechanism is essential for predicting sea level changes.

The analyses of short-term sea level changes records have shown possible evidence of various solar and tidal cycles. Fairbridge and Hillaire (1977) visually identified a summer storminess cycle of about 45 years from an uninterrupted time series that goes back more than 8,000 years, obtained from 183 uniform beach ridges of the Hudson Bay region. Apparently, the visually inferred cycle is nearly comparable to double Hale solar cycle (22 years) suggesting that the underlying mean frequency of 45 years in the emergence of beach ridges glacial ices retreat of the Hudson Bay during the past 8000 years might have its origin in solar sunspot related phenomena.

Global sea level changes have also indicated longer cycles of about 16–20 successive low and high peaks during the past 8000 years (Fairbridge and Hillaire, 1977). Visual inspection of this record shows a 400–500 years cycle (Fig. 8.1a). The maximum entropy method (MEM) spectral analysis of sea level changes curve reveals three stable periodicities around 1083, 634 and 464 years (Negi and Tiwari, 1987) (Fig.8.1b). Stability of these three periodicities is remarkable. It is interesting to note that cycles of 1083 and 634 years in sea level changes record correlates fairly well with the known planetary conjunction cycles of 1133 and 567 years respectively as discussed in chapter 3.

Astronomical variations of 550, 1100, 1650 years are known as periods of earth/moon orbital motion. It is noted that maximum perigal spring tides repeat

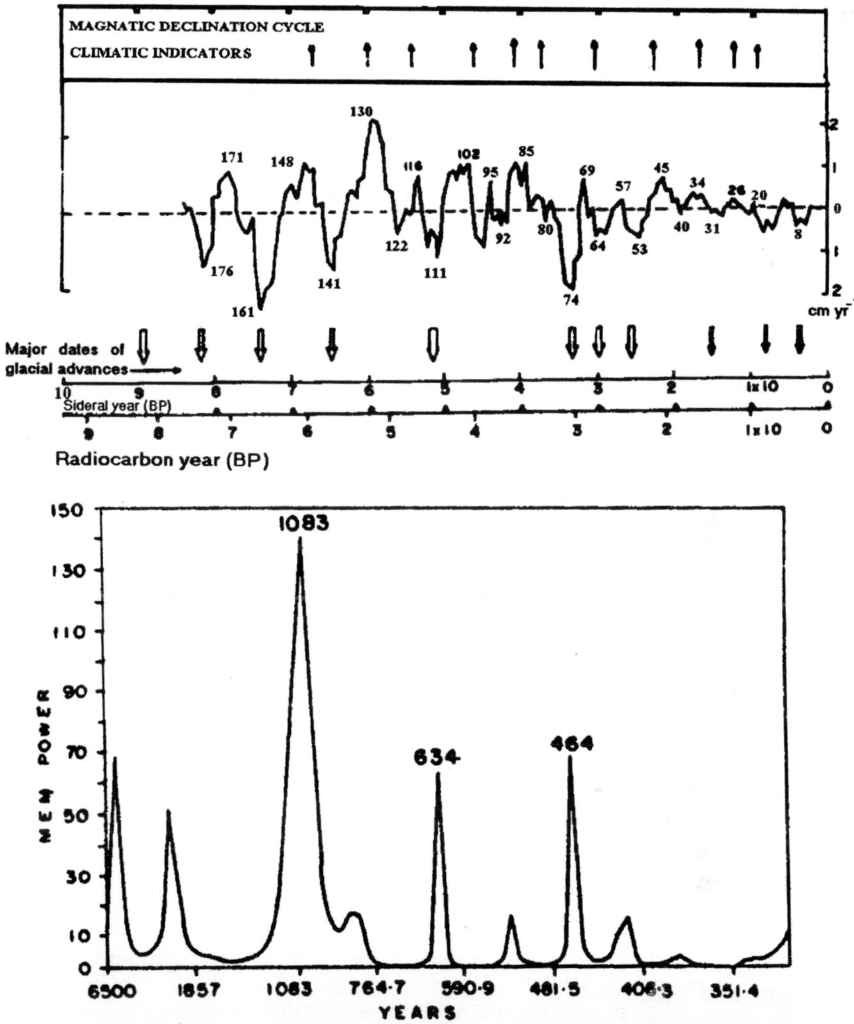

Fig. 8.1 (a) Time Series of Rate of Holocene Sea level Change Relative to Present Coast of Hudson Bay. (Fairbridge and Hilliare, 1977). (b) Maximum Entropy Power Spectra of Sea level Changes During the Past 6500 Years

(*Source*: Negi and Tiwari, 1987)

approximately after 1668 years when moon is in perigee and earth at perihelion (Fairbridge, 1984 and reference therein). A 434-year cycle in climate and sea level changes record is approximately 1/4 harmonic of 1668 years cycle. The planetary conjunction cycle of 567 years and 1100 years mark the times of conjunction of major planets and also coincides well with high sunspot activity and many other major geological events (Fairbridge and Hilliare, 1977). Pisias (1978) analysed a long Holocene core from the South Barabara Basin of

California (microfossils record) and demonstrated reversal cycles of warm and cool phases with complex periodicity of the order of 350–1000 years. They further asserted that curve is modulated by 350–1000 years trend that seems to correlate with global climatic record. A similar periodic oscillation was also observed in other microfossils (indicator of warm and cold cycles) record.

The short-term time series analysis of sea level variations distinctly show solar and tidal cycles. Maximum entropy spectra of tide gauge records spanning the late 19th and 20th centuries yielded a possible persuasive evidence for the 18.6 and 6.3 years cycles in sea level (Currie, 1976). Lisitzin (1974) had also detected a 18.6-year cycle (Lunar-nodal term) corresponding to the precession of the moon's nodes. A similar cycle has also been reported in the Japanese air pressure data (Currie, 1982). Accordingly, a solar cycle variation of 1 mb in atmospheric pressure would induce the observed 9 mm signal in sea level. Lamb (1972) has summarised departures of mean pressure at sea level associated with different phases of the sunspot cycle and the departures on typical 1 mb. However, Woodworth (1984) has examined global tide gauge records and found evidence of the 11 years solar cycle in mean sea level. Although comparable solar and tidal cycles are present in sea level records, its direct physical impact is yet to be established. Furhter, in many cases, however, it is difficult to distinguish impact whether the underlying phenomena is caused by solar cycle or tidal cycle.

Climate Change

The data on climate variability are available in two forms:

(a) From the instrumental records of temperature, rainfall and atmospheric pressure variations etc., and

(b) Proxy data decoded from sources such as sediments, glaciers, varves and tree rings etc.

Instrumental Records of Temperature Variations

There have been considerable and careful efforts to establish reliable and accurate records of surface air temperatures for a century or more. Appropriate analyses of such instrumental records of historical temperature variability are useful to understand the basic nature of global warming trend. Ghil and Vautard (1991) used singular spectral analysis to analyse one of such annually averaged time series of global surface air temperatures for the past 135 years, spanning the period from 1854–1988, to decipher significant modes and secular warming trends in the data. (Fig. 8.2) Their analysis revealed inter-decadal oscillations of 21 and 16 years, and inter annual periods of 6 and 5 years (Fig. 8.3). They have suggested that inter-annual oscillations are probably related to global aspects of the El Nino-Southern Oscillation (ENSO) phenomenon. The inter-decadal oscillation could be associated with changes

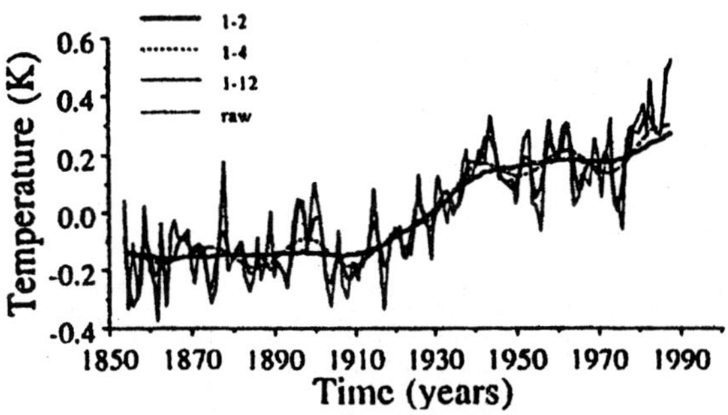

Fig. 8.2 Reconstruction of the Trend and Filtered Part of the Global Temperature Time Series for Northern Hemisphere (NH) (Heavy Solid and Heavy Dots) and Southern Hemisphere (SH) (Light Solid and Dots) Temperatures

(*Source*: Ghil and Vautard, 1991)

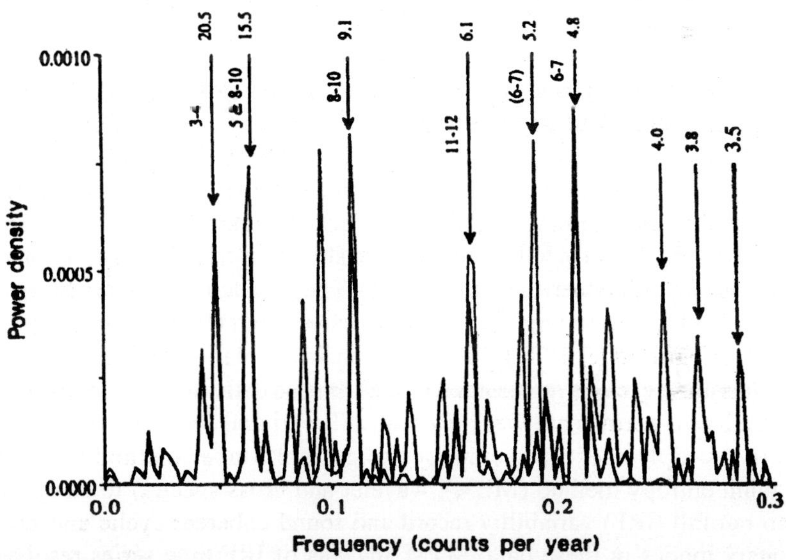

Fig. 8.3 Multitaper Power Spectra of the Temperature Time Series: All the Peaks Marked by Arrows have Identical Frequencies in Both Hemispheres, and Pass the F-test at the 97% Level in Each Hemisphere Separately

(*Source*: Ghil and Vautard, 1991)

in extra-tropical ocean circulation. It is interesting to note, however, that a 21-year periodicity is also very close to the known sunspot solar cycle (Hale cycle) and solar constant changes of upto 0.1% have been observed for the 22 years Hale cycle (Eddy et al., 1982; Ghil and Vautard, 1991) Although a possible link between climate and solar variability may be expected (Kelly and Wigley, 1990), conflicting evidence, however, exists for such link. Several other studies also reveal solar/sunspot cycle in temperature records (Lassen and Friis-Christensen, 1995) but the origin of these cycles is, at present, debated. For instance, Ghil and Vautard (1991), discounted the solar origin of 21 years signal in temperature record on the following grounds: firstly, according to them any insolation effects of the sun's 22-year activity cycle should be dominated by the rectified 11-year period, which they did not find in their analysis. Secondly, the multi-channel singular spectral analysis of North American temperature data reveals no coherence with the 11-year insolation signal. Thirdly, although the 0.1% amplitude of solar irradiation cycle would be sufficient to cause variations of 0.1°C in atmospheric temperatures, it is definitely not sufficient to heat the upper ocean down to the penetration depth of a decadal signal. Hence, the underlying inter-decadal oscillations could involve water masses below the seasonal thermocline and outside the tropics. According to Ghil and Vautard (1991), reference and references therein, recent findings show that temperature and salinity change down to the permanent thermocline on this scale.

Rainfall

The rainfall variability is also one of the phenomena closely connected with the tropical climate system and onset of the modern state of ENSO. Several studies have suggested that the most obvious cycle is the 18.6 year lunar-nodal cycle in the instrumental record of precipitation or in its inverse, the drought incidence in semi arid region to marginally humid regions (Currie and O'Brien, 1988). This cycle has been demonstrated in tree rings and in other data series of North America (Currie, 1981). The long-term Chinese historical records of rainfall at Beijing have given the same result (Hameed et al., 1983; Currie and Fairbridge, 1985). In connection with solar cycles, the 78 – 80 year Gleissberg cycle is represented in the Chinese data, but the signals from 11 and 22 year periods are weak. Recently, Tiwari and Rao (2004) used the modern multiple spectral techniques (e.g., multi-taper method (MTM), maximum entropy method (MEM), wavelet and cross spectra) to study the Indian rainfall (IRF) variability record and found coherent cyclic and non-stationary modes in these data. MTM analysis of IRF time series resolves statistically significant variability (>90% CI):

(i) at multi-decadal (66–70 years) scales related to the well-known global temperature variability of internal atmospheric-ocean origin,
(ii) relatively weak signal at 13 and 22 years (solar cycles), and
(iii) the 2.5 to 7.5 years cycles associated with ENSO frequency band.

MEM analysis confirms the stability of spectral peaks. Wavelet spectral analysis of the above time series reveal non-stationary 'localised modes' of ENSO evolution corresponding to 2–7 years and higher order terms. Matching periodicities are present in these records, but cross-spectral analysis of IRF and NINO3 temperature records exhibit significant 'coherency' (>80% CI) only at periods 5.4 years and 2.7 years, suggesting the significant role of ENSO dynamics in organising the subtle Indian monsoon at these frequencies.

Long-term Rainfall Proxy Record

The significance of global millennial scale climate shifts during the last glacial is now well recognised. Besides the instrumental record of historical rainfall variability, the geologic proxy data also record pronounced rainfall and climate variability on different time scales ranging from annual to millennial time scales. Hong et al. (2001) examined a 6000-year record of changes in drought and precipitation in northeastern China based on a $\delta^{13}C$ time series from peat cellulose (Fig. 8.4). The variability of precipitation shows periodicities of around 70, 80, 90, 107, 110, 123, 134, 141, 162, 198, 205, 249, 278, 324, 389, 467, 584, 834 and 1060 years (Fig 8.5). They suggest that on time scales

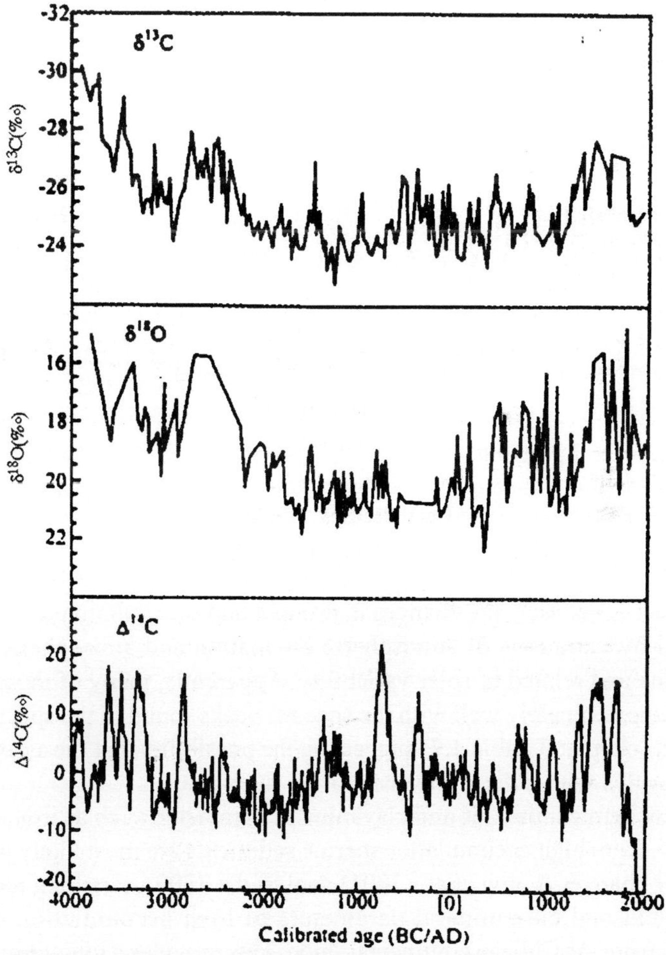

Fig. 8.4 A Comparison Between $\delta^{13}C$ and $\delta^{18}O$ of Peat Cellulose (this Study) and $\Delta^{14}C$ of the Tree-rings Plotted Against the Same Time Scale

(*Source*: Hong et al., 2001)

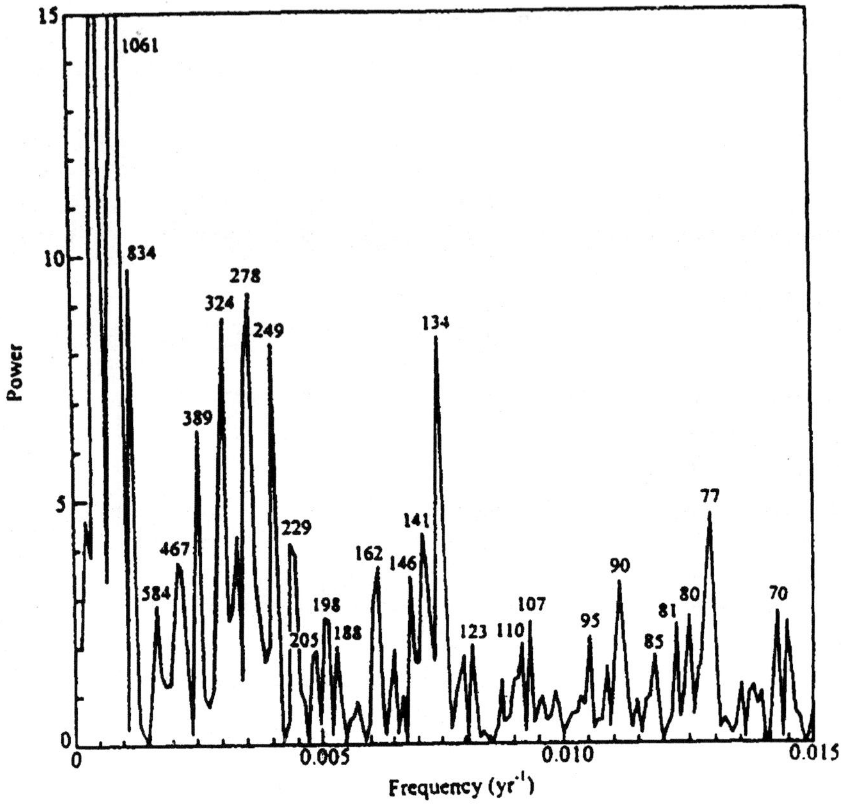

Fig. 8.5 Power Spectrum of the $\delta^{13}C$ Time Series of Peat Cellulose from Jinchuan. Numbers Above Peaks Indicate the Corresponding Periodicities (Years)

(*Source*: Hong et al., 2001)

of decades to centuries, the changes in drought and precipitation are likely to be caused by variations of atmospheric circulation and atmospheric-ocean interactions and related to solar variability. Apparently, many of these peaks correlate approximately well with the spectral peaks found in the spectrum of ^{14}C record (chapter 3, table 3.7) suggesting the possibility that the association of these cycles with solar cycles cannot be ruled out.

The geochemical proxies and clay mineral parameter, such as iron content variability, from high accumulation marine sediments are most likely induced by rainfall changes. Lamy et al. (2001) analysed a 7700 years long record of geochemical and clay mineral parameters of high accumulation marine sediments from the Chilean continental slope. This provides a long-term record of rainfall variability in southern Chile related to the position of southern Westerlies. They used the iron content of the bulk sediment in combination with clay mineral data as indicators for quantitative changes in the terrigenous sediment input. Spectral analysis of this data reveals dominant periodicities of about 1500 ± 200 year and 850–1050 year. A similar periodicity of the order of 1500 years have been also suggested by Sarkar et al. (2000) in the Holocene monsoon record of eastern Arabian sea. The possible cause of near 1500-year periodicity has been associated with an internal oscillation of atmosphere ocean system involving the deep-water circulation in the North Atlantic and possibly to the formation of Southern ocean deep waters. (Bond et al., 1999). Variations in solar activity or internally forced oscillations of the ocean atmospheric system (Van Geel et al., 1999) have been associated with the 1500 years cycle. Another cycle with an average period of about 900 years similar to the above cycle (Paul and Schulz, 2002) has also been interpreted either as an astronomically induced (Loutre et al., 1992); or as internal oscillation of ocean–atmospheric system, which might possibly be triggered by salinity anomaly in the North Atlantic or ENSO (Heusser and Sirocko, 1997; Groots et al., 1997). The significant unique role of internal ocean atmosphere variability (global thermocline circulation), (ENSO) or orbital variations can presently not be decided however (Lamy et al., 2001). In addition to this another climate periodicity of the order of 2500 years is associated with changes in solar activity (Stuiver and Reimer, 1993).This indicates that the Holocene climate records are dominated by quasi-periodicities varying from several hundred years to thousand years giving rise to broad variety of possible forcing mechanisms.

In addition to these, Campbell et al. (1998) reported a 1500 years periodicity alongwith several weaker century and millennial scale cycles in a palaeo-climate proxy record of Western Canada. Their results suggest that post glacial climate anomalies such as the Little Ice Age and Younger Dryas are periodic phenomena rather than unique, aperiodic event. Based on their result, Campbell et al. (1998) have drawn important conclusion that even in the absence of anthropogenic climate forcing, a natural warming trend will continue until AD 2400.

It is well-recognised that the millennial scale changes in the climate system is also a result of major abrupt oscillations and transitions from one state to another. One such example is abrupt circum-North Atlantic climate oscillation associated with the Younger Dryas event of Europe. Different views exist on the possible causes of these events. Based on some high quality data one school suggests that the atmospheric cold event was associated with abrupt shifts in atmospheric ocean circulation (Bond et al., 1992; Clark et al., 1999; Heinrich 1988) while the other group believes that the circum-North Atlantic Younger Dryas event resulted from the effects of glacial melt water on ocean thermohaline circulation and dramatic decrease of northward heat transports. It is now recognised that many such major abrupt century scale warm-cold oscillations took place in the circum-North Atlantic region during the last glacial period (Ruddiman, 1977; MacManus et al., 1998; Alley and MacAyeal, 1994). The climate system went through rapid changes during the last glacial stage. These so-called 'Dansgaard-Oeschger abrupt events' are well documented in the North Atlantic area and Greenland ice cores. These events appear to be apparently related to reorganisation of the ocean–atmospheric system and seem to group into longer oscillations, each terminating in massive discharges of iceberg into the North Atlantic known as Heinrich event. There are two types of abrupt changes:

(1) Heinrich events with a periodicity of ~6–7 k.y. (Bond et al., 1992; Heinrich, 1988). These events are followed by consecutive abrupt warmings on the Greenland continent.

(2) Climatic oscillations in the Greenland ice cores records oscillations with periods of 1–3 k.y. (Dansgaard et. al., 1993).

Evidence suggests that ice cores from Greenland ice sheet reveals substantial (15°C) variations in temperature (Jouzel, 1999) and atmospheric dust concentrations that recurred with periods of a few thousand years between 115 and 10 k.y. BP (Dansgaard et al., 1993). These oscillations correspond to considerably high air temperature shifts between a stadial (cold phase) and an interstadial (warm phase). These variable climatic states, or stadial/interstadial temperature oscillations, over the Greenland ice sheet are referred to as Dansgaard–Oeschger (D–O) cycles. Similar changes are documented in sea surface temperature (SST) reconstruction, percentage of carbonate, magnetic susceptibility, planktonic foraminifera oxygen isotopic composition and foraminiferal assemblage records from the North Atlantic Ocean (Cortijo et al., 1995).

There is evidence of so-called Heinrich events that are correlative with major changes in the global climate dynamics from the North Atlantic to the Antarctic. These events have possible implications for the East Asian winter monsoon (Porter and Zhisheng, 1995). These environmental cycles range in a period of 1–10 k.y. (Clark et al., 1999) and are discovered in short-term episodes of ice rafting (Ruddiman, 1977; Heinrich, 1988). The horizons rich in ice rafted debris are indicative of the so-called Heinrich events. Bond et al. (1992) and Broecker et al. (1992) have suggested that these events lasted perhaps less than a thousand years and occurred at intervals of 7–13 k.y. during the last thousand years. The average period of the Heinrich event is about 6–7 k.y.

The Greenland temperature oscillations coincided with Heinrich events, as well as changes in the sea-surface temperature, with periods close to 1.5 k.y. and 1–3 k.y., (Bond et al., 1993; McManus et al., 1994; Fronval et al., 1995; Bond et al., 1997; Grootes and Stuiver, 1997) from 20 to 110 k.y. BP. In the low-latitude areas, the phytoplankton activity quantified by primary productivity is correlated to the wind stress dynamics on the sea surface (Nair et al., 1989). The primary productivity have frequency close to the D–O cycles and indicate a teleconnection of the east Asian winter monsoon with Greenland climate. According to these workers, the East Asian winter monsoon oscillates with periodicities of ~6, 4.2–3.4, 2.3–1.5 k.y. The 1.5 k.y. cycle exhibits a strong stationary signal suggesting the millennial-scale dynamics of primary productivity with the East Asian winter monsoon (Garidel-Thoron et al., 2001).

Proxy Records of Climate Indicators

Proxy records of climate variability preserved in ocean sediments carry significant information about the interactions and evolutionary history of the geo–bio–cryo–hydrosphere and atmosphere. These 'proxy' records occur in many forms (e.g. tree ring, varve, glaciers and deep sea sediments etc.) and provide indirect but very useful information of physical behaviour that reflect the influence of the earth's climate.

Dendro-climatology

Tree rings are high resolution proxies of past climate variability and are being routinely used for better understanding of the climate system. The power spectral density of radio carbon variations, using an obsolete chronology based on tree ring count, exhibits spectral lines at number of periods: 2300, 964, 753, 717, 493, 413, 357, 229 and 208 years suggesting a possible link between terrestrial climate and sun's activity. In view of a common driving mechanism, this might suggest that radiocarbon variability is consistent with a model of atmosphere ocean resonance underlying the radio-carbon periodicities (Sonett and Finney, 1990). In another study, Castagnoli et al. (1989) provided phenomenological evidence of an identical periodicity of about 206 years in quite different records of solar activity, such as tree ring, radio-carbon, the sunspot numbers and the thermoluminescence profile of sea sediment possibly suggesting a link between terrestrial and solar driven processes. However, they note that the dominant 206-year periodicity in these records may not necessarily represent a basic frequency of the sun-earth system but may be produced as a low frequency modulation, by the beat of two high frequency oscillations. Such findings are interesting and quite useful for understanding some of the temporal variations of solar terrestrial system and also for shedding new light on the mechanisms, which drive the terrestrial climate rhythms.

Stuiver and Braziunas (1989) analysed a 9,600 years long high precision ^{14}C. Variable production rate of cosmogenic isotopes such as ^{14}C in atmosphere provide a time history of changing conditions of Sun's surface. Trees, through CO_2 assimilation, lay down a ^{14}C record which provide clues towards the causes underlying ^{14}C production-rate changes. The maximum entropy spectral analysis of above timeseries revealed a fundamental oscillatory mode of 420 years along with several other harmonics. They emphasised possible link between sun and climate with a particular cycle of 140 years (Fig. 8.6a, b).

In another study, Scuderi (1993) analysed temperature variability, decoded from tree ring, of the past 2000 years. Interestingly his analysis also revealed a 125 year periodicity that may be linked to solar activity reflected in radio-carbon and auroral records. (Figs. 8.7 and 8.8). In addition to this, there are

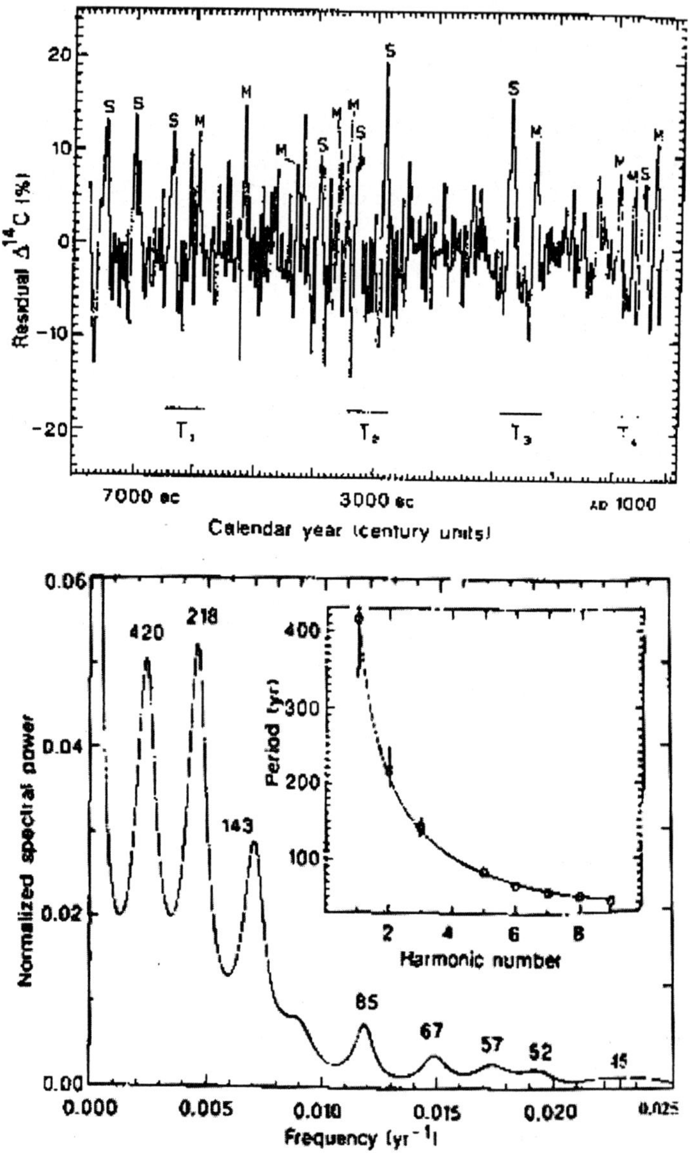

Fig. 8.6 (a) Residual Atmospheric $\Delta^{14}C$, Obtained by Deducting the Long–term $\Delta^{14}C$ Trend, M and S Denote Maunder–and Sporer–type Events (Stuiver and Braziunas, 1989). (b) MEM Power Spectrum (AR=20) of $\Delta^{14}C$ Production Rate Q. The Periods are Listed with the Individual Peaks. The Hyperbola in the Inset Shows the Harmonics Expected for a Fundamental Frequency of 1/420 per Year

(*Source*: Stuiver and Braziunas, 1989)

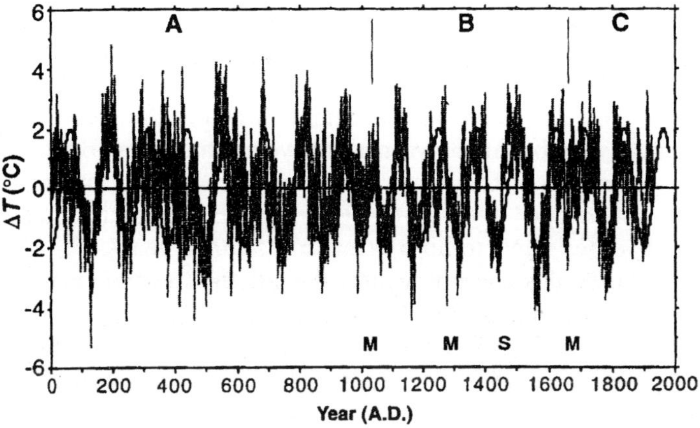

Fig. 8.7 Reconstructed June Through January Temperature Anomalies for Southern Sierra Nevada Relative to the A.D. 1 to 1980 Mean

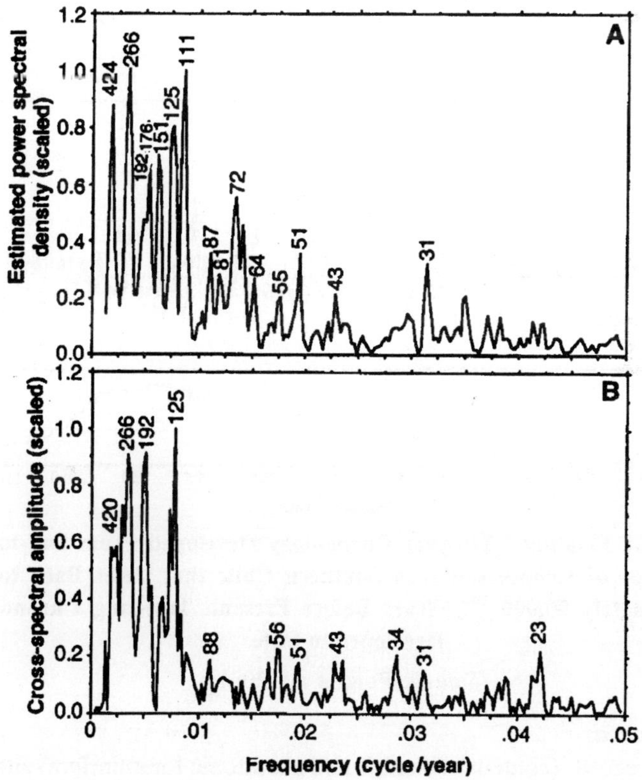

Fig. 8.8 (a) Power Spectral Density of the Cirque Peak Temperature Sequency (1050 BC to AD 1981). (b) Cross-spectral Analysis of the Cirque Peak Ring Width Record and $\Delta Q/Q_t$ (production rate of ^{14}C) (1050 BC to AD 1870)

(*Source*: Scuderi, 1993)

also several weak periodicities particularly at around ≃ 420, 266 and 192. Although Scuderi 1993 did not discuss the physical relevance of these periodicities, it is interesting to note that a 180–190 years periodicity called King Hele cycle does exist of solar origin. Similarly, 266 years periodicity is approximately double of 130 years solar cycle (Stuiver and Braziunas, 1989).

Some researchers (Roig et al., 2001) have reported a floating 1229 years chronology developed from sub-fossil stumps of E cupressoides in the southern Chile that dates back to approximately 50,000 years (Fig. 8.9) These researchers used this chronology to calculate the spectral characteristics of climate variability and obtained oscillation at periods of about 150–250, 87–94, 45.5, 24.1, 17.8, 9.3 and 2.7–5.3 years (Fig. 8.10). Interestingly, some of these periodicities (87–90), 45, 24 and 8.7–9.3 years match quite closely with solar cycles discussed in chapter 3 while others in period of 2.7 to 5.3 years with well-known ENSO modes. From the above discussions it is seen that spectral analysis of tree rings record provide evidence of periodicities ranging from the modern ENSO activities to the past solar activity. This might suggest the combined role of solar cycles as well as oceanic-atmospheric processes on climate variability.

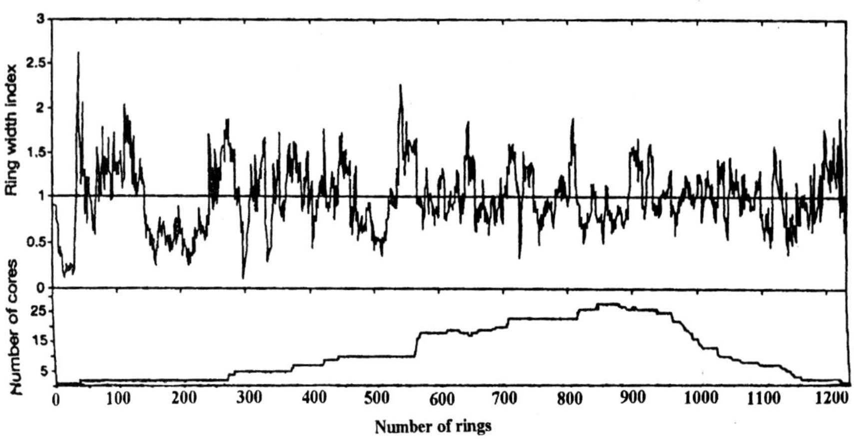

Fig. 8.9 A Floating 1,229-year Chronology Developed From Sub-fossil Stumps of F.cupressoides in Southern Chile that Dates Back to Approximately 50,000 ^{14}C Years Before Present. Tree-ring Chronology Developed to Date

(*Source*: Roig et al., 2001)

Faunal Records

The faunal records (content of benthic and planktonic foraminifera) and stable oxygen isotope provide a systematic development of palaeo-environmental conditions. Spectral analysis of Holocene faunal records from North Sea for the Troll 8903/28-03 core identified the major peaks corresponding to periods of 1400, 975 and 800 years (Klitgaard-Kristensen et al., 2001) suggesting the periodic evolution of Norwegian sea surface temperature. A periodicity of 1400

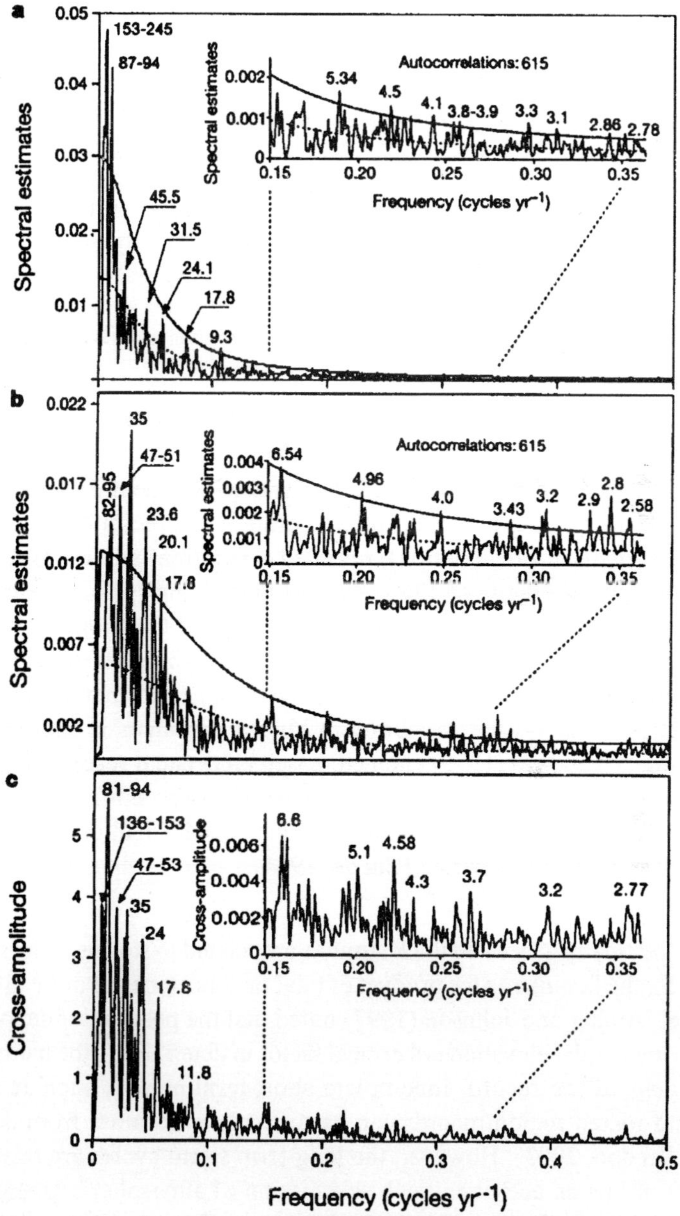

Fig. 8.10 The Spectral and Cross-spectral Characteristics of the Sub-fossil (a) and Living (b) F, Cupressoides Woods. The Cross–spectral Relationships Between Subfossil and Modern Chronologies. Expanded Details of Each Power Spectrum are Shown in the Insets

(*Source*: Roig et al., 2001)

years is, particularly, important since similar periodicity is also reported in several other geophysical records (Bond et al., 1997). The other cycles are also comparable to frequencies previously recorded in various climate records, Although the origin of these periodicities are not well known, several researchers have suggested that the global occurrence of the 1500 years periodicity might indicate either an external forcing (solar output) to the earth's climate or is related to the earth's deep ocean circulation. It may be noted that similar explanations have been proposed for several other Holocene records where these have been correlated either with solar variability and/or possibly linked with variations in formation of North Atlantic Deep Water. The details of the above relationship between the sun, the ocean and the global climate still remain to be understood fully (Campbell et al., 1998; Stuiver and Braziunas, 1993).

Drought Cycle

Hodell and his colleagues (2001) have analysed lake sediment cores from the Yucatan Peninsula, Mexico, to reconstruct the climate history of the region over the past 2600 years. Time series analysis of sediment proxies, which are sensitive to the changing ratio of evaporation to precipitation (oxygen isotopes and gypsum precipitation), revealed a recurrent pattern of drought with a dominant periodicity of 206 years. The underlying cycle is identical to a similar 208 years period documented in records of Cosmogenic nuclide production (carbon-14 and beryllium-10) that is thought to reflect variations in solar activity (Castognoli, 1989). They have suggested that a significant component of century-scale variability in Yucatan droughts is explained by solar forcing. Yu and Ito (1999) claimed, based on decadal resolution data, a 400-year signal of drought in North America, which is also similar to 420 years solar cycle. There is also evidence for 18.6 lunar-nodal (tidal origin) drought cycle in Western North America (Currie, 1984).

Storm Cycles

Although the tidally generated thick laminae are related to storms, (Murakoshi et al., 1995), the laminae as the product of tidal factors alone are quite difficult to interpret. Archer and Johnson (1997) noted that the position of deposition relative to mean tidal elevation is a critical factor in determining the nature and completeness of the record. Incomplete short-term period, such as semi-diurnal and mixed region records can be difficult to separate from diurnal records (Weedon, 2003). However, the long-term storm cycles are related to the impact of human activities, such as emission of atmospheric greenhouse gases, which may lead to increases in the frequency of severe storms, in certain regions of the globe. The existence of natural variability in storminess confounds reliable detection of anthropogenic effects. Noren and his colleagues (2002) have dated terrigenous inwash layers in cores from 13 lakes from the northeastern United States. Visual inspection of the temporal records shows four peaks centered around 2600, 5800, 9100 and 11900 years in storminess during the above period suggesting that the frequency of storm related floods has varied in the regular cycles during the past 13 k.y. The multi-taper spectral analysis show a statistically significant characteristic time scale of about 3 k.y. Solar influence has been suggested as one of the prime candidates for such a long-term natural climate variability (Bray 1968, 1971; Bond et al., 2001). It would be interesting to clarify whether the tidal generated imprint varies with the level of solar variability. However, it is well known that oceean-atmospheric modes, such as the Atlantic oscillation and El Nino/Southern Oscillation (ENSO), account for significant fractions of modern climate variability on inter-annual time scales also. Hence, such dominant atmospheric modes may also be modulated on longer time scales, and may, therefore, account for much of variability in the palaeoclimate records (Noren et al., 2002).

Cycles in Varve Sediments, Ice Cores, Glaciers

The formation of varved sediments display certain characteristic which can be used to distinguish spring, summer and winter laminae. The spring and summer laminae are coarser-grained, lighter in colour and richer in carbonate; whereas the winter laminae are darker, and contain clay and organic substances (Weedon, 2003). Lakes have the most ideal environment for the generation of varves. However, they can also be found in sheltered inland marine waters.

Several researchers have found convincing evidence for tidal periodicities in varve sediments. Annual cycles are the most typical astronomical cycles. The seasonal changes are sometimes so large that they affect and have affected present and ancient sedimentation (Weedon, 2003). Tidally generated strata are found from the Archaean to the recent (Eriksson and Simpson, 2000). Proterozoic Elatina Formation laminitis is one of the best examples of such an ancient formation. Williams (1981b) and Sonett and Williams (1985) interpreted this formation initially as glacial lake varve with implied annual laminae thickness controlled by solar cycles. However, in his later studies (Williams, 1988 and 1989), he used contemporary sediments with high accumulation rates to demonstrate tidal control. Several researchers (Schaaf and Thurow, 1997; Stuiver, 1997) have found evidence for the 18.6-year solar–lunar nodal cycle in varved (annually laminated) sediments and suggested that it might be related to climate and other meteorological phenomena. Berger et al. (2004) analysed varves archive for different time period in the sediments of Santa Barbara Basin. In particular, they analysed the record since AD 1350 for varves, AD 1120 for total organic carbon (TOC) accumulation and AD 1200 for fish scales. Their analysis revealed significant periodicities at about 169, 93, 51.4 and 17.7 years in varves, 124 years, 56 years and 37.4 years in TOC and 103 years, 128 years in fish scales. According to these researchers, varves and TOC can be interpreted as governed by a common forcing mechanism (tidally generated) i.e., in terms of the 4.425 year perigee period (describing the changing distance between earth and moon) and the 18.6 year lunar nodal period (which pertain to line up earth-moon sun). According to the record of Santa Barbara, basin is partly a result of pulsed sedimentation aligned with tides.

Varve sediments are ideal study material for determining not only annual, but also longer cyclicity (Weedon, 2003). The two-and three-year cycles occur frequently matching with similar cycles observed in the changes of climate. Five- and six-year cycles are also frequent. Very prominent are, however, the 10–15-year cycles which apparently correspond to the sunspot cycle of average 11 years. The 11 years are only a statistical average. Cycles corresponding to the 11-year or a similar period or its multiple, have been found in nearly all varved sediments and can hardly be accidental. Sonett and Williams (1985) have shown some evidence for solar forcing in varve record of modern preglacial lake in Alaska. The early Holocene varve in the USA contain evidence for weak signal of the Hale (22 year) and Gleissberg cycle (90 year). There is also evidence for solar cycles of 90 year and 11 year in late Pleistocene-Holocene varve from the Santa Barbara Basin, California (Schaaf and Thurow, 1997) and the Hale cycle (22 year) in late Pleistocene varve from New England (Rittenour et al., 2000).

Bray (1968) has proposed linking of recent glaciation with solar activity. Bray has documented several glacial episodes separating with time interval of 2400–2700 years. He also proposed a solar cycle of about 2600 years. The proposed 2,400–2,700 year solar cycle was tested against late-Pleistocene chronology by projecting periods back from the Little Ice Age of 2,400 to 2,700 years at 50-year intervals. Combinations of geo–bio–ocean–atmosphere-cryosphere and geochemical variations and primary productivity support the idea of a 2600-year solar cycle (Leventer et al., 1996). The results of the projected solar activity periods showed that an interval of 2,600 years gave the best fit with the compiled climatic and other environmental data. The closeness of the relationship between climate and predicted solar activity is quite strong. This suggests probable role of long-term solar cycle in coupled earth's system processes.

Cyclic Pattern in the Asian Monsoon System

The monsoon system is considered as one of the main determinants of global environmental conditions. Hence understanding the monsoon history and the nature of its palaeo-climate dynamics is essential both from academic and economic point of view. Since the Indian Ocean monsoonal system provides one of the main linkages to the global weather system, the variations in intensity would have broad oceanographic and economic implications. This, however, requires very accurate data of monsoon indicators and adequate techniques of analysis, that can unveil its correct deterministic, periodic and/or random nature.

Spectral analysis of a high resolution Late Quaternary upwelling record from the Cariaco Basin has shown coincident periodicities between the upwelling index, G. bulloides indicator of palaeo-environmental condition and atmospheric ^{14}C (Peterson et al., 1991). Naidu and Malmgren (1995) carried out high-resolution study of fluctuations in upwelling intensity along the Oman Margin, Western Arabian Sea through the last 19,130 years based on several marine

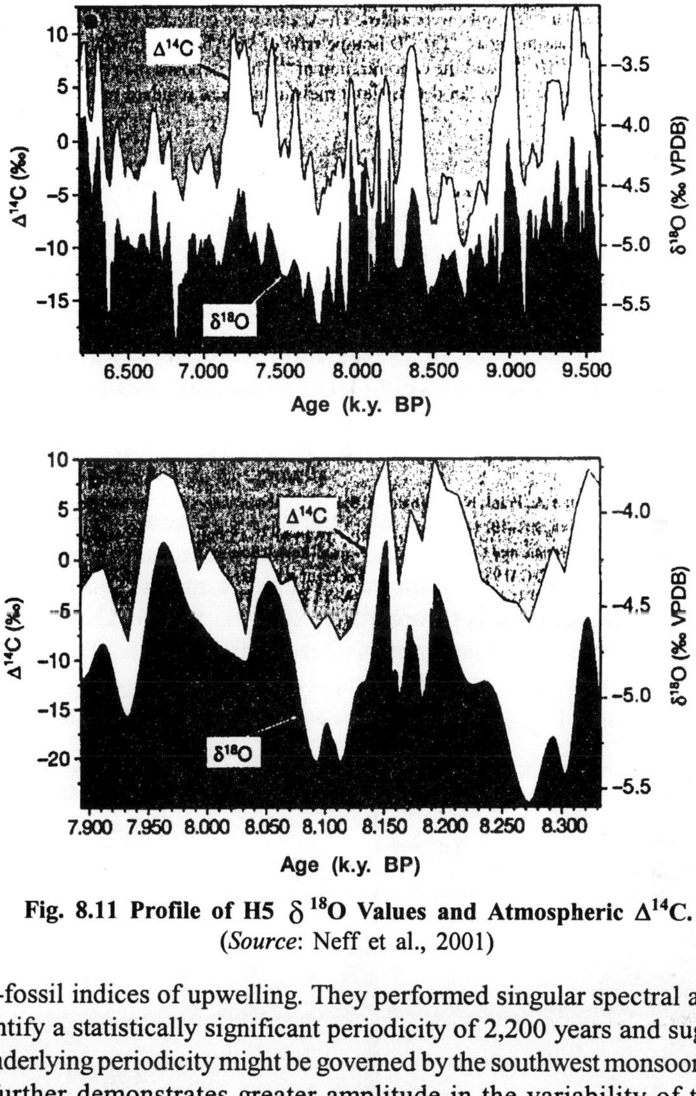

Fig. 8.11 Profile of H5 δ^{18}O Values and Atmospheric Δ^{14}C.
(*Source*: Neff et al., 2001)

micro-fossil indices of upwelling. They performed singular spectral analysis to identify a statistically significant periodicity of 2,200 years and suggested that underlying periodicity might be governed by the southwest monsoon. Their data further demonstrates greater amplitude in the variability of the SW monsoon during the Holocene than during the last glacial periods. They have noted two implications of reported periodicity:

(a) oceanic circulation changes partly influenced monsoon strength at sub-Milankovitch cycles, and

(a) similar periodicity monitored in atmospheric ^{14}C might be induced by oceanic circulation changes.

Sarkar et al. (2000) have also reported a 1500 years producity in Holocene monsoon of east Arabian sea. On a still shorter time scale, singular spectrum analysis of four global temperature records identified temperature oscillation of the order of 65–70 years. They argued that the oscillation arises from predictable interval variability of the ocean atmospheric system (Schlesinger and Ramankutty, 1994).

The possible link between solar variability and ^{10}Be variation and climate on the above time scale has been suggested. Bond et al. (1992) have further shown that there is strong similarity between the variation production rate of ^{10}Be and indrift ice extent, which are also associated with North Atlantic circulation. These variations have striking similarity at the time scale of hundred to few thousand years and may be related to solar variability extending globally beyond the North Atlantic attesting the global teleconnection of monsoon system and external forcing phenomena (Neff et al., 2001 and Bond et al., 2001). The spectral analysis of high-resolution profile of oxygen isotope of stalagmite H5 from Hoti cave (Fig. 8.11) show statistically significant periodicities centred at 1018, 226, 28, and 9 years (Fig. 8.12). In addition to this, two broader cycles are centred on 101–90, and 35–26 years. These cycles are close to the periodicities of the tree ring ^{14}C record (206, 148, 126, 86, 26, and 10.4 years), which are known to arise from the solar modulations (Neff et al., 2001). Further, it is interesting to note that cross–spectral analysis of the tuned H5 $\delta^{18}O$ record for the entire studied samples versus the $\delta^{14}C$ record analysis reveal peaks at around 977, 206, 91, 57, 46, 33 years which are significant at 95% confidence limit (Neff et al., 2001) (Fig. 8.13). Oscillation of the order of 206 years is particularly important because there is an evidence of a near 208-year ^{14}C cycle as a result of solar activity. This cycle has been

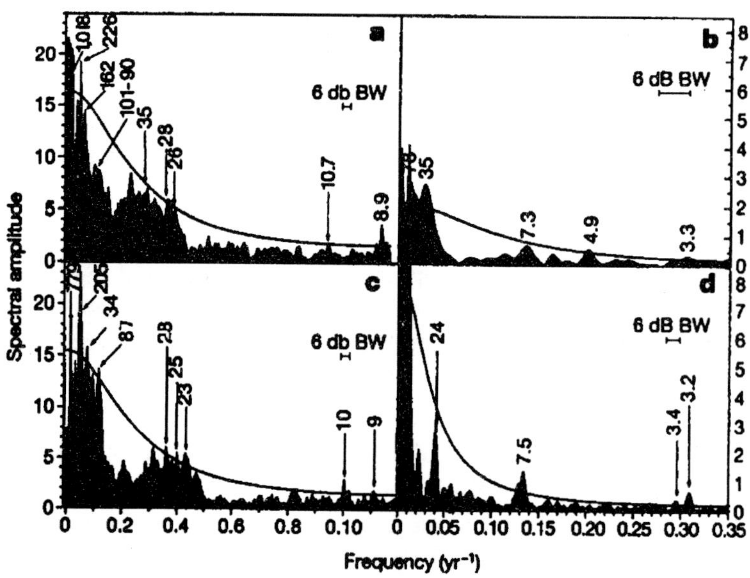

Fig. 8.12 Frequency Analysis of the H5 $\delta^{18}O$ Record for the Entire Sample for the High Resolution Interval, a, b, Untuned; c, d, Tuned Peaks are Labelled with their Period (in years).

(*Source*: Neff et al., 2001)

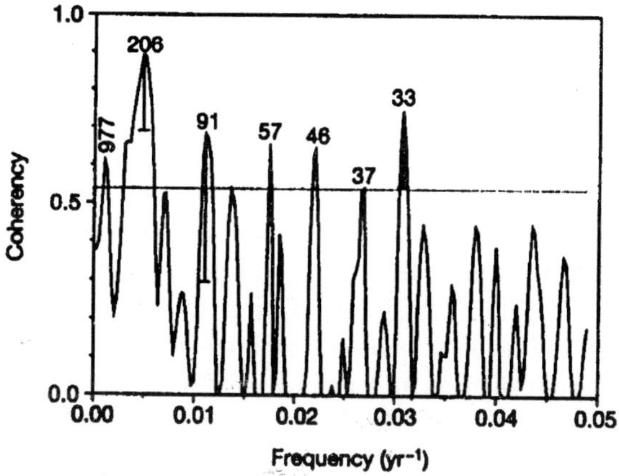

Fig. 8.13 Cross-spectral Analysis of the Tuned H5 $\delta^{18}O$ Record for the Entire Studied Samples Versus the $\delta^{14}C$ Record Analysis were Used. Peaks Common to Both Time Series are Labelled with their Period (in Years), Error Bars are Given as Lower 95% Bounds, Horizontal Line is 5% False-alarm Level

(*Source*: Neff et al., 2001)

recognised and identified globally and hence solar modulation may be a possible forcing mechanism. It is interesting to note that not only this frequency but there is also strong similarity between the smoothed ^{18}O and ^{14}C time series, both in temporal and spectral characteristics, even on millennial scale. It is tempting to expect, therefore, that a close correspondence between these phenomena and their parallel evolution seem very unlikely to have occurred by chance (Neff et al., 2001).

In addition to this, they have also found periodicities of the order of 24, 7.5, 3.2 years in high-resolution interval of the record, which are similar to the North Atlantic Oscillation cycles. This then might suggest possible link between monsoon rainfall in the Hoti cave area and North Atlantic Oscillation.

Laminated Sedimentary Formations

Astronomical cycles on still shorter time scale have prominent impact on the earth's climatic processes due to different tidal and solar cycles as can be seen in reviews by Burroughs, 1992 and Weedon, 2003. Time series analysis of laminated marine and lacustrine sedimentary records revealed significant periodicities at around 50 years, 22–24 years and 11 years. The 11 years and 22–24 years cycles are similar to solar cycles and are not uncommon in various climate records (Pike and Kemp, 1997). Climate cycles of around 50–70 years have been widely reported in various oceanic and atmospheric phenomena: e.g. tree-ring, flood and drought records, laminated marine and lake sediments, fish scale deposition/population, global mean surface temperature measured in North America etc. The underlying cycle is linked to the changes in atmospheric and oceanic circulations. Some researchers, however, also believe that the 11years, 22–24 years, and ~50 years cycles are all suggested to be influenced by solar cycles (Pike and Kemp, 1997).

During the last twenty years, numerous scientific studies have been conducted and new climate data of these studies have been quite reliable and produce statistically significant set of weather cycles (Burroughs, 1994). The analyses of various kinds of precise geophysical/geological data indicated cyclic pattern governed by external forcing. These cyclic behaviour in various earth system process and their association with sunspot (solar) and tidal influence seem so high that it is difficult to discount their possible impact. It is interesting to note that there is evidence of rhythmic sedimentation in Precambrian (600–700 m.y.) old Elatina formation in Australia (William 1981b, 1989). Scientists have identified fine-grained sandstones and Siltstones with rhythmic laminations, which were correlated well with sunspot cycles of periods of 9–14, 22–25 and 90–110 years. It also indicates cyclic pattern of periods of 180–200 years and 300–400 years. These advances and improved understanding of cyclic behaviour of long-term and short-term global climate system has facilitated in revealing various aspects of physical interconnections between the extra terrestrial phenomena and all the quasi-periodic geo-processes.

CO_2 and Climate Variations

Fluctuations in the level of CO_2 concentration are also one of the main causes of global climate change and temperature fluctuations. High resolution carbon and oxygen isotope records from the tropical Western Pacific, Sulu sea, over

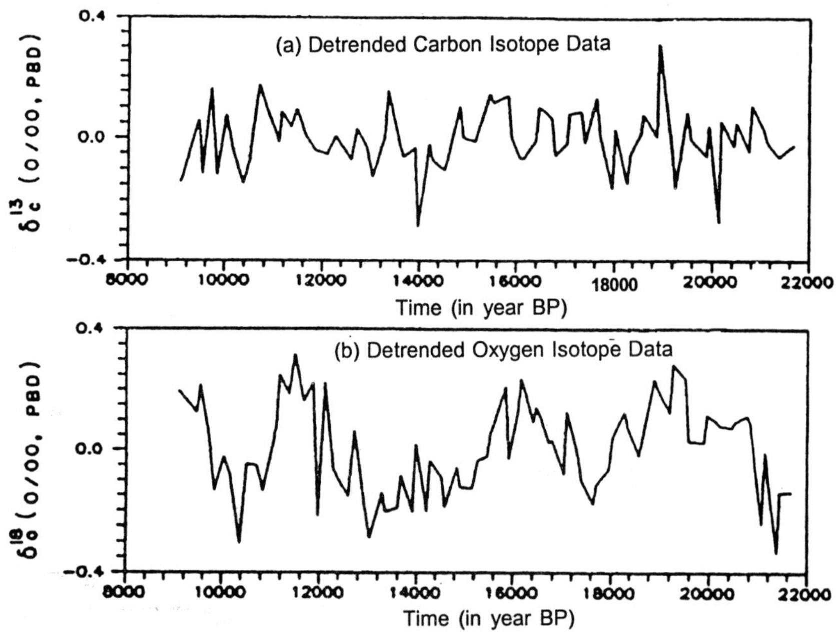

Fig. 8.14 Detrended (a) Carbon Isotope Data, (b) Oxygen Isotope Data

(*Source*: Linsley and Thunell, 1990)

the past 22000–9000 years (Fig. 8.14) has been presented by Linsley and Thunnel (1990). Clean spectral analysis of these records reveal statistically significant periodicities of 2980, 690, 322, 250, 174, 140, and 1100, 533, 425, 183, 151, years respectively (Tiwari and Rao, 1999). These peaks fall into solar climate frequency bands suggesting intricate physical link between solar and climate cycle (Figs. 8.15 and 8.16). Table 8.1 provides detailed explanation of these cycles.

Table 8.1 Cycles Obtained from the Clean Spectral Analysis of High-Resolution Deglaciation Record References and its Comparison with ^{14}C as Primary and Beat Solar Periodicities

Split periods f_1, f_2 *in years*	$f_1 + f_2$		*Beat periods (years)* $f_1 - f_2$	
140, 153	73	Global climate cycle (Schlesinger and Ramankutti, 1994)	1644	Moon–earth tidal cycle (Fairbridge and Hillaire, 1977)
174, 187	90	Gleissberg cycle	2200	Primary solar cycle Sonett and Finney(1990)
459, 447	233	Primary cycle in radiocarbon climate records (Sonett and Finney, 1990)	15898	Precession split cycles
611, 596	302	Radiocarbon spectra	1044	Primary cycle in radio carbon spectra (Sonett and Finney, 1990)
691, 757	361	Radiocarbon spectra 'super cycle' (Sonett and Finney, 1990)	7949	Higher order climate cycle (1/3 of precessional period)

(*Source*: Tiwari and Rao, 1999b)

As we have seen in Chapter 3 (Table 3.7), there is evidence of longer solar cycle of order of 2300 years in the production of ^{14}C (Mayewski et al., 1997; van Geel et al., 1999). Then the above-discussed longer climate and sedimentary cycles might also be linked to similar long solar cycles. Several researchers, however, have noted that the measured concentration of ^{14}C in tree rings might actually have been controlled by ocean circulation (Stuiver and Brazinuas, 1993 and Stuiver et al., 1997). Bond et al. (2001), however, have maintained that the similarity of variations in ^{10}Be and ^{14}C production rate to the extent of North Atlantic drift ice in the North Atlantic during the Holocene strengthens this link.

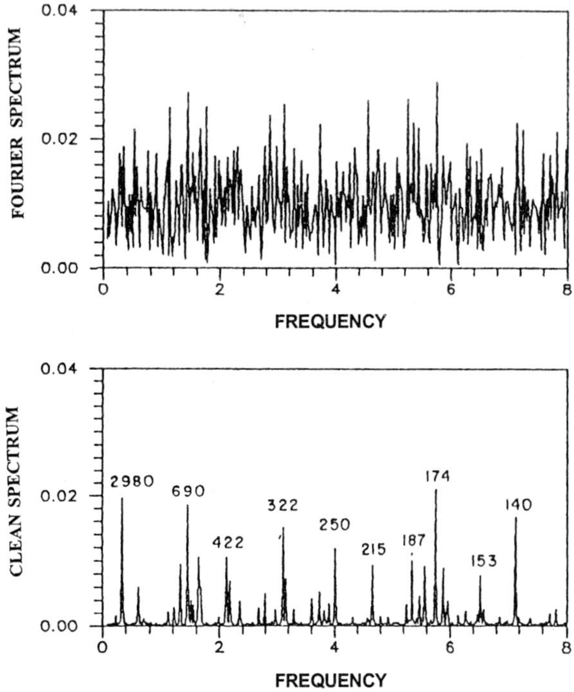

Fig. 8.15 Top: Fourier Spectra of Carbon Isotope Data. Bottom: Clean Spectra of the Same Data (Spectral Peaks are in Years)

(*Source*: Tiwari and Rao, 1999, 2000)

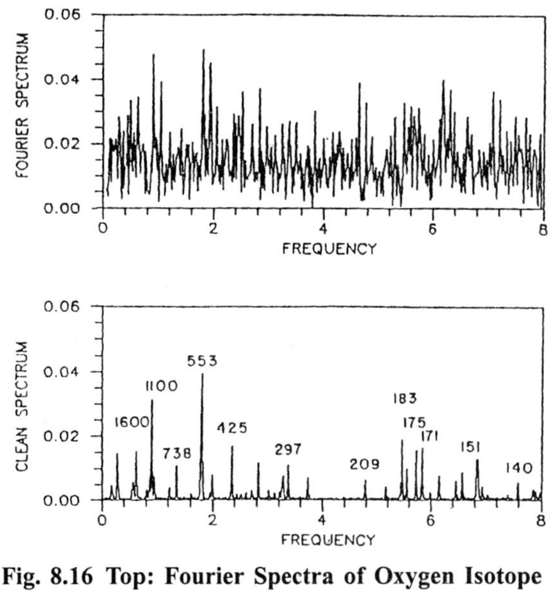

Fig. 8.16 Top: Fourier Spectra of Oxygen Isotope Data, Bottom: Clean Spectra of the Same Data (spectral peaks are in years)

(*Source*: Tiwari and Rao, 1999, 2000)

Secular Variation of Geomagnetic Field

Much effort has been made to identify harmonic/periodic components in the secular variation of earth's magnetic field (Mackereth, 1971; Creer et al., 1972; Thompson and Turner, 1979; Stuiver, 1978). Records decoded from sediments core particularly lake deposits, generally yield a continuous time sequence and are thus well suited for spectrum analysis.

A sinusoidal variation in virtual dipole moments is supported by the discovery of westward drift of the non-dipole field with a period of about 2000 years (Bullard et al., 1950; Bucha, 1967, 1969 and Cox, 1968). Some researchers have also shown regular swings in declination in lacustrine records. Mackereth (1971) initially provided some strong evidence for discrete periodicities in the geomagnetic signal. Analysis of results from around various parts of the globe show a tendency for periodicities to fall into the following bands: 60–70, 400–600, 1000–3000, and 5000–8000 yr (Barton, 1983). The latter two bands are often, though not always, associated with clockwise and anticlockwise variation of the magnetic vector respectively, when viewed from tail to tip (Barton, 1983). The first three bands are generally attributed to non-dipole field effects and the last to the dipole field. Since many of the records are not much longer than 10000–15000 years, periods in the range of 5000–8000 year are poorly defined and are strongly dependent on the type of detrending technique employed for data analysis (Courtillot et al., 1977).

Fourier and MEM analyses of palaeomagnetism time series for the past 5000 years indicated periodicities of the order of 1250 years, 700 years for declination time series and 1650 and 300 years for inclination time series (Bonifay et al., 1987). These researchers have also identified longer periodicities of the order of 4600 years, 1700 years, 1250 years, 725 years and 1000 years in longer time series inclination data. The periodicities of the order of 4600 years, 1250 years and 1000 years are common in both inclination and declination data (Bonifay et al., 1987).

Jin and Thomas (1977) analysed the geomagnetic dipole field from 1901 to 1969 using maximum entropy spectral analysis. The values of the geomagnetic dipole field from 1901 to 1969 were determined from the Gauss-Schmidt coefficients resulting from a physical harmonic analysis of the annual mean magnetic data of various geomagnetic observations (Jin, 1973). The analysis indicates dominant spectral line corresponding to periodicities of 66, 33, 22, 17, 13, 11 and 9 years. Further, similar analysis of the rate of change of geomagnetic field indicates similar periodicities. In addition to this MEM analysis of rate of change of LOD (length of day time series) (1865–1961) reveal identical periodicities. According to them, the existence of common spectral peaks with period of 66 and 33 years in LOD and geomagnetic activity is clearly established. The existence of higher harmonics with the period of 22, 17, 13, 11 and 9 years is somewhat uncertain and may be related to the obvious characteristic of line splitting in MEM analysis.

Furthermore, the 22-year magnetic cycles and the solar cycle are unambiguously higher harmonics of 66-year period, the authors argue.

Common periodicities in the spectra of different phenomena may be used to infer some direct or indirect causal relationship between them. In the field of geomagnetism the concept has been used to demonstrate a relationship between field fluctuations and the solar cycle (Currie, 1973; Courtillot and Le Mouel, 1976; Alldredge, 1976, 1977), the length of day (Jin and Thomas, 1977), and variations in varve thickness, which must be climatically controlled. In any case, the fundamental period of 66 years may be related to the motion inside the fluid of the earth core during past 1000 years. Several researchers (e.g. Vestine and Khale, 1968; Yukutake, 1973; Currie, 1973; Braginsky and Fishman, 1977) also suggested a significant concentration of power in the geomagnetic spectrum at about 60–70 years and is attributed to torsional oscillations in the magneto-hydrodynamics dynamo (Braginsky, 1970). This cycle has important implications for climate studies, since similar periodicity also appear in temperature record discussed earlier and there are suggestions which link magnetic field variations and climate.

Currie (1968) has analysed mean daily values of magnetic data for 27 observations by using Fourier spectrum techniques and discussed the physical significance of periods greater than 40 days. The coherence and phase for the continuum indicate that the entire continuum spectrum from 40 days to 3.7 years is primarily due to storm modulation. A satisfactory theory of the internal origin of the field must correctly predict the form of the continuous geomagnetic power spectrum. Based on observatory records, this has been well defined for periods between 1 and 100 years (Currie, 1968; Allredge, 1977), but has been left to the realm of speculation for longer periods. Since palaeomagnetic records of the secular variation are generally directional only, the choice of time-dependent variable is important. Interpretation of the results must be made in the light of the following limitations (Courtillot et al., 1977):

(i) The low frequency ends of spectra, particularly for observatory records, will be strongly dependent on the detrend applied.
(ii) The high frequency ends of spectra will underestimate the true spectrum due to loss of coherence of short period variations over long record lengths.
(iii) The deep-sea records are apparently dominated by noise.

The above works conclude that no distinct boundary exists between the inferred frequency domains of dipole and non-dipole field effects.

Short-term Seismic and Volcanic Cycles

Subduction zones, oceanic ridges and major faults are the main focal points/sites of large earthquake occurrences and volcanic eruptions. The plate motion mainly drives these activities. However, other factors such as tides can also trigger earthquakes (Kilston and Knopoff, 1983) and volcanoes (Hamilton, 1973; Kasahara, 2002). The signatures of tidal effect have long been implicated in their generation, although evidence has been very sparse. As discussed in chapter 3, gravitational forces exerted by the moon and the sun cause ocean tides in the ocean and earth tides in the solid earth, with diurnal and semi-diurnal periods. However, the observations indicate that the elastic strain resulting from earth tides is too small to trigger earthquakes and volcanism. Nevertheless, the idea that tides may influence these geological and geophysical activities have been discussed long back when an interesting earthquake sequence was observed during an earthquake swarm east of Ito on the Izu Peninsula, central Japan (Glasby and Kasahara, 2001.

There is some evidence of quasi-periodicities of the order of seasonal and 11-year solar cycles. The possible link between the earthquake activities and these external cycles have been suggested through the variation in magnetic field and earth rotation. The analysis of LOD proxy for earth axial rotation yields in the high frequency range period of 27 days, 13 days, down to 5.6 days caused by lunisolar tides and also irregular periodic variations between 40–15 and 182 days. There is also evidence for tidal and seasonal signal in micro-earthquakes. It is suggested that seismic seasonality might manifest periodicity of crustal strain accumulation (Murakami and Miyazaki, 2001). The possible influence of tidal effects on the periodicity of the earthquake activity has been suggested in diverge geological settings (Glasby and Kasahara, 2001). They have suggested that the tidal stresses are very small to initiate earthquakes themselves. However, tides might trigger earthquake activity in tectonically active (increasing tectonic stress) areas.

An apparent volcanic cycle of 7–8 years has been suggested in volcanic dust veils index (VDI) by Kelly (1977). DVI is an annual measure of the volcanic dust in stratosphere, and correlate with various aspects of climate change on time scales of 50 years and more (Chance and Kelly, 1979). According to Kelly, the volcanic cycle of 7–8 years and associated climate change on this scale may be forced by earth's rotation. Several researchers suggested that variations in seismicity, earth's axial and polar motion and atmospheric circulation may be linked (Kelly 1977 and references therein). It is interesting to note that an identical periodicity of 7–8 years has been suggested in earth's rotational speed (Maksimov et al., 1970). It is also suggested that interaction between the annual component of the polar motion, forced by seasonal redistribution of the atmospheric mass, and the Chandler Wobble period of 14 months gives rise to an approximately 7–years fluctuation in the polar motion (Kelly, 1977 and references therein). It has been also suggested by Maksimov (1970) that the polar motion could have a tidal influence on atmosphere. The ensuing discussion suggests that volcanicity, seismicity, the atmospheric circulation and earth's rotation, possibly form a complex interactive loop with 7–8 years cycle (Kelly, 1977). Besides this, Ammann and Naveou (2003) shared evidence of multidecadal and a 5 years cycle in tropical volcanic eruption.

SUMMARY AND CONCLUSIONS

In addition to the various cycles that we have discussed in this chapter, there are several other weak evidences of periodicities in different earth-atmospheric and ocean phenomena. We, however, have focussed here only on those cycles, which are globally persistent, remarkably regular and well established. Table 10.2 in chapter 10 provides the details of such annual, inter-annual, decadal, inter-decadal and century and millennial scale cycles and their possible interpretations (Raymo, 1998; Poore et al, 2003).

Presence of identical periodicities in various climate and earth phenomena and solar and tidal cycles of external origin may suggest that the two phenomena are related in some way directly or indirectly. However, coming to the degree of possible impact of linkages of these various cycles is a debated issue Burroughs, 1992, Weedon, 2003 and reference therein). It is not yet clear, however, as to what extent external forcing, variations in solar activity and tidal cycles affect the climate and temperature variability and other related terrestrial phenomena. For instance, the measured concentration of ^{14}C in tree–rings record, that display a number of climatic periodicities including larger perodicities of approxmetly 3000 year and 2200 year has been associated with the solar origin. Several others, however, believe that the observed concentration might have been controlled by oceanic circulation. Some recent results based on similarity of variations in ^{10}Be and ^{14}C production rate, however, uphold and strengthen the idea that sun–climate relation does exist at least to the extent of North Atlantic drift ice in North Atlantic during Holocene. The remarkable stability and the global presence of millennial scale cycles, however, seem to be apparently triggered by some external forcing. It has been argued that the variations in solar irradiance necessary to cause the observed changes in ^{14}C are estimated of the order of a few tenths of one per cent (Fligge and Solanki, 1999). According to them, such minor variations are believed to have only a very small direct impact on causing significant differences in sensible heating for the climate change. However such changes can lead to changes in atmospheric or oceanic circulation that amplify this initial input.

Some recent model study of the relationship between the 11-year solar cycle and climate has suggested that maximum solar irradiance leads to increased easterly trade winds and a broadened Hadley cell. This might emphasise the fact that solar effects on as interdecadal and century's climate may not be negligible. It has also been suggested in various recent studies that several periodicities in climate records on other different time scales also are identical to periodicities identified in proxy records of solar variability (Poore et al., 2004). This might suggest that atleast some of these are due to solar forcing.

Considering the global warming of the recent century, however, several researchers have argued based on quantitative considerations that solar effects are small relative to the anthropogenic release of greenhouse gases, primarily carbon dioxide (Hoffert et al., 1999). According to several researchers, the available data suggest that temperature increase of 1 K would increase equilibrium CO_2 by less than 5%, which is much less than 30% increase recorded since the beginning of the industrial revolution. Contrary to this, the solar brightening itself may be responsible for both the observed century surface temperature and increased carbon dioxide. It has been argued that a carbonated drink expels most of its CO_2, if warm (Parker, 1999).

There is strong evidence of longer 10,000–12,000 years cycle, which has been invariably associated either with second harmonics of precessional cycle or non-linear response of climate response. Similarly, the well-known Heinrich cycle of the order of 6000–7000 years corresponds to third harmonic of the precessional cycle. It is also a matter of debate as to what extent sub-Milankovitch cycles are related to internal ocean–atmospheric dynamics and global thermohaline circulation. It appears that ENSO are the ultimate cause of global decadal interannual or millennial scale changes in earth processes in general and climate variability in particular. Alternatively, the fact that non-linearity plays some role in modulating and amplifying climatic signals cannot be discounted.

In addition to quasi-periodic/irregular Dansgaard-Oeschger-Bond cyclicity and associated Heinrich events, a more pervasive 1470–1480 years periodicity has been reported from North Atlantic ice rafted detritus records (Bond et al., 1999) and in several other ocean and climate records. Wunsch (2000), however noted that reality of this periodicity in various climate record is doubtful, which might arise due to 'aliasing'. Several independent and careful analysis of various climate records, however, appears to provide considerable credential of its reality and physical relevance (Schulz, 2002). Spectral analyses of sea level changes, tree ring temperature from different parts of the world reveal three dominant periodicities around 1668 ± 100, 1130 ± 100, 570 ± 60 and 464 ± 50 years. Interestingly, at least three of these oceanic and climatic periodicities match (within palaeo-sea level estimation errors) with known harmonics of planetary cycles of 1668, 1133 years and 567 years. However, it is not well-understood that to what degree external factors, like solar and tidal variability have been related to the origin of 1500 years cycle or 1134 years and 567 years or 420 years climate cycles. It is suggested that although tidal effects of planetary conjunction is quite small, it speculated that these periodic variations may be due to short-term effects on the basic energy flow from the sun due to changes in orbital angular momenta and disturbances in orbital configuration of all planets. The climatic mechanism involved, however, require further investigation. If some of these periodicities do really exist and have global appearance, it will have significant implications for the developing sea level theory of anthropological cycles in global human history. For instance, sea level cycles of 550 years might suggest that Dwarka coast (west coast of Arabian sea) must have experienced six oceanic transgressions and regressions during the past 3500 years (Negi and Tiwari, 1989).

Berry and Barker (1975) distinguished fortnightly and monthly cycles in fossils and varves. Weedon (2003) has discussed in detail the effect of tides on various sedimentary processes. It has been suggested that origin of many varved deposits however, do not correspond to an annual cycle, but to a shorter one (for instance to two rainy seasons per year).

Although there have been many suggestions and arguments in favour of and against the ideas of linking of long and short term climate and other environmental and earth's cycles with the solar and tidal cycles, it appears that there is enough scope to revise these suggestions using more precise data and statistical techniques. This will provide better opportunity to ponder over physical mechanisms and prediction of these intimately associated societal phenomena.

9

El Nino-Southern Oscillations (ENSO)

INTRODUCTION

One of the most well-known short-term quasi periodic variation of the earth's climate is the El Nino that originates in the tropical latitudes of Pacific Ocean. It is the strongest source of variability in the earth's climate system and its climatic impact has been realised globally through the disruption in major rainfall systems and occurrences of droughts, catastrophic storms and floods in various parts of the world. Since these consequences are directly related to the societal value, the study of El Nino behaviour has recently emerged as one of the major topics of modern climate research. The available observations indicate that the El Nino dynamics is non-random phenomenon but displays a very complex and chaotic nature. Predicting the El Nino behaviour in the midst of chaos is, therefore, a challenging task for the climatologists. This is due to the fact that the oceans absorb and release heat with distinct modes of oscillations, which are apparently manifested in the form of temperature variability. These oscillations vary on multilateral time scales from years to decades and centuries and hence are influenced by several other factors such as long-term fluctuations in cloudiness, sustained changes in precipitation, in the extent of ice pack and rates of evaporation. The apparent dynamical changes at high latitudes take decades or centuries before influencing the temperature or upwelling of cold water at low latitude and thereby set a long-term oscillatory pattern in the system (Burroughs, 1992). The ocean and atmosphere are also constantly exchanging heat and water in the tropic by virtue of their coupling. Any changes in atmospheric conditions, therefore, can lead to changes in the ocean's surface, which in turn can alter the climate pattern. These ocean-atmospheric feedback mechanisms have the potential to set up oscillatory behaviour on time scales of weeks to months, thus producing periodicities or quasi-periodicities in the weather (Burroughs, 1992). Non-linearity and hence chaos in the climate and weather pattern might stem either from the underlying internal/external mechanisms or from both. It is, therefore, quite difficult to understand precisely such a complex and chaotic behaviour of oceans and atmospheric phenomena.

There are several unresolved and exciting issues related to complex ENSO dynamics (Philander and Fedorov, 2003), for instance, it is to be explored if El Nino is one phase of a continual, self-sustaining natural mode of the coupled ocean–atmosphere that has the complementary phase, or if El Nino is a temporary departure from 'normal' conditions triggered by random disturbances, such as burst of westerly winds. Also, it needs to be found out if El Nino is governed solely by internal causes or there is external trigger. The posible causes and irregularities in the ENSO oscillations are at present a matter of great debate and focus of active research. During the past two to three decades, a lot of research work has been carried out which shed light on the various aspects of ENSO phenomenon. In this chapter, we discuss the current status of ENSO as re-interpreted in the light of modern research and its impact in terms of both human activity and the environment. We also discuss the possible prospects of its modelling and prediction.

EL NINO, LA NINA, SO AND ENSO

The term 'El Nino' historically refers to the massive warming of the coastal waters off Ecuador and Peru that takes place irregularly/quasi-periodically at intervals of 2–7 years and alternates with an opposite cold phase, generally called as La Nina. The 'El Nino' (in Spanish El Nino means the 'Christ child') begins around Christmas (Neelin and Latif, 1998). However, the El Nino related warming in the Pacific Ocean does not occur in isolation. It is now clearly recognised that the atmospheric counterpart of the El Nino is the Southern Oscillation. The term 'Southern Oscillation' (SO)' refers to large-scale associated changes in atmospheric circulation. Scientists coined the acronym ENSO for El Nino and Southern Oscillations to uniquely describe this large-scale, interannual 'Pacific climate phenomenon'. Theoretical and numerical modelling studies demonstrated that there is strong coupling between the atmosphere and tropical ocean, and that the crucial interactions give rise to catastrophic climate changes with several human consequences, such as drought, flood as well as distinctive teleconnected patterns of climate anamolies. The ENSO coupling also have significant impact on the Asian monsoon system that drives surface ocean seasonality in the Indian and Western Pacific Ocean and the Atlantic Inter Tropical Convergence Zone (ITCZ), a near equatorial band of intense rainfall and converging trade winds that interact with the surface ocean (Cole, 1994). SO was discovered in 1923 by the British climatologist, Gilbert Walker, who has suggested that there is a quasi-periodic standing oscillation in the atmospheric surface pressure, now widely known as 'Walker Circulation', that span the Pacific from east to west. Jacob Bjerknes (1969) showed more convincing connection between the two

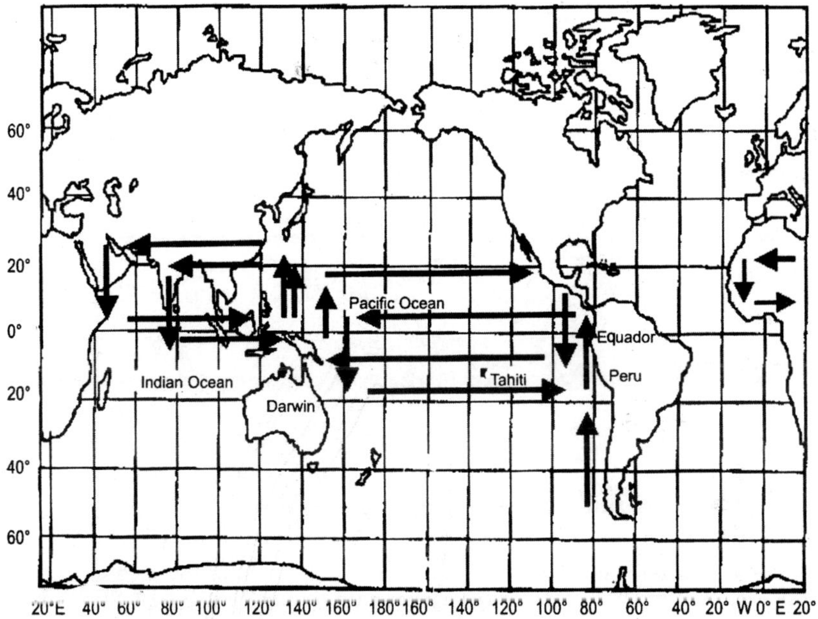

Fig. 9.1 World Map Showing Walker Cell and the Important Sites of El Nino Occurrences.

(*Source*: Das, 1988)

phenomena and also proposed couplings between the atmosphere and ocean. He further proposed that the changes in atmospheric heating associated with tropical Pacific Sea Surface Temperature (SST) anomalies cause changes in the mid-latitude circulation patterns. This teleconnection is consistent with global nature of Walker circulation. Figure 9.1 shows Walker cell and the major sites of ENSO activity on the global map. Arrows indicate Walker cells. The solid arrows indicate the normal years while dotted ones are for the abnormal years (Das, 1988). The Southern Oscillation Index (SOI) is defined as the changing difference in atmospheric pressure between Tahiti, French Polynesia and Darwin, Australia. This is a measure of the strength of the trade winds, which have a component of flow from regions of high to low pressure. Trade winds blow from east 30° at high latitude also called 'preventing easterlies'. These are actually surface winds of Haddely cell. The changing pressure difference is associated with changing SST distributions across the equatorial and tropical pacific. In the eastern Pacific region, high SOI (large pressure difference) is associated with stronger than normal trade winds corresponding to El Nino and low SOI (smaller pressure difference) is associated with weaker than normal trade winds and correspond to La Nina (Philander 1989; Einfleld 1989a,b).

There have been several catastrophic ENSO events in history. These events are most energetic and extreme in the sense that the temperature of equatorial waters from the South America coast to the date line goes up by an average of 2°C with the warming along the coast exceeding 6°C. The climatic consequences of these catastrophic events were often devastating. Figure 9.2 shows the global effects of El Nino 1982/1983 (Molnar and Cane, 2002). As clearly seen in Fig. 9.2, the ENSO impact is worldwide and dominated either by flood and heavy rainfall or by dry and wet conditions.

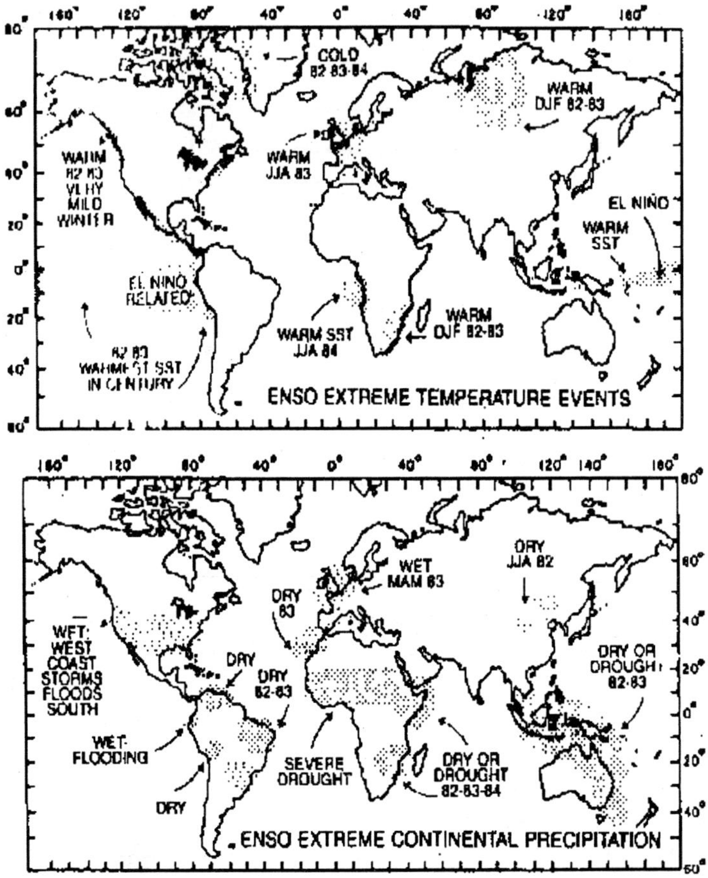

Fig. 9.2 Maps of the Climatic Anomalies Attributed to 1982–1983 El Nino.

(*Source*: Cane, In predicting El Nino and The Southern Oscillation; 1985)

A commonly used measure of ENSO strength is the SST anomaly averaged over the eastern equatorial Pacific. The SSTs (NINO3 region defined as 5°S to 5°N and 150°W to 90° W) index are strongly linked to climatic processes and hence it can be used to monitor climatic change (Evans et al., 2000). Observational evidence indicates that there is anti-correlation between SST and surface pressure gradient (Neelin and Latif, 1998). The negative phase of the SO correlates with El Nino episodes, indicating abnormally high air pressure over Indonesia and the western tropical Pacific and abnormally low air pressure over the eastern tropical Pacific. Conversely the positive phase of the SO correlates with La Niña episodes, indicating abnormally low air pressure over Indonesia and the western tropical Pacific and abnormally high air pressure over the eastern tropical Pacific.

Webster and Palmer (1997) demonstrated some distinctive features of SST anomaly of the eastern Pacific El Nino and La Nina. Accordingly Fig. 9.3 'A' shows the interannual oscillations in the SST at the equator in the eastern Pacific (averaged over the area 5° S to 5° N, 80° to 120° W) on the background of the decadal fluctuations obtained by means of a low pass filter after removal

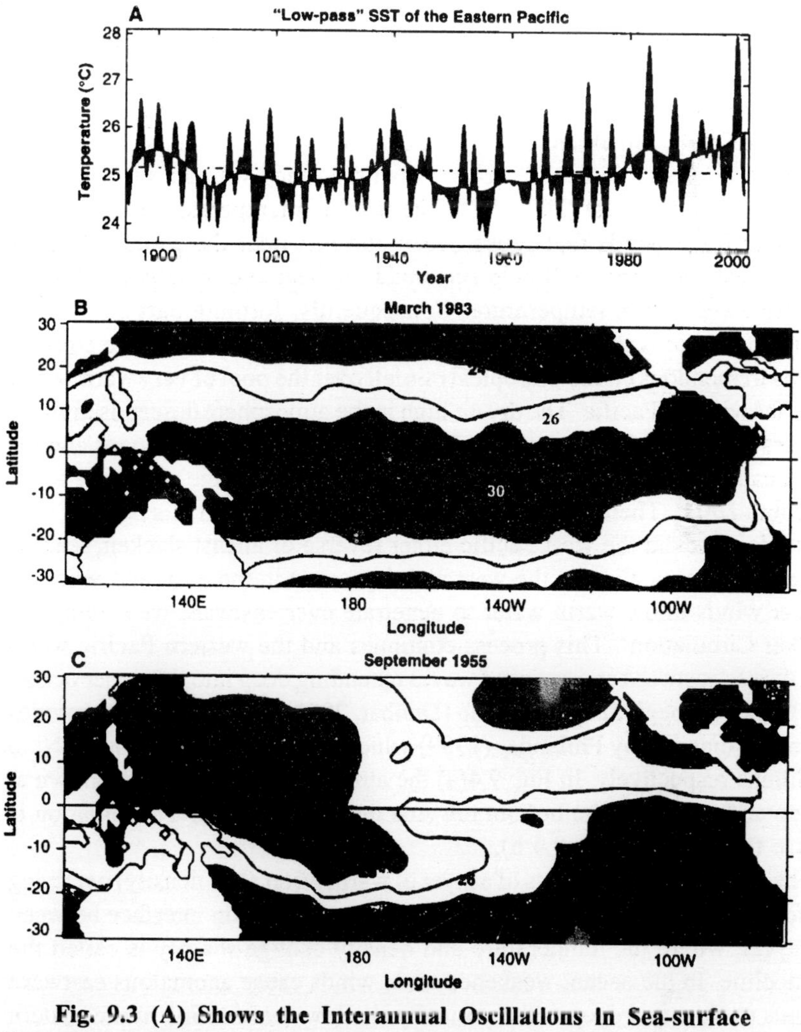

Fig. 9.3 (A) Shows the Interannual Oscillations in Sea-surface Temperature (SST) at the Equator in the Eastern Pacific (Averaged Over the Area 5° S to 5° N, 80° E to 120° W). For details see text.

(*Source*: Webster and N. Palmer, 1997)

of the annual cycle and higher frequency variability. The horizontal dot dashed line is the time average for the record. The SST (°C) at peaks of (B) El Nino in March 1983 and (C) La Nina in September 1955. Some of the differences between 'B' and 'C' (high temperature that extend far north in September and far South in March) are attributable to the seasonal cycle.

Evolution of ENSO

The ENSO cycle arises as a coupled ocean–atmospheric interaction in which anomalies in SST in the Pacific cause the trade winds quasi-periodically to strengthen or weaken. This pattern drives the change in ocean circulation that produce anomalous SSTs. Ocean-atmosphere feedback can amplify perturbations in either the equatorial SST or the Walker Circulation–the thermodynamic circulation of air parallel to the equator (Neelin and Latif, 1998).

The essence of the ENSO dynamics is described in the following way (Neelin and Latif, 1998; Philander, 1989; Dunbar, 2000). Accordingly, the eastern Pacific is normally cooler (22–24°C) than the much warmer surface-water temperature in the west (29–30°C). The resulting temperature gradient sets up strong trade winds that cause ocean currents near the equator to flow mostly westward. This will help accumulating heat and thereby reinforcing the westward rise in temperature. Consequently, forming part of 'Walker circulation', that is thermodynamic circulation of air parallel to the equator, there is tremendous centre of tropical rainfall over the pool of very warm water in the far western Pacific. The dry air high in the atmosphere descends towards cool sea surface in the eastern Pacific. At this point, something snaps and the winds can no longer hold the wall of water up towards the western Pacific (Weedon, 2003). There are changes in situation, however, at this point of time and trade winds in the west Pacific either reverse or atleast slacken, leading to elevated thermocline in the west and depressed in the east, and whereby weaker winds allow warm water to penetrate ever eastward weakening the 'Walker Circulation'. This process continues and the western Pacific warm pool moves eastward as a series of waves extending deep into the upper ocean. This marks the beginning of El Nino (Dunbar, 2000). Figure 9.4(a,b) provides schematic diagram by Philander (1989) which explains La Nino and El Nino conditions respectively. In Fig. 9.4(a) the atmospheric condition is shown at the top and oceans at the bottom towards the end of a year. The situation is reverse for El Nino (Fig. 9.4 b).

The tropical ocean consists of a layer of warm water (low density) overlying a much deeper layer of cold water (high density). The thin interface between the layers, where the temperature and density change sharply is called the thermocline. In the ocean, weakened trade winds cause anomalous eastward currents to flow and deepen the normally shallow thermocline in the eastern basin. Figure 9.5 shows the interactions that sustain the ENSO cycle. Figure 9.5a shows the normal state of affairs whereas Fig. 9.5b depicts the thermocline condition during the El Nino events.

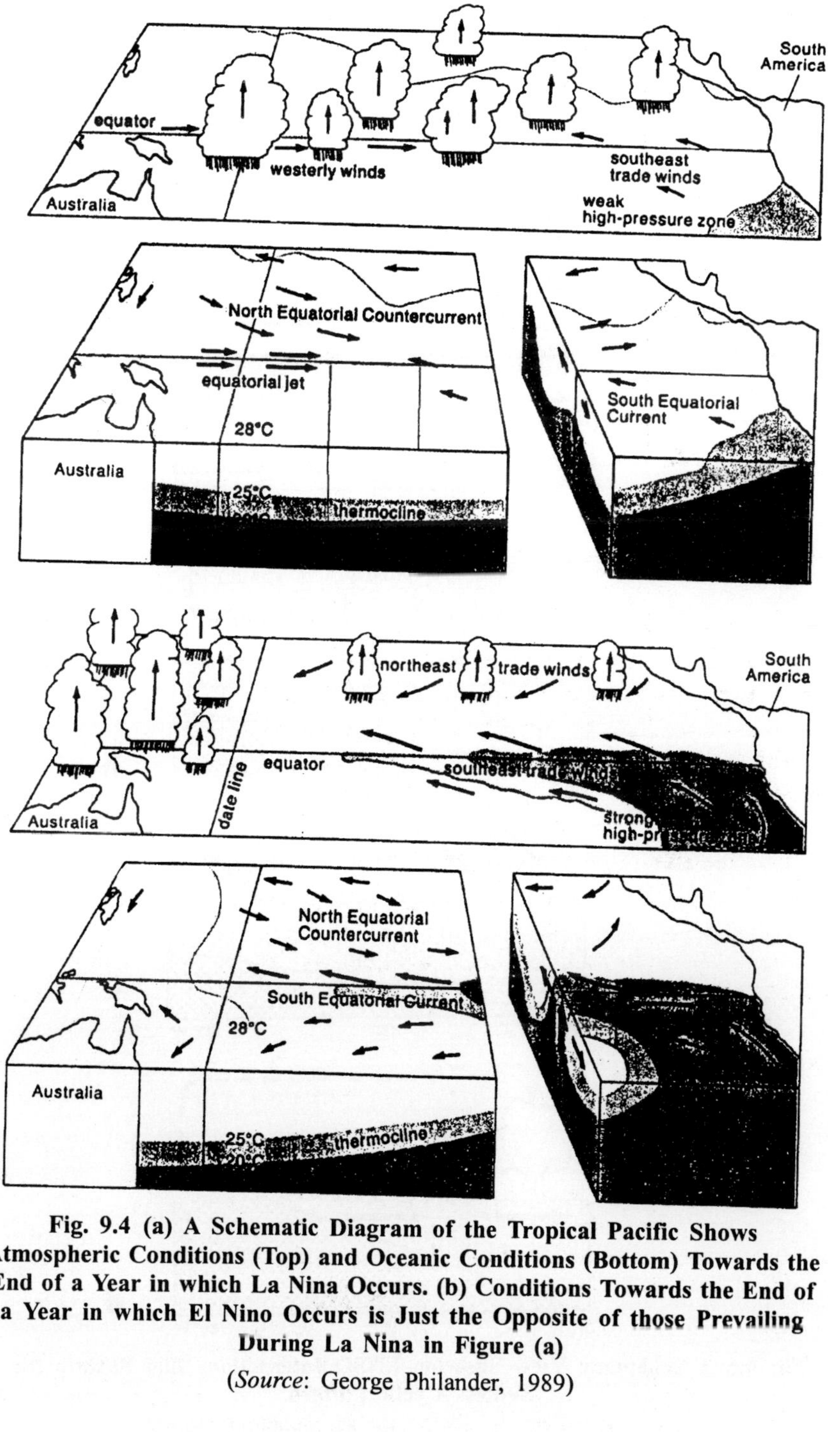

Fig. 9.4 (a) A Schematic Diagram of the Tropical Pacific Shows Atmospheric Conditions (Top) and Oceanic Conditions (Bottom) Towards the End of a Year in which La Nina Occurs. (b) Conditions Towards the End of a Year in which El Nino Occurs is Just the Opposite of those Prevailing During La Nina in Figure (a)

(*Source*: George Philander, 1989)

As discussed above, El Nino is one extreme of a year-long oscillation in the entire Pacific basin and the atmosphere above it. To this chain of interactions, there must be added a suitable mechanism for flipping the system from El Nino to subsequent La Nina. Like El Nino, La Nina produces extreme weather and abnormal conditions in the western Pacific similar to those El Nino produces in the east. The change in the thermocline condition as shown in Figs. 9.5c and d is based on the idea of Professor Cane and his modelling group at Lamont

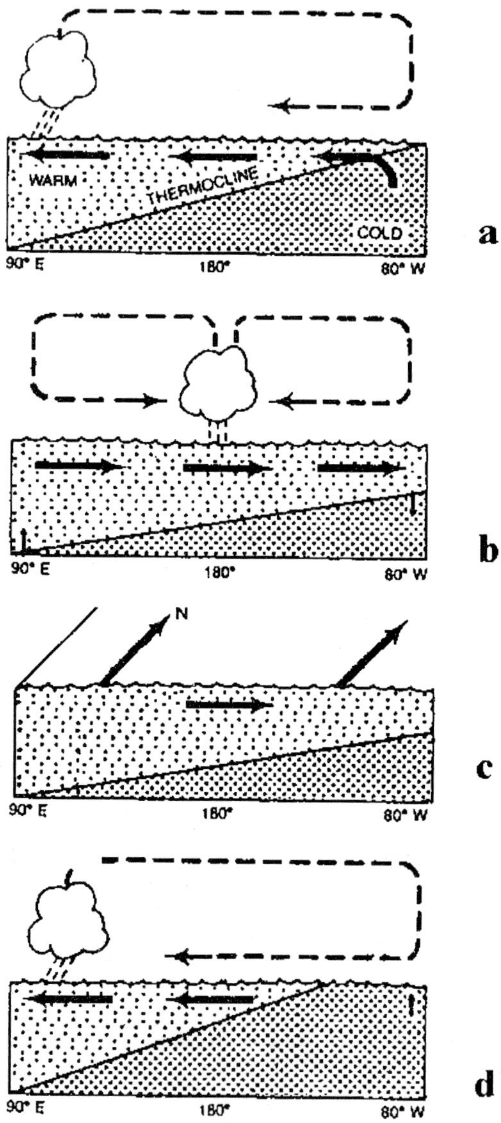

Fig. 9.5 A Schematic View Showing ENSO Interactions that Sustain the Endless Cyclic Pattern.

(*Source*: Mark. A. Cane and his modeling group)

to demonstrate the swing of ENSO cycle. Accordingly, at the peak of an El Nino event water moves not only from west to east but also poleward, emptying the reservoir of warm water at the equator (Fig. 9.5c). After some time, there is no longer enough warm water to sustain above normal surface temperature in the east Pacific and El Nino begins to decay. Fig. 9.5d shows the aftermath (La Nina)condition where the eastern Pacific is cooler and easterlies stronger than normal. This also shows the process of transitions from one phase to another and continue in endless cycle.

There are, however, different views and theories about what exactly causes El Nino and its swing, the truth is, no one knows what really causes catastrophe (Fedorov and Philander, 2000; Cane, 1986). A quasi-periodic oscillation about every 3–7 years is a reverberation back and forth between two states apparently revealing the non-stationary characteristics and variability in the amplitudes. One of the possible explanations for such behaviour could be that ENSO signals might reflect a chaotic system due to non-linear resonance of natural variability of the Pacific ocean with the annual cycle (Jin et al., 1994; Tziperman et al., 1994). It seems somewhat still unclear that why not a perpetual El Nino or La Nina instead of every 3 to 5 years flips from one to other? It is believed that the ocean carries certain 'memory' by virtue of its slower time scales of adjustment. Perhaps this memory carries oscillation phase-to-phase. The slow adjustment of the ocean layer between the thermocline and the surface layer causes the depth of the equatorial thermocline to vary slowly thereby carrying the oscillation between the warm and cold phases (Neelin and Latif, 1998). Whatever may be the cause of quasi-periodic variations of EI Niño, the ensuing discussions suggest that an interaction between the ocean and the atmosphere was necessary to explain the ENSO swings because the trade winds have implication for creating shallow or deep thermocline. Model study shed light on breaking the entrenched ocean-atmosphere mode and bringing transition from normal to El Nino and back. The study suggests that the underlying cause may perhaps be linked to the development of slow eastward and westward moving oceanic waves respectively (McPhaden and Yu, 1999; Bigg, 1996).

COUPLED OCEAN-ATMOSPHERIC MODEL

We have discussed two phenomena above: El Nino, an oceanic phenomena and Southern Oscillation, an atmospheric phenomena. The successive interval of El Nino episode varies from approximately three-to-seven years and the intensity of different episodes also changes. Based on the spectral analysis, the characteristic time scale for the Southern Oscillation has been estimated to be approximately four to five years. The coupling between the two phenomena (i.e., exchange of ocean and atmosphere) therefore leads to irregularities in the system dynamics. Cane (1992) and Philander and Fedorov (2003) discussed the different ENSO modes and irregularities in the coupled dynamics using the coupled ocean-atmospheric model (Neelin, 1990). The main feature of this model is that it has an atmospheric component that produces surface winds strictly in response to sea-surface temperature and the atmospheric noise is absent. The model sheds light on how changes in intensity of ocean-atmospheric interactions affect the southern oscillation. In Fig. 9.6 (left panel) ocean-atmosphere interactions are weak and the simulated SO is damped. This highlights the imposed initial burst of westerly winds that lasts for a month and induce El Nino condition that persist for a while and then decays. The central panel depicts more intensive ocean-atmospheric interactions that result in a self-sustaining SO. Further intensification is shown in the right-hand panel, which appears to cause the SO to grow to such large

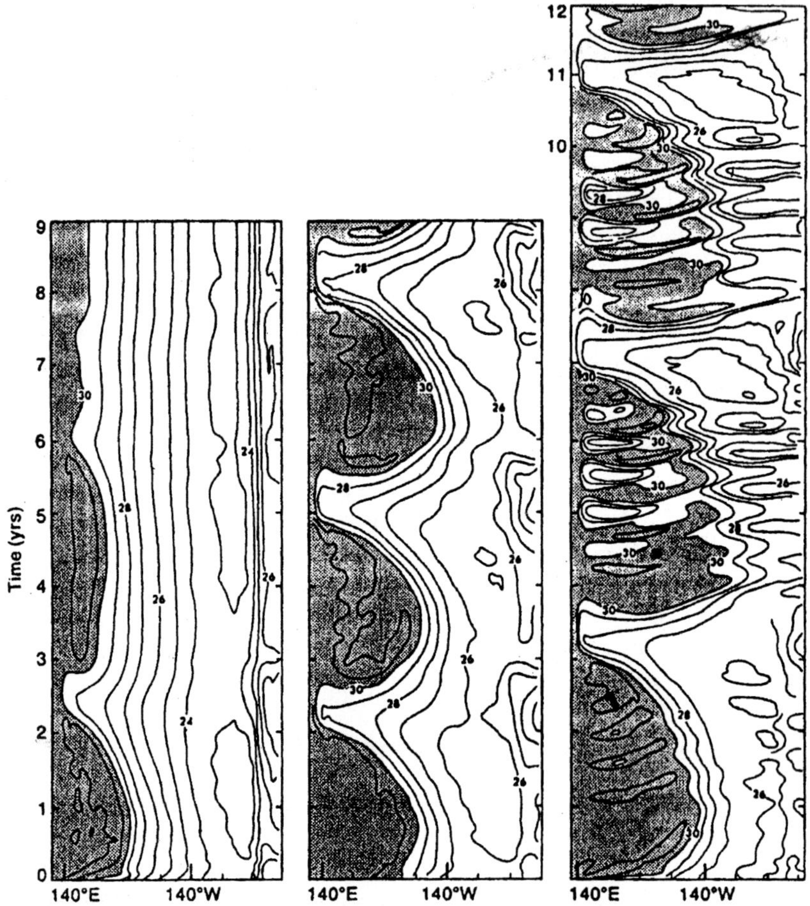

Fig. 9.6 Evolution of Sea Surface Temperature Along the Equator in a Coupled Model (Neelin, 1990). The Strength of the Coupling Between the Ocean and Atmosphere Increases from (a) to (b) to (c). The Evolution Over a Nine Year Period is Given in (a) and (b) and that Over a 12 Year Period in (c). Shaded Portion Show the Temperature Higher than 30°C

(*Source*: Philander and Fedorov, 2003)

amplitude that secondary instability develops. The ENSO phenomena of coupled ocean-atmospheric system appear to be unique, of its kind having response of mixed oscillatory signals of cyclic/stochastic and chaotic nature.

Neelin and Latif (1998) critically reviewed the significance of irregularities in different ENSO models. The ENSO's cycles are inherently irregular which is possibly a manifestation of low dimensional deterministic chaos arising from the internal non-linear interaction of the Pacific ocean–atmospheric system. Alternatively, some others argue that the majority of irregular ENSO variability may be a form of correlated Gaussian noise and can be explained with a simple linear model driven by external stochastic forcing. The noise may, for example, be a product of short term local weather fluctuations including solar radiation anomalies, changes in CO_2, salinity anomalies, precipitation events, westerly wind bursts and other wind anomalies besides natural random background variation (Enfield, 1989a,b). All these various factors lead to stochastic fluctuations in SST and thermocline depth and may, therefore, be important driving forces behind the ENSO phenomena. Since ENSO irregularity limits its predictability differently, this particular issue has practical as well as theoretical implications.

Deterministic Chaos

The observation of global ENSO connection appear to imply that the oceanic and atmospheric anomalies of the equatorial Pacific are quasi-periodically forced non-linear systems. Several researchers (Battisti and Sarachik, 1995 and references therein) have examined non-linear atmosphere model and found that irregular interannual variability can result from the non-linear coupling of the atmosphere and ocean. Various spectral analysis reveal that variations in amplitude and period of ENSO are its inherent characteristics. The possible explanation suggested by some researchers (Jin et al., 1994; Tziperman et al., 1994) is that ENSO variability is nothing but a chaotic system. As the system is quasi-periodic, the chaos might ensue due to non-linear resonance of natural variability of Pacific ocean with annual cycle. In general, two parameters pave the quasi-periodicity route to chaos. One parameter affects the strength of non-linearity and the other one affects the inherent frequency of the ENSO oscillation relative to the annual cycle (Neelin and Latif, 1998). As non-linearity increases in the system, so does the tendency of ENSO cycle to lock its frequency on rational fractions of the annual frequency. This hypothesis is closely associated with the non-linear interaction of ENSO with the annual cycle of seasons. Chaotic ENSO behaviour has been observed in several models. In several other models also the transition to chaos takes place primarily along the quasi-periodicity route, which is found in periodically forced non-linear systems. Chaos can ensue as the system jumps between the various sub-harmonic resonances (Neelin and Latif, 1998).

Weather Noise

The possible cause for irregularity of ENSO has been attributed to noise internal to the atmosphere or ocean. Several researchers have discussed the impact of the atmospheric weather from synoptic to inter-annual (Battisti and Sarachik, 1995 and references therein). Much of the atmospheric variability associated with weather has relatively short decorrelation times. Because this variability appears essentially random on the long timescales associated with ENSO, it can be treated as a stochastic noise process, termed as 'weather noise'. Several recent studies have underlined the importance of weather noise in promoting the irregularity of the ENSO cycle (Neelin and Latif, 1998). For instance, adding weather noise to an otherwise periodic coupled model produces irregularity that is generally consistent with observed ENSO signals. In the ENSO power spectrum, weather noise broadens and flattens the roughly four-year peak. This spectral signature typically differs from that of models whose irregularity is induced purely by chaos.

Changing Background State

In the study of inter-annual timescales, it is common to make the approximation that an average over a longer interval can be treated as a fixed background state (Neelin and Latif, 1998; Urban et al., 2000). Modelling study suggests that the period and amplitude of the stimulated ENSO are sensitive to changes in the basic-state to the mean depth of the thermocline, thus variations in the climate system may affect the way the ENSO cycle varies on inter-decadal and longer timescales.

Several researchers have proposed that disturbances propagating into the tropical Pacific from the extra-tropics or the Indian ocean can trigger or modify the ENSO cycle (Barnet et al., 1981 a,b). Some other investigations suggested the possible impact of solar variability and changing atmospheric CO_2 concentrations on the tropical Pacific coupled inter-annual variability (Graham and White, 1995).

During early eighties and mid-1980 George Philander and his colleagues and Mark Cane and Stephen Zebiak develeped hierarchy of ENSO prediction models ranging from statistical forecast models to fully coupled ocean–atmosphere models. Information about the evolving sub-surface ocean anomalies are introduced either directly, as sub-surface temperature data or indirectly as estimates of the sub-surface initial field from an ocean model driven over several previous years by the observed wind stress. However, question arises that to what extent these models could be able to predict day-to-day ENSO. Views are emerging that 'Pacific Phenomena'/coupled ENSO dynamics is chaotic, a property of many natural systems which unfold in mysterious and complex ways.

RECORDS OF ENSO ACTIVITIES

El Nino has left signatures of its past appearances throughout the globe and corals have recorded these signatures as a tape recorder. Coral (Fig. 9.7) refers to a general name for any of the large group of bottom dwelling sessile, marine invertebrate organism, (polyps) that belong to the class *Anthozoa* (phylum *Coelenterates*) that are common in warm intertropical modern seas and abundant in the fossil record in all periods later than the Cambrian that produce external skeleton of calcium carbonate and that exist as solitary individual or grow in colony (Bates and Jackson, 1980). Corals can live for several centuries and because of their rapid growth—as much as two centimeters per year—analysis can be performed at a resolution of weeks to months. Accurate chronologies can be developed from annual banding or geochemical signals.

Coral record is a spectacular archive of past climate variability and provides vital information on factors governing behaviour of climate-related processes. One can have continuous records of past climate history from corals in the

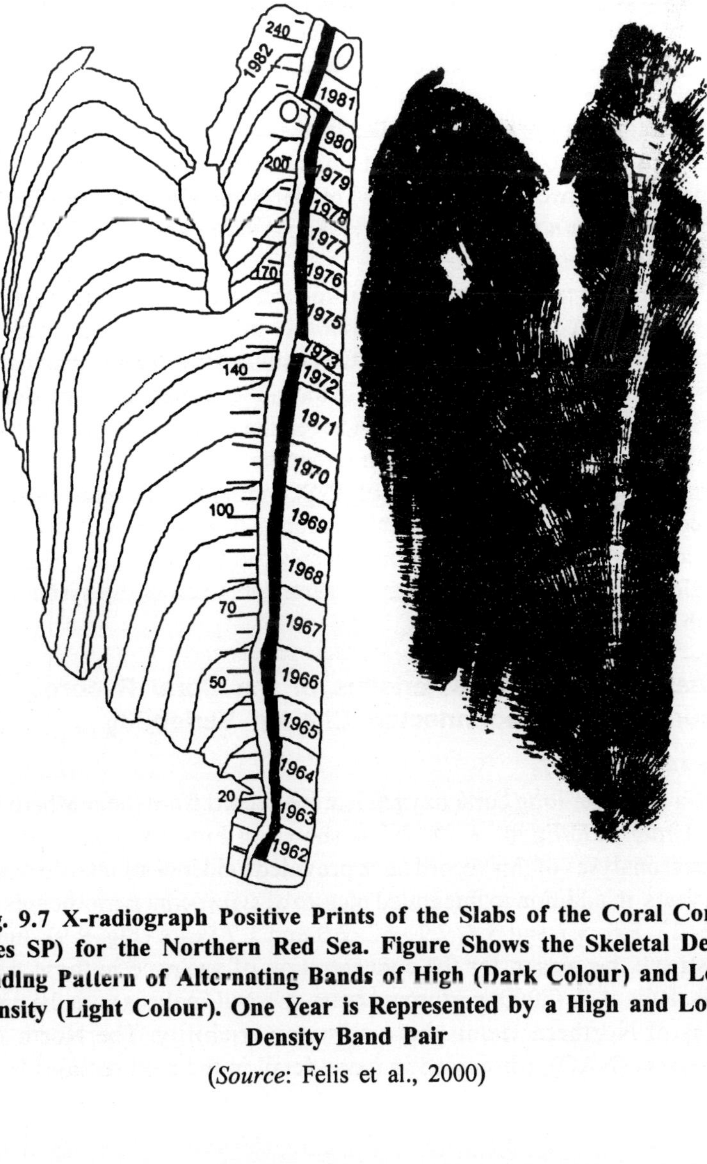

Fig. 9.7 X-radiograph Positive Prints of the Slabs of the Coral Core (Porites SP) for the Northern Red Sea. Figure Shows the Skeletal Density Banding Pattern of Alternating Bands of High (Dark Colour) and Low Density (Light Colour). One Year is Represented by a High and Low Density Band Pair

(*Source*: Felis et al., 2000)

form of sub-annually collected isotope ratio and geochemical measurements apparently related to sea–surface temperature (Cole and Fairbanks, 1990; Cole, 1994). The proxy records of stable isotope and chemical tracers in corals provide the climate history. Geophysical tracers within coral skeleton aragonite, such as stable isotopes, trace metals and radiocarbon, accurately monitor key features of tropical ocean–atmosphere variability. They track parameters such as sub-tropical SST, rainfall, nutrient availability, river run off, and the upwelling and advection of distinct water masses (Boiseau et al., 1998). Thus, corals are uniquely well suited to provide a new network of quantitative climate reconstruction through the tropical oceans. There are some difficulties, however, faced in climate reconstructions concerning chronological dating. For instance, oxygen isotopes in corals are affected by both temperature and salinity of the seawater in which the corals grow, resulting in an inherent ambiguity (Trenberth and Otto-Bliesner, 2003). The second problem is based on trace element ratio normalised to calcium (such as Sr/Ca and Mg/Ca) in the skeletons of corals and shells. It is assumed that the trace metals are incorporated into the skeleton at concentrations that depend on growth temperature. However, as noted by Trenberth and Otto-Bliesner (2003) since living organisms are complex, their response varies and empirical calibrations are always necessary.

One of the reasons to study coral records were motivated by a desire to understand the complete natural variability in the ENSO system and related regional phenomena. Although ENSO constitutes the dominant signal of inter-annual climate variability over much of the globe, its natural variability remains poorly characterised beyond the past several decades (Philander and Fedorov, 2003). However, coral records from across the equatorial Pacific are revealing new insights into the natural modes of ENSO variability, including the previously unrecognised links across seasonal, interannual, and interdecadal time scales (Linsley et al., 2000), recent trends in specific variables (i.e. rainfall, SST); and global spatial variability in these parameters across the system's domain (Cobb et al., 2001).

Global Spectral Characteristics of the Coral Records Documenting Teleconnected Climate Variability

Inter Decadal Cycle

Here, a 245-year long coral oxygen isotope record from the northern Red Sea (Ras Umm Sidd/Egypt, ~ 28° N) is shown in Fig. 9.8 (Felis et al., 2000). Spectral analyses of this record have provided evidence of interdecadal signal ~ 70 years in addition to integrated picture of significant periodicities peaking at 23–22, 8.6, 5.7 and 3.4, 2.7, 2.2, 2.0 and 1.7 years (Fig. 9.9). In order to investigate the reason for the concentration of variance in these frequency bands, Felis et al. (2000) performed cross-spectral analysis with the prominent modes of Northern Hemisphere climate variability. The North Atlantic Oscillation (NAO), (discussed in more detail in the next section) has strong

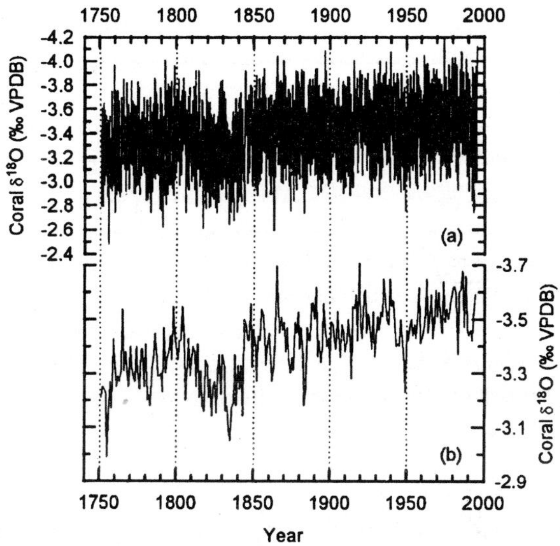

Fig. 9.8(a) A 245-Year Coral Oxygen Isotope Time Series from Northern Red Sea in Bimonthly Resolution. (b) Mean Annual Time Series

(*Source*: Thomas Felis et al., 2000)

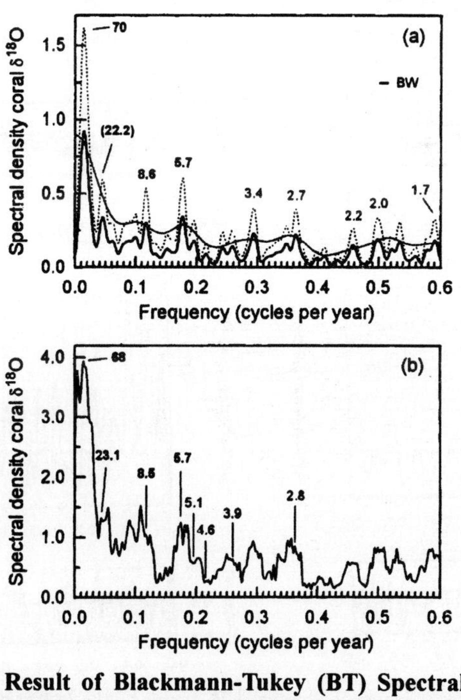

Fig. 9.9 (a) The Result of Blackmann-Tukey (BT) Spectral Analysis of the Bimonthly Record Shown in Figure (9.8b). (b) The Result of Multitaper Method (MTM) Spectral Analysis of the Same Time Series. For Details See the Text

(*Source*: Thomas Felis et al., 2000)

influence on large-scale variations in the atmospheric circulation over the North Atlantic and its surrounding continents controlling regional distributions of surface temperature and precipitation (Hurrell, 1995). The cross-spectral analysis of the winter NAO index with the mean Ras Umm Sidd coral oxygen isotope indicates that the two time series are coherent at periods of 22.6, 7.6, and 5.7–5.8 years suggesting that the prominent periods of 22–23 years and 5.7 years evident in these analysis are related to the NAO (Felis et al., 2000). However, the two phenomena do not seen to be connected at interdecadal scale. The similar cross-spectral analysis of NAO index with various other indices (Fig. 9.10) shows the result of cross spectral analysis (a) between mean annual Ras Umm Sidd coral $\delta^{18}O$ and the winter (December–March) North Atlantic Oscillation (NAO) index (Hurrell, 1995) for the interval 1864–1994, (b) between the bimonthly variations in coral $\delta^{18}O$ and the extended

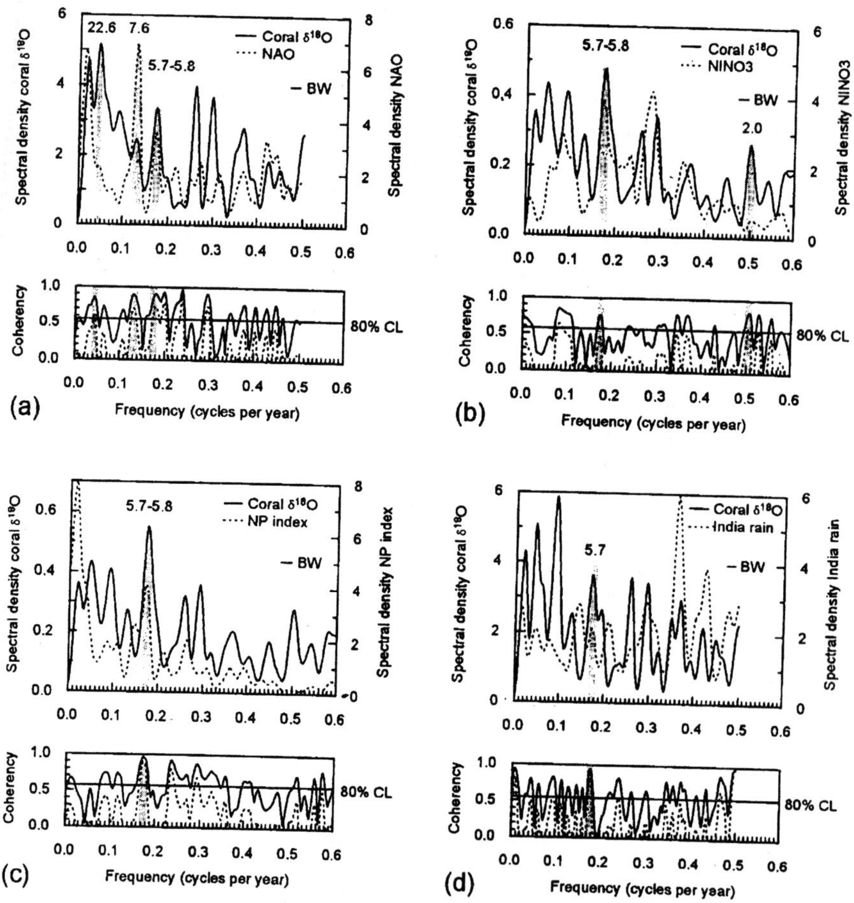

Fig. 9.10 The Results of Blackman-Tukey Cross-spectral Analysis of NAO and Various Other Indices)

(*Source*: Felis et al., 2000)

NINO3 SST Anomaly Index (150°–90°W, 5°N–5°S) (Kaplan et al., 1998; Reynolds and Smith, 1994) for the interval 1856–1995, (c) between bimonthly variations in coral $\delta^{18}O$ and the North Pacific (NP) SST anomaly index (150°E–120°W, 50°–30°N) (Kaplan et al., 1998) for the interval 1856–1991, and (d) between mean annual coral $\delta^{18}O$ and the all-India summer (June–September) monsoon rainfall index (data) for the interval 1844–1992. Figure 9.10 also shows dominance of power at 5.7-year period indicating the role of ENSO dynamics in these processes. These results are consistent with the singular spectrum analysis. It is interesting to note, however, that a similar cycle of the order of 22–23 years are also found in solar-sunspot cycle and the relation between the sun and weather cannot be discounted. More discussions on the issue are given in a separate section.

Inter-Annual Signals

Several studies have also shown that ENSO encompasses most prominent signal of climate anomalies on inter-annual time scale centred across the equatorial Pacific and perturbed rainfall anomalies globally via disruption of normal atmospheric circulation pattern (Cole et al., 2000). On inter-annual time scale, several frequencies are apparent: namely 5–5.5, 3.5–4 and 1.7–2.4 years. The amplitude of these inter-annual modes, however, varied substantially over the last 150 years. The energies of 1.7–2.4 years band was apparently modulated by decadal cycle. (Charles et al., 1997)

Decadal Cycle

Decadal climate variability in the tropical Pacific is the subject of continuing debate. Decadal scale spectral modes in the ENSO variability have been interpreted as resulting from slow movements of Pacific water from high and mid to low latitudes in the Pacific (Gu and Philander, 1997). The most prominent cyclical mode has a decadal periodicity of 11.8–12.3 years and is relatively constant in amplitude throughout the record. Charles et al. (1997) suggested that there is a clear distinction between inter-annual ENSO and decadal monsoon influences. Cole et al. (2000), however, found that decadal variance in both the Malindi and Seychelles record is more coherent with ENSO indices than with Indian or East African rain indices. Statistical studies of coral records and Pacific indicators suggest that the Indian ocean decadal variability reflect decadal ENSO like variability originating in the Pacific. If the decadal variability in the coral records represents monsoon forcing, then the coherence between the coral and NINO3.4 records require a mechanism by which decadal variance in the monsoon generates or is greatly amplified by Pacific decadal variability (Charles et al., 1997). It may, therefore, be suggested that decadal variability in the coral records is primary response to Pacific influences, on the basis of:

(a) the statistical hierarchy of stronger correlations with NINO3.4 than with monsoon rain indices, and
(b) the well-documented capability of the tropical Pacific to amplify and propagate climate anomalies, particularly in monsoon region. (Cole et al., 2000).

Several studies have suggested that observed differences between the recent spectral signature of ENSO (1960–1998) and the average spectral signature (1856–1976) might be due to the possible influence of anthropogenic greenhouse forcing on post-1976 variability.

ENSO Signature in the Indian Rainfall

The possible link between the ENSO and Asian/Indian Rainfall (IRF) has been the subject of longest standing debate. Searches have been made to find any possible links with the external and/or internal forcing (Kailas and Narsimha, 2000). For this, the IRF data have been updated time-to-time and various spectral analysis and filtering techniques were used to unravel the nature of Indian monsoon variability (e.g., cyclic, stochastic or chaotic). Possible causes include external factors (Dunbar, 1994), solar and sunspot cycles, increasing greenhouse concentration and predictable El Nino and unpredictable noise. Some researchers have identified the role of solar/sunspot cycles while others emphasised evidence for ENSO signal in the Indian rainfall data. In addition to this, several others have suggested the possible tele-connection of IRF with internal ocean–atmospheric processes on the decadal–multidecadal and inter-annual (ENSO and NAO related) time scale (Charles et al., 1997).

The updated instrumental annual rainfall index (Singh and Sontake, 1996) spanning over a period from 1826 to 1994 (Fig. 9.11, a) are area weighted

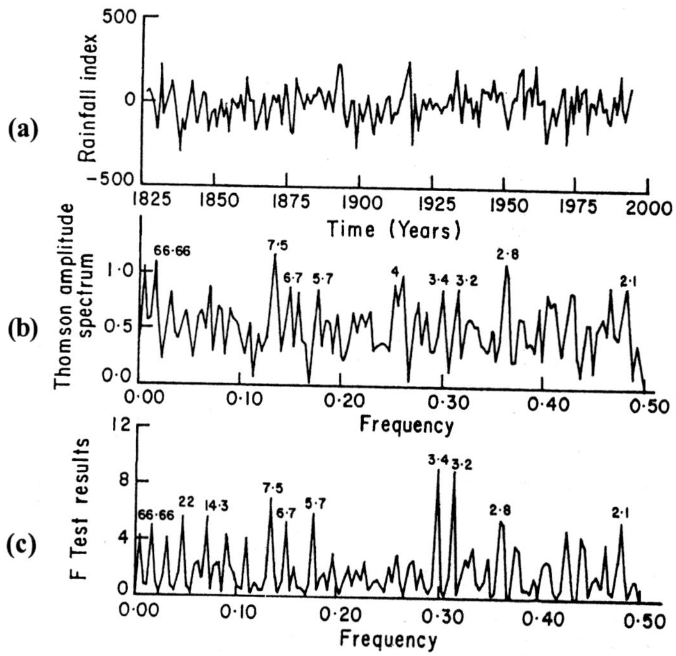

Fig. 9.11: (a) Updated Indian Rainfall (IRF) Index for the Period 1926–1994. (b) Thomson Amplitude Spectra; (c) F Significance Test (Tiwari and Rao, 2004)

and homogeneous for the whole country. As can be seen from Fig.9.11(a), the important feature of these records is its distinct changes in the character of variability. The IRF time series exhibits small oscillatory fluctuations superimposed on apparently long-term decadal and inter-decadal variations, which can be revealed through detailed spectral analysis.

Multi-Taper Spectra (MTS)

MTS analysis of IRF data (Tiwari and Rao, 2004) (Fig. 9.11(b)) exhibits significant peaks (at > 90% confidence level according to F-test), (Fig. 9.11(c)) that occur around 66.66 years, 20–22 years and 13–14 years and centred at 7.5–5.7, 3.4–3.2, and 2.8–2.1 years. MEM spectra confirm the stability of these peaks (Tiwari and Rao, 2004). The possible sources for the 66.66 years rainfall cycle could be either due to:

(a) random forcing, or
(b) related to some external physical processes such as external solar forcing and
(c) internal ocean–atmospheric system.

In view of consistency of spectral power near the 65–70 years period, inspite of variable bandwidth and tapers, it seems highly unlikely that this peak has appeared by chance. Hence, we discount the first possibility of random forcing. It is difficult to resolve the second question at this stage. However, a statistically significant cycle at 66–70 years in the IRF time series is interesting because it is similar to the well-known inter-decadal periodicity of 50–70 years reported in the global mean temperature record (Schiesinger and Ramankutti, 1994; Minobe, 1997). It also coincides with the typical time scale of anthropogenic climate modification (Stocker, 1994). The evidence for this sort of 65–70 years oscillation is also convincing in the instrumental observations, proxy climate records and in the climatic models (Kerr, 2000). The spectral power at 22 years cycle though relatively weak, has the same frequency as the Hale-solar magnetic cycle, which has been argued to have a detectable influence in modern temperature records (Kerr, 2000; Tammy et al., 2000). The third peak corresponding to 12–14 year period may be related to the decadal climate–temperature cycle or solar cycle (Dunbar, 1994).

The MTS also revealed statistically significant peaks (at > 90% CI) in modern ENSO-related climate variability. The signatures of ENSO quasi-cycle have been recognised globally. The possible coupling between the IRF inter-annual variability and SST anomaly in the Arabian Sea is interesting for the study of the Indian monsoon. The Kutch region of Arabian Sea provides a sensitive location for Indian monsoon from which one can monitor the state of the ENSO system, as intense rainfall accompanies enhanced connection in the region during ENSO warm extremes. In the Arabian sea, migration of south west (SW) low convective zone system associated with intense winds causes upwelling of cool, nutrient rich water in several areas like the Somali basin and west coast of India leading to significant changes in IRF. Hence, analysis of new proxy record of temperature variability (coral growth rate) from sensitive locations like Arabian sea and West Pacific regions combined with IRF time series may shed some light on the degree of coupling between the ENSO and Indian tropical monsoon.

The history of annual coral growth record (Chakraborty and Ramesh, 1998), analysed here, comes from a live coral head *Favia speciosa* which was collected in June, 1990 off the coast of Pirotam island (22.6° N, 70° E) situated in the Gulf of Kutch (Northern Arabian Sea) at a water depth of ~ 1 metre during low tide. This 42-year long crude record with annual resolution, spanning (1948–1989 AD) has been constructed carefully to assess its potential for retrieving past environmental changes in this region (Chakraborty and Ramesh, 1998). This coral growth rate record provides key features of ENSO including SST, rainfall, upwelling and trade wind reversals.

MTS analysis of relatively shorter coral growth rate record reveals peaks at 4.2, 2.8 and 2.2 years, which are also statistically significant according to the F-test (Tiwari and Rao, 2004). A smaller peak of order of 12 yrs also appears, but in view of relativly short time span of coral record, it is prudent to restrict our discussion only to the main ENSO band. Similar power spectral analysis of SST averaged over the east central Pacific at the equator for 1950–1991 and normalised surface pressure difference between Tahiti, in the mid-Pacific, and Darwin Australia (a measure of the pressure gradient across the tropical Pacific) exhibit broad but robust spectral peaks centred at approximately 4 years and 2 years period (Hanson et al., 1989; Enfield, 1999a). MEM spectral analysis is known for resolving hidden periodicities in shorter geophysical time series. Tiwari and Rao (2004) analysed coral growth rate time series using the MEM with variable length of prediction error filter (30–60%) to confirm further the stability of periodicities found in the MTS spectra.It is interesting to note that MEM spectra of coral time series reveal peaks at 12.5 years, 6.3 years, 4.2 years and 2.8 years and confirm the stability of these peaks.

Non-Stationary Spectral Modes in ENSO

The evolutionary spectral analysis (Urban et al., 2000) of new coral record suggests that before 1890, the spectra show persistent decadal variability centred at 12.5 years and thereafter a transition begins which can be characterised by a shift of variance from decadal towards shorter period (5–7 years). Also, there is spectral variance centred on 2.9 years, which emerges around 1880 and persists until about 1920. At this point, the 2.9-year variance disappears and inter-annual variance weakens generally, except for weak persistence at the 5–7 year period. Spectral result further indicates that around 1955, the variance shift towards broad concentration centred at approximately 4 years period. This cycle persists and intensifies through the end of the twentieth century. It is also interesting to note that the low-frequency variance concentration in the final year of coral record reflect the step function change in 1976 (Urban et al., 2000). Based on these results it has been argued that observed changes in time and frequency domains in coral records do not correlate linearly with global radiative forcing (solar, volcanic or greenhouse gas concentrations). However, significant shift to warmer and wetter condition in 1976 observed in several coral records coincide with shift to warmer conditions that appear clearly in hemispheric and global temperature indices. This change has been attributed to the intensification of the tropical hydrological cycle by anthropogenic greenhouse forcing, which has strong regional signature in the central equatorial Pacific consistent with the substantial oxygen isotopic shift.

Wavelet Spectra of IRF and NINO3 Records

Wavelet spectral analysis of non-stationary IRF and NINO3 time series provide vital clue to understand the physical linkages, as how the different significant harmonics in IRF record evolve in time. Figure 9.12 shows the wavelet spectra of IRF series revealing the significant changes in the contributions of spectral variance at various periods matching with ENSO frequency bands (Tiwari and Rao, 2004). There are some important observations:

(1) A dominant cyclic mode of 55–70 years continuing till 1960.
(2) 2.8–3.5 years mode continuing till 1960.

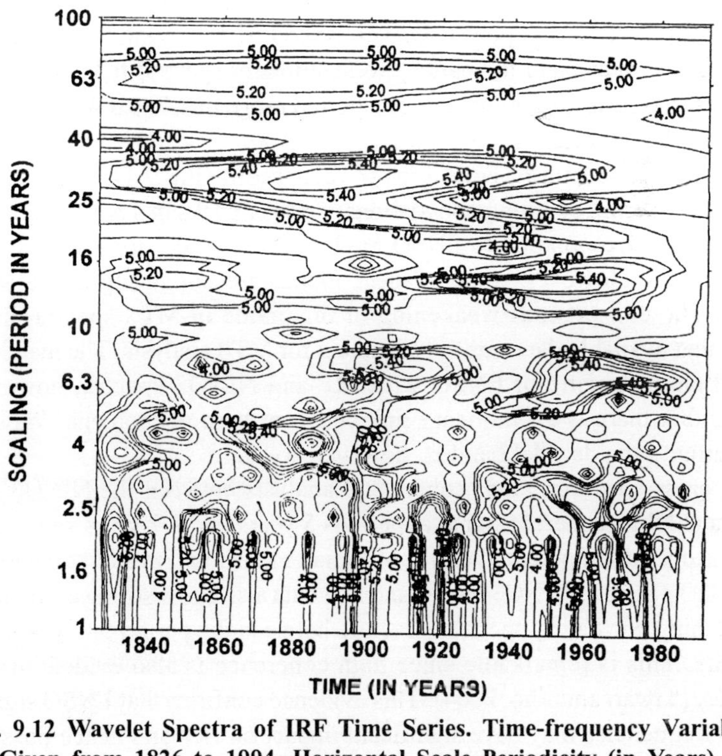

Fig. 9.12 Wavelet Spectra of IRF Time Series. Time-frequency Variability is Given from 1826 to 1994. Horizontal Scale Periodicity (in Years) and Vertical Scale (Years)

(3) A non-stationary mode of approximately 13–16 years mainly dominated between periods 1925–1980 and 1820–1850.
(4) 6–8 years (1890–1980).
(5) 4–6 years till 1900 and then shifting to 3–4 years mode from 1930–1980.

The energy around 2.3–2.7 years frequency band was apparently modulated by the decadal cycle. During this period, a general weakening of southern oscillations is suggested within and beyond the Pacific ocean (Cole et al., 2000).

The wavelet spectra of NINO3 time series (Fig. 9.13a) (Kaplan et. al., 1998) are displayed in Fig. 9.13b. Apparently, Fig. 9.13b shows high power concentration mainly below 20 years cycle bands and clearly reveals non-stationary characteristics. Comparison of wavelet spectra of IRF time series (Fig. 9.12) and NINO3 time series spectra (Fig. 9.13(b)) shows no one-to-one match. However, the comparison indicates that wavelet spectral power of NINO3 data varies substantially during different time windows and match nicely with IRF activity only at certain frequencies. For example, high wavelet power of IRF data between the periods 3–7 years exhibits more or less a clear match with the wavelet spectra of NINO3 data for the identical period testifying the link between these two phenomena. For certain period of time (e.g. from 1880–1920) there is high power seen for period >12–15 years also and then shifted to 5–7 year band thereafter. Non-stationary spectral behaviour with significant power is also dominant for a period range of 1.7 to 3.4 year period similar to IRF spectra.

The wavelet spectra of coral growth rate record (Fig 9.14) also clearly indicates clustering of strong non-stationary power peaks in a period range of 2 to 7 years corresponding to ENSO modes. These peaks are also stable in MEM spectra. However, some of these modes are not strong in the MTS spectra. The conspicuous weakening of this mode in MTS spectra might suggest that available data was not sufficient for MTS analysis. The matching spectral characteristic of ENSO with IRF and NINO3 spectra, however, suggests that there is evolutionary link between these phenomena. Wavelet power around the decadal band is, however, weak.

The 'cross-spectrum' and 'coherency' analysis between the NINO3 and IRF data exhibits coherent signal at periods 5 years and 2.7 years suggesting ENSO as one of the potential driver of Indian monsoon. Further comparison of wavelet spectra of NINO3 record and IRF data also show strong correlation and matching oscillatory spectral structure between the period 5–7 years and 2–3 years. This is remarkable since high coherence is also evident at this frequency (Tiwari and Rao, 2004). This evidence confirms that ENSO signals at these frequencies may have significant impact on IRF and hence possible global link of the Indian monsoon with ENSO.

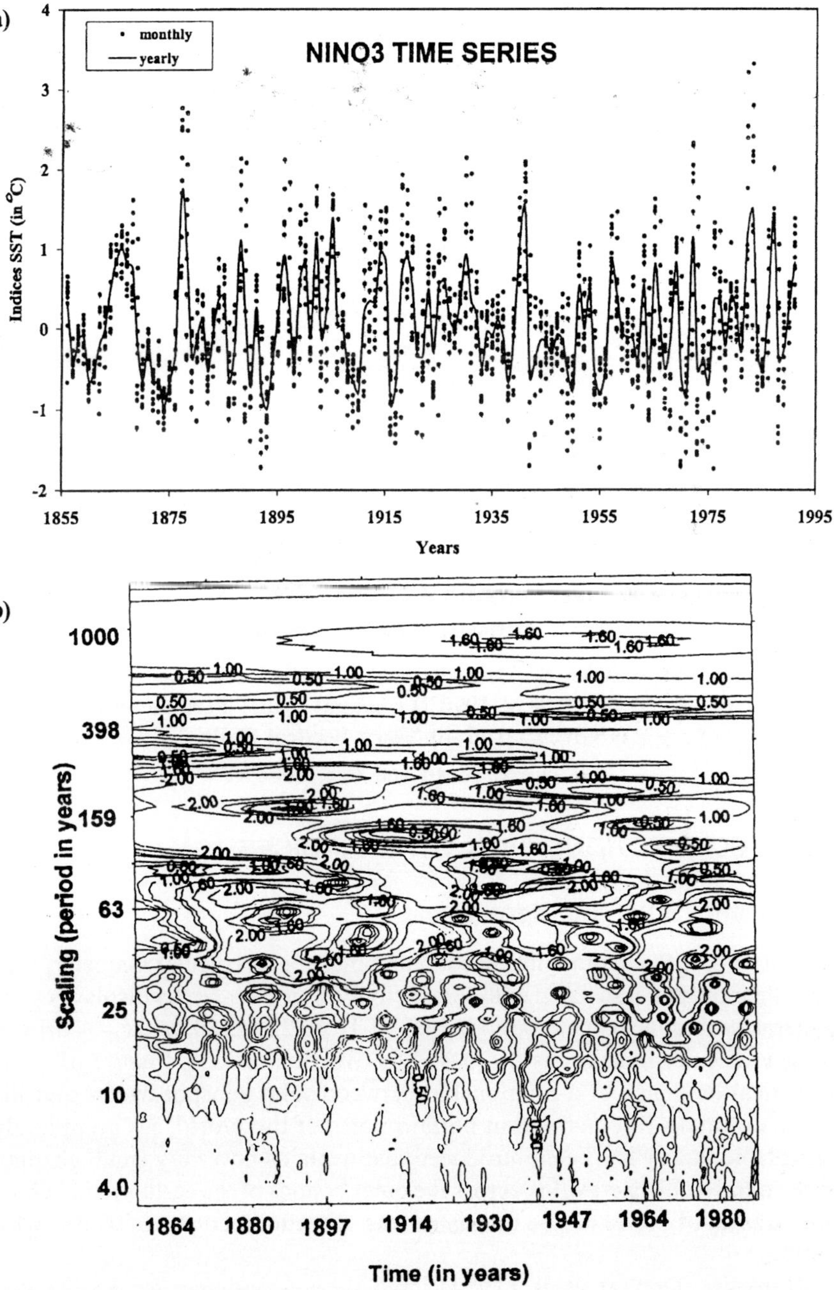

Fig. 9.13 (a) NINO3 Time Series Representing ENSO Related Temperature Variability. (Kaplan et al. 1998); (b) Wavelet Spectra of NINO3 Time Series. Vertical and Horizontal Scales Represent Periodicity (in Months) and Time Respectively

(*Source*: Tiwari and Rao, 2004)

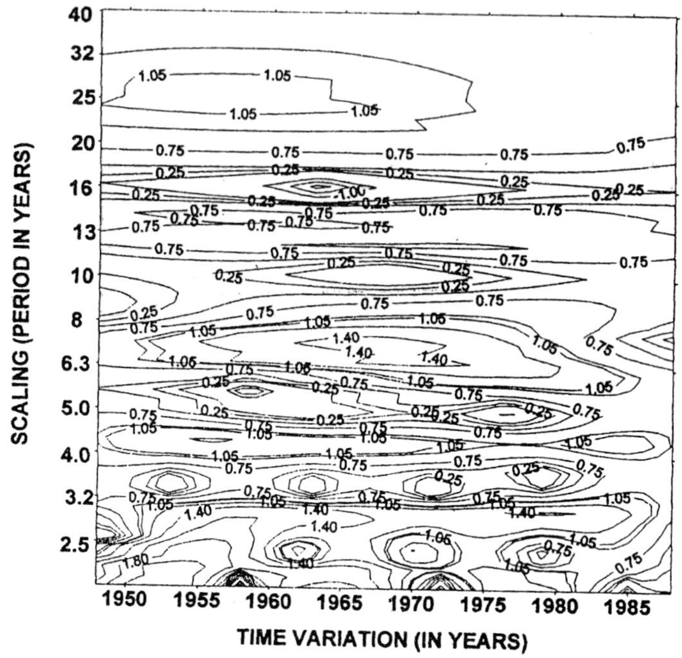

Fig. 9.14 Wavelet Spectra of Coral Growth Rate Time Series. Time-Frequency from 1948 to 1990 Horizontal Scale; Periodicity (in Years) is Given up to 40 Years Vertical Scale

(*Source*: Tiwari and Rao, 2004)

ENSO CONNECTIONS WITH OTHER INDICES

ENSO and Solar Sunspot Cycle

Global climate fluctuations that occur with a more or less regular frequency may have their origins associated with solar activity. Sunspot activity displays cyclic patterns with approximate periodicities of 11 yr, 22 yr, 80 yr etc. A critical review of such linkages is discussed by Weedon (2003). Inspite of some statistical evidence of a relationship between solar sunspot cycles and the earth's climate fluctuations in certain parts of the world, no completely acceptable theory has been introduced that explains how very small changes in the ultra-violet energy flux across the outer bounds of the earth's atmosphere due to sunspot cycles can be translated into climatic fluctuations (Burroughs, 1992).

However, Dunbar et al. (1994) have shown evidence for 83, 23 and 11 year periodicities in an oxygen isotope record from a Galapagos Island coral. In view of similarity of these periodicities with solar cycles, they suggested the control of SST by solar activity, which would affect the frequency of ENSO events recurrence. Rittenour et al. (2000) found evidence for a statistically significant 22-year cycle, along with shorter cycles of ordering 3–5 years buried in the layered bottom sediments of New England lakes. Intriguingly, the 22-year oscillation is a significant number to researchers looking for sun–climate links. As discussed in chapter 3, the 22 years cycle is twice the 11-year period at which sunspot abundance and, solar brightness varies, and it equals the length of the cycle in which the sun flips its magnetic poles back and forth. Kerr (2000) discusses the relevance of these cycles in climate record, in detail. Accordingly, when an 11- and 22-year periodicity shows up in climate records, suspicion falls on the sun, although it's never been clear exactly how a feeble change in solar brightness or the flipping of the sun's magnetic field would trigger measurable climate change (Kerr, 2000).

Although these researchers (Rittenour et al., 2000) were a little skeptical about robustness of 22 yr periodity but after testing its statistical significance against 'red background noise' they found 22-year cycle to be a real feature. It is not a dominant signal, but is always there. From the physical link point of view, researchers are still very hesitant to say that sunspots are behind any 11-or 22-year climatic periodicity (Kerr, 2000). It is suggested that the link between the sun and climate would be strengthened if, rather than just pointing out solar like climate periodicities, researchers could demonstrate that their 11- or 22-year climate variations were in step with the variation on the sun (Kerr, 2000). Another set of the latest set of temperature data since 1958 displays an 11-year variation in the Northern Hemisphere troposphere that is in step with sunspots over four solar cycles.

A number of models and data analysis claim a multidecadal oscillation – within the range of 35 to 70 years in various climate records. But such climate oscillation seems to be debated and it is difficult to believe more firmly over these controversies. However, evidence for this sort of 50–70-year oscillation is accumulating in the instrumental observations, proxy climate records, and the climate models as discussed in Schlesinger and Raman Kutti (1994); Minobe (1997); Delworth and Mehata (1998).

Spectral analysis of relatively longer climate indicator records and atmospheric ^{14}C production rate show common periodicities with 88%–95% confidence intervals (Wang et al., 2003). The frequencies significantly above the red noise indicated three long-term common periodicities of 800–1000, 450–550, and 350–390 years. It is interesting to note that 800 and 450 year cycles were not only reported in the atmospheric ^{14}C production rate, but also in ^{10}Be flux in the Greenland ice layers and North Atlantic marine sediments. as suggested by Bond et al. (2001), these two periodicities reflect persistent solar influences in the North Atlantic region.

Model studies suggested that the solar forcing warmed the equatorial stratosphere and strengthened the 'Hadley Cell circulation', a large belt of circulating air that lies between the equator and 30° latitude with north south, in the troposphere (Haigh, 1996; Shindell et al., 1999). Here air rises at the equator and moves towards the pole and sinks at 30° to flow back towards the equator. Alternatively, the El Nino-Southern Oscillation (ENSO) variability has been considered as a driver that amplifies a weak solar input by the stochastic resonance (Ruzmaikin and Feynman, 2002; Lean and Rind, 2001). The Mississippi River Valley loess-palaeosol record suggested that during the late Wisconsin glaciation, ENSO variability might have amplified solar forcing to a great extent. The resultant heat and moisture transport from eastern tropical Pacific, Atlantic and Gulf of Mexico have played a significant role on millenial (800–1000 year) to centennial (450–550 year) climate cycles over the northern hemisphere (Wang et al., 2003).

ENSO and QBO

The Quasi-Biennial Oscillation (QBO) is a result of reversals of stratospheric wind direction (Labitzke and van Loon, 1998). This mode recurs approximately biannually. It has often been seen that proxy records which used to decode the evolutionary history of ENSO variability also involves spectral component similar to QBO. This creates, therefore, some confusion regarding the origin of this cycle (Burroughs, 1992). Several researchers have suggested a possible link between the QBO and the ENSO (Dunbar, 1994 et al.; Rittenour et al., 2000). A shift in the pulse of ENSO at approximately 2.2 years bands in the different spectra may either be related to the inherent non-linearity in the internal oceanic mechanisms or due to its possible coupling with QBO term (Tiwari and Rao, 2004). It has also been suggested that the QBO in the atmosphere, which acts as a 'pacemaker' for the ENSO phenomena is also, known to have strong link with solar activity (Enfield, 1989b) and SST in the east central equatorial Pacific (Rasmusson et al., 1990). The stratospheric QBO is also coupled to a regionally intensified biennial oscillation (BO) in the western equatorial Pacific low level wind and Pacific equatorial SST shield, both phase locked to the annual cycle of Indo Pacific region (Yasunari et al., 1989). Available data indicate that solar activity was strong prior to 1900 and relatively weak after this. Interestingly, these results match with the suggestions (Enfield, 1989b and references therein) that when solar activity is strong, there is well-developed stratospheric QBO, linked to a seasonally phase-locked BO in the Indo-Pacific region, tending to favour ENSO development at intervals closer to four years, i.e., a multiple of QBO periodicity. However, when solar activity and QBO are weak, ENSO is free to develop at a shorter periodicity closer to 2–3 years that is after 1900, which is set by the southern oscillation time scale (the latter is inherent feature of the coupled ocean atmospheric system governed by the propagation rates for planetary waves in the Pacific ocean).

CO_2 and El Nino

Recent studies based on theoretical models and data analysis have shed light on controversies concerning the possible effect of global warming on El Nino over the past few decades and in the future. Based on preliminary investigations, several researchers (Meehl et al., 1993: Graham and White, 1995 and Kunston and Manabe, 1994) have suggested the possible effect of changing atmospheric

CO_2 on the tropical pacific interannual variability. Some other researchers, however, believe that such observations are simply an artefact of numerical and methodological calculations and relevance of the results from these studies to nature is uncertain (Battisti and Sarachik, 1995). In a more recent spectral study, however, it is suggested that random disturbances maintain neutrally stable ENSO mode whose properties depend on the background (time-averaged) climate state (Urban et al., 2000). The past observations indicate that the intensity and frequency of ENSO events have changed, with strong events at 1880–1920, 1939–1942, 1954–1955, 1982–1983, 1989–1990 and 1997–1998. Because ENSO events have strong influence on regional and global temperatures and rainfall, the possibility that CO_2-induced global warming is linked with El Nino cannot be excluded (Urban et al., 2000). Apparently, there is a tendency for more frequent El Nino and La Nina events since the late 1970. According to these researchers, this has also been linked to decadal changes in climate throughout the Pacific basin. Statistical studies, however, suggest that there is a possibility that the ENSO changes may be partly caused by the observed increase in greenhouse gases. The relationship between El Nino and anthropogenic climate changes has emerged as a key goal of modern climate research. There is unprecedented evidence of changes in amplitude and phase of past El Nino – since 1970, which appear to be outside the range of natural variability (Urban et al., 2000). Some recently published results also provide strong evidence that external forcing such as increase in greenhouse gases concentrations are influencing the recent behaviour of El Nino. The separation between the natural and forced variability remains central to understanding global warming. Understanding of the mechanisms that control and affect the El Nino is, therefore, crucial to a better understanding of climate variability and change.

ENSO Teleconnections

Although El Nino's mass warm water wandering phenomenon is regional, mostly restricted to tropical Pacific, the effects this has on climate truly stretch around the globe, as wandering ocean currents and atmospheric winds exert its monumental power also out of the tropics and around the planet (Fig. 9.2) (Cole, 1994). During the severe ENSO event 1982–83, when the sea surface off Peru was warmed by more than 7° C, it was found that there are strong links to weather in other regions as, for instance, floods in California and intensified drought in Africa (Cane, 1985). An unusually warm equatorial ocean over a large zonal extent would create anomalous zonal and meridional SST gradients over large space scales. The resulting gradients would provide an enhanced input of thermal energy to the direct atmospheric circulations, especially the meridional circulation of the atmospheric 'Hadley cells' in that quadrant of the globe (Barnet, 1981a, b). This would, in turn, increase the poleward flux of angular momentum to the winter hemisphere jet streams through a more efficient meridional circulation, finally strengthening the mid-latitude westerlies and affecting weather patterns downstream (to the east) of the disturbance. According to several researchers, the impact of Western Pacific Warm pool is not limited to the surface of the globe, but also appears to have significant effect on core mantle boundary and hence on earth's rotation. This became the basis for the much widely used notion of 'teleconnections' by which El Nino warming can project climatic anomalies to remote regions of the globe.

Several, researchers have suggested that in a statistical sense 'teleconnections' simply mean link among the various climatic variables separated by large distances. For instance, in the Indian ocean, the SST variability reflects obvious influence of El Nino-Southern Oscillation and both forces and feedback on the monsoons of India and Africa. Important observational studies led to empirical evidence that linked the tropical ENSO events to seasonal climate anomalies over Australia, South America and into the tropical Indian ocean, and to anomalies in the mid-latitude northern hemisphere climate, especially over the north Pacific ocean and over central and north America during winter (Molnar and Cane, 2002; Cobb et al., 2001, Bojariu and Gimeno, 2003). A number of oceanic and atmospheric phenomena such as North Atlantic Oscillations, North Pacific and North American climate oscillation exhibiting/causing common quasi-cyclic pattern decadal/inter decadal time scale revealing the possibility of 'teleconnection' of their phenomena (Minobe, 1997; Fealis et al., 2000; Schlesinger and Ramankutti, 1994; Stocker, 2002).

The strongest and most reliable effects occur in the tropical Pacific ocean. Other parts of the world, especially in the middle latitudes and the climate between tropical precipitation fields and large-scale circulation patterns of extra-tropical regions far away from the Pacific may also be affected by 'teleconnections' (Charles et al., 1997). The case of ENSO connection with precipitation over India is quite intriguing. It is presumed that there are precursor climate anomalies (other than El Nino) to produce a normal monsoon. However, the modern spectral analysis suggests that ENSO component at around 4–5 years have strong coherence with average precipitation anomalies over India (Tiwari and Rao, 2004). This might hold the view that Indian monsoons should be predictable in terms of ENSO signals. However, the Indian monsoon and ENSO signal seems to have a common chaotic component associated with an irregular alternation between the periods of excessive rain or drought. By virtue of a common chaotic property, El Nino would influence the probabilities of these rainy and drought regimes. It may not be possible to make exact prediction.

NORTH ATLANTIC OSCILLATIONS (NAO) AND ENSO

World climate can be visualised as integrated dynamical climate system. The well-known teleconnections linking the NAO and ENSO are the major focus of present research. NAO refers to another major source of interannual variability in the atmosphere. Important features of NAO and its various linkages are reviewed by Appenzeller (1998) and Hurrel (1995), Bojariu and Gimeno (2003) and Wanner et al. (2001). Accordingly, the fundamental dynamics of NAO are likely to arise from atmospheric processes: the basic features of the basin-scale atmospheric circulation determine its position and spatial structure. The NAO is characterised by changes in surface westerlies across the North Atlantic onto Europe (north-south difference in pressure). The changes in circulation associated with changes in the NAO index are determined from the differences in sea-level pressure between winters with an index value greater than 1 and those with an index value less than –1. A low-pressure region is centred near Iceland and a high-pressure region in the subtropics near the Azores. This pressure contrast drives the surface winds and wintertime storms from west to east across the North Atlantic. When the pressure is lower than the normal near Iceland, it tends to be higher than the normal near the Azores and vice versa; this out-of-phase relation defines the NAO index. The NAO is the dominant source of the Northern Hemisphere climatic variability. There is growing evidence that the NAO has profound impact on both marine and terrestrial ecosystems (Cullen and deMenocal, 2000). Hence, understanding the nature of variability of NAO during the past centuries is of considerable importance in general for society and in particular for palaeoceanographic community.

There is evidence for interannual variability in NAO, in addition to several other periods. Spectral analysis of the 1900–1983 winter NAO index showed prominent spectral peaks at around 20 years and 7.3–8.0 years (Rogers, 1984). Hurrell and van Loon (1997) identified spectral peaks at 2 to 3 years, 6 to 10 years and inter-decadal period bands using a much longer winter NAO index data (1865–1994). Schlesinger and Ramankutti (1994) performed singular spectral analysis using temperature record to show a quasi-periodic cycle of around 70 years in north Atlantic region. Allen and Smith (1996), however, have questioned this cycle. Attempt has also been made to reconstruct long term NAO index. Appenzeller et al. (1998) performed wavelet spectral analysis using a 350-years ice core record from West Greenland. This analysis has revealed highly nonstationary characteristics of NAO index with most variability centred on less than 15 years. There is also some apparent variability around 80–90-year variability restricted to the last 150 years. Cullen et al. (2001), however, used ice core and tree ring data and confirmed that the longer period oscillation is not stable through time but some periodicities of the order of 7–8 and 12 years are dominant during certain time intervals (1759–1979). Further study by Black et al. (1999) also reveals 12–13 year variations in the planktonic foraminifera Globogerina bulloides in laminated sediments from the Cariaco basin for the last 825 years. They have linked this cycle to surface productivity and upwelling. A better understanding of the temporal relationship between the NAO and ENSO is essential in elucidating the Northern Hemisphere mid-latitude climatic variability. Several recent studies have shown possible common inter-decadal link between the NAO and ENSO (Huang et al, 1998; Dima et al., 2001; Gallego et al., 2001). Cross-spectral analysis between these two time series reveals a large significant power at 5.7 years (Rogers,

1984). Cullen et al. (2001) has presented multiproxy reconstruction of the NAO spanning over a period of 1750–1979 and 1701–1979 (Fig. 9.15) for the Hurrell (1995) NAO_{SLP}. Multitaper spectral analysis results of the Hurrell (1995) are shown below in Fig. 9.16 which clearly reveal only one certain significant peak (at 99% confidence level) which may be related to the QBO (Saravanan, 1990). In addition to this, there are some other peaks in a period range of 7– 8 years and 12.5 years, which are also significant (95–99% confidence level). The low frequency (70 years) mode is, however, not present, as suggested by Schlesinger and Ramankutti (1994), in these records (Felis et al., 2000). The 7–8 years peak may be interpreted as a possible link between Pacific and Atlantic. The 12.5 years peak may be related to a dominant mode of Atlantic variability, believed to be driven by coupled tropical ocean-

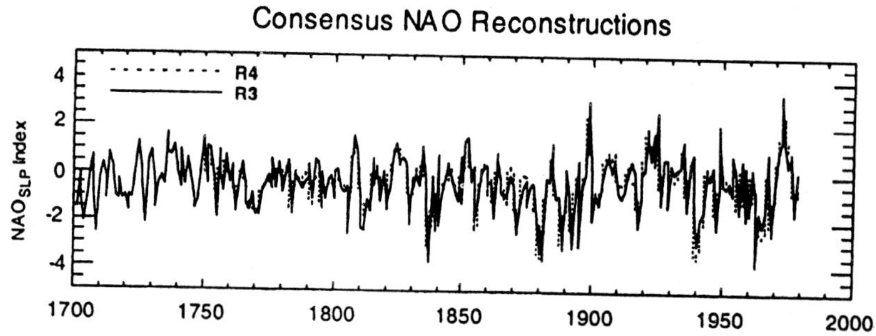

Fig. 9.15 Shows NAO Time Series
(Source: Cullen et al., 2002)

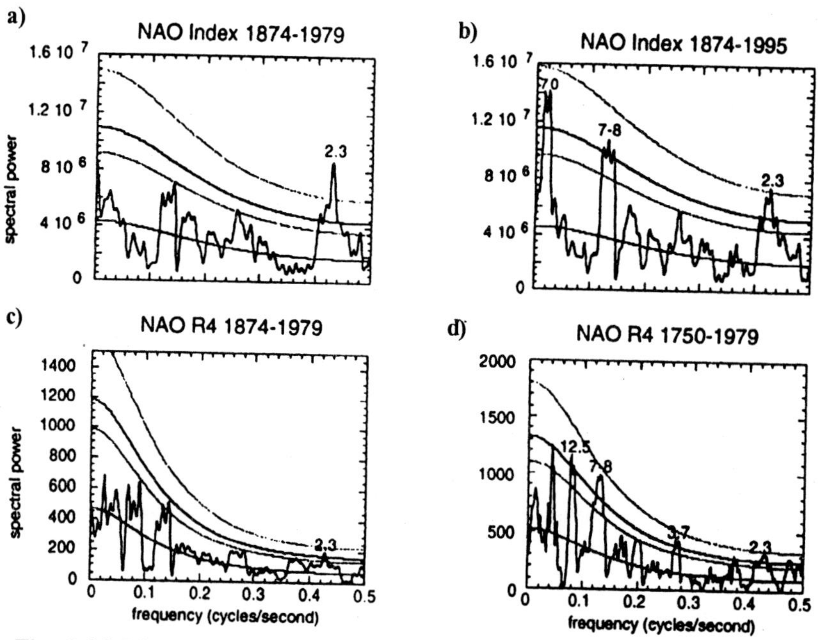

Fig. 9.16 Multitaper-Spectra of NAO Time Series Shown in Fig. 9.15
(*Source*: Cullen et al., 2002)

atmosphere dynamics (Cullen et al., 2001). However, several studies suggest that there is significant solar influence on the NAO (Gimeno, 2003; Kodera, 2003). Kodera (2003) has shown that during the solar activity winters, the NAO signal in the sea level pressure is confined in the Atlantic sector, while during the high solar winters, NAO related anomalies extend over the northern hemisphere, in particular, over the polar region and Eurasian continent. These studies suggest that major modes of climate variability such as NAO or ENSO could modulate the relationship between solar activity and global temperature (Gimeno et al., 2003).

The atmosphere, the ocean and the coupled atmosphere-ocean system have all been suggested as main driving forces for the NAO. However, it is not very clear what drives these climate fluctuations. The atmosphere generally has a short memory compared to the oceans, where signals can be propagated over years or even decades. It is somewhat difficult to decide which of the two is more influential in determining characteristic time scale for climate oscillations (Uppenbrink, 1999). The NAO's impact is more restricted than the ENSO's. In North America, Europe and North Africa, precipitation patterns and wintertime temperatures can be attributed to its phases. The NAO may well be more influential than ENSO in regions surrounding the North Atlantic, where its climate shifts can affect food production, energy consumption and other economic factors (Cullen and Demenocal, 2000). Changes in marine and terrestrial ecosystems have also been attributed to the NAO (Bojariu and Gimeno, 2003).

IMPACT OF ENSO PHENOMENA ON SOCIETY

El Ninos are a regional phenomenon but have become synonymous with larger-scale, teleconnected, climatically significant, warm events having their broad influence on global economy and everyday life. It has been found that large scale unusual warming has severely disrupted local fish population every few years (Webster and Palmer, 1997). ENSO phenomena are usually accompanied by severe drought over Australia and Indonesia, together with a weakened summer monsoon rainfall over South Asia. Catastrophic events such as flooding often occurs along the Pacific coast of South America. During the ENSO period, ocean upwelling containing high-nutrient cold water diminishes leading to severe impact on fish stocks. El Nino also influences tropical cyclones, reducing their frequency in the Atlantic, but increasing it in parts of the Pacific.

The direct societal impacts of El Nino can be understood in the following two ways (Webster and Palmer, 1997):

(a) Homes and crops might be destroyed due to flooding, and fisheries ruined. It is realised that climate variations associated with the El Nino have a wide range of effects on the global ecosystem including the large-scale distribution pattern and population of fish and shellfish, the production of zooplankton, plant phenology and growth of reproduction and demography of many terrestrial animals.

(b) Distinct disease patterns follow the waxing band waning of an event, through contamination of water supplies by flooding and the creation of increased breeding areas for vectors such as mosquitoes.

It has been suggested that different kinds of water and vector-borne diseases such as hepatitis, dysentery, typhoid, cholera, malaria, dengue, yellow fever, plague, hantavirus and schistomiasis are associated with ENSO phenomena and follow El Nino cycles (Webster and Palmer, 1997).

CONCLUSION

As we have seen, many uncertainties remain to be resolved regarding the ENSO behaviour. However, several recent multiple statistical and spectral analysis of new global time series based on coral of SST and rainfall time scale related to ENSO (Coral record) activity and other related proxies indicate a robust multi-decadal cycle of order of 65–70 years and even larger may possibly be related to the well-known time scale of orbital variations (Clement et al., 1999) or internal ocean-atmospheric system dynamics (Boiseau et al., 1999). The spectral analysis also reveal apparent, quasi-periodicities in a period range of 2–7 yrs. This spectral band exhibits global characteristics. However, particularly for the Indian monsoon, the cross-spectral analysis between the IRF and NINO3 time series reveal coherent peak at 5.4 yrs and 2.7 yrs modes suggesting the role of ENSO signal in Indian monsoon only at these frequencies. The new coral growth rate data for Arabian sea, although short, reflect the role of coupling of ENSO dynamics with Indian monsoon. The wavelet time-frequency analysis of these data reveals the non-stationary character e.g. periodicity changes with time. It is to be noted, however, that the non-stationary behaviour of ENSO cyclic pattern leaves enough scope for further study and understanding of ENSO dynamics in more detail using the precise global records. This is particularly important for checking the stability and temporal continuity of the cyclic pattern that in turn will be useful for the prediction of ENSO phenomena and global monsoon system.

Another important point to make here is that a 2–2.2 years cycle apparently seen in ENSO record is similar to QBO (Burroughs, 1992). There is some indication that QBO coupled with ENSO phenomena may play a subtle role in triggering inter-annual variability in the Indian monsoon. The non-stationary character of QBO in temperature record manifests variable amplitude and may be linked to QBO. However, before concluding such physical linkages, it would be prudent to understand link between QBO and ENSO in more clear way. If such link is strengthened further, these results may provide some significant implications for the modelling of not only Indian monsoon but also global climate dynamics.

The world climate is being increasingly recognised as integrated system. Hence so called 'teleconnections' linking the NAO to other climate systems and ENSO appears to be the main focus of future research.

The researchers have always been skeptical about the apparent link between the sun's variability on the earth's climate. It has been quite difficult to explain the impact of small changes in the sun's input (solar variability) that might change climate on the earth and nobody had come with a convincing mechanism to explain how tiny changes on the sun might change climate on earth. But associations between sun and climate keep cropping up (Kerr, 2000). Now, two correlations— 22-year climate cycle recorded in glacial sediments and tracing of an 11-year cycle from the stratosphere into lower atmosphere— may be robust enough to give the sun–climate link a touch of more respectability (Dunbar et al., 1994; Rettenour et al., 2000).

10

Summary and Conclusions

Long-term geo–bio–ocean–atmospheric records are episodic in nature. In many cases, there are large gaps and hiatuses, in these records, which hamper appropriate time series analysis. Sedimentary and stratigraphic records entail enough red (noise) signals and hence frequency analysis of such record displays deceptive peaks in the spectrum. Searching for evidence of a statistically significant periodicity in episodic record is quite a difficult task and appropriate non-parametric method, as one proposed by Stothers (1979) is required. The method of time series analysis also requires appropriate procedure for judging the statistical reliability of detected peak. In addition to this, these on shorter time scale environmental records also suffer from several other problems such as bioturbations and discontinuous sedimentation rates. It is necessary, therefore that prior to spectral analysis and proposing physical interpretations and linkages, these points must be considered seriously. Further, in order to avoid aliasing and statistically non-significant spectral peaks to appear in spectrum, appropriate care should be taken to preprocess the record under study as discussed in chapter 2.

Various time series and spectral analyses of the geo-records suggest that most of the earth system processes are quasi-periodic and/or chaotic in nature on the different time scales and there is nothing like randomness in the earth system processes, however, some exceptions cannot be ruled out. Further, several studies also indicate that the earth system processes are coupled and interlinked themselves in some way or other on different time scales. However, detailed physical mechanism is not yet clearly understood. Such orderly dynamic system seems to be governed by some common physical mechanism/ mechanisms. Various independent studies have suggested that underlying forcing mechanism may be either of external and /or of internal origin. Hence, the main focus/purpose of this book/monograph is two-fold: (i) to present evidence of periodicities and (ii) discuss the various views of exogenic and endogenic linkages in the different earth system processes on different time scales.

Several studies have deduced that long-term geo–bio–ocean and atmospheric records show deterministic and quasi-periodic pattern. Modern time series analysis of the worldwide episodic as well as continuous long term geo-records pertaining to the various tectonic activities, mantle (convection and major plumes activities) and core processes (geomagnetic reversals) suggest that at least two long-term quasi periodicities of order of 280–320 m.y. and 33+5 m.y. in almost all the earth system records exist. There is also evidence of weak quasi-periodicities of the order of 137–155 m.y., 80–90 m.y., 50–60 m.y. and 5–10 m.y. in some geological records. In particular, a 145+10 m.y. periodicity is suggested in climate and glaciation records. Some studies have also suggested even longer cycle of the order of 600–800 m.y. in climate, tectonic and magmatic intrusive records. Evidence of periodicities indicates orderly behaviour of long-term earth's system processes.

A 30 m.y.-periodicity in particular, in the various records of earth system has been one of the hotly debated issues amongst the geoscientists from statistical as well as from physical point of view. However, various studies suggest that 30 m.y. is remarkably stable, statistically significant and invariably exist in many earth system processes (Rampino and Caldeira, 1997a,b). Table 10.1 provides a list of geo–bio–ocean–atmospheric and geo-chemical processes where a dominant periodicity of order of 35+5 m.y. is reported. A long-term common characteristic time scale in the earth's system (geo–bio–oceanic and atmospheric) processes might suggest that these processes are inter-linked and may be driven by some common physical mechanism.

Table 10.1 Review of 30 ± 5 m.y. Periodicity in Various Earth Processes (Based on various literature quoted above)

Period of record in m.y.	*Geological events*	*Dominant periodicity in m.y.*
0–570	Geomagnetic Reversals	32 ± 3
0–250	Flood basalts/mantle	33 ± 3
0–570	Orogenic events	30.6
0–180	Sea floor spreading	26.7, 18.4
0–200	Sea level changes	28 ± 2
0–245	Ocean Anoxic events	26.1, 39.7
0–258	Evaporite deposits	28 ± 2
0–160	Phosphorus burial rate	33 ± 3
0–140	Fluctuations of dolomite	24 – 26
0–120	Variation in the diameter of CTBF	35 ± 5
0–250	CO_2 variations	33 ± 3
0–260	Mass extinctions	26 – 32
0–260	Impact cratering	32 ± 3

Given that a 30 ± 3 m.y. periodicity exists in various geological records and mass extinction, one might obviously ask: why do the long-term terrestrial records of geo–bio–ocean–atmosphere show a periodicity of this order? And, if an underlying periodicity of about 30 ± 3 m.y. does really exist in these records, what could be the possible driving mechanism of such a coupled complex system? In general, one might link this underlying cycle with the earth interior processes, in which periodic plumes and associated plate tectonic cycles with their characteristic time scale, possibly also control oceanic and surface processes such as weathering, climate and CO_2 variations. However, since most of the geological processes seem to be catastrophic in nature, it may be somewhat difficult to explain that such gradual tectonic and environmental stresses, in the absence of any critical catastrophic external forcing/triggering, could have had caused major geological upheaval and abrupt extinction. Hence, an alternative external forcing mechanism has been proposed to explain the physical validity of this cycle in view of a similar astronomical cycle of solar system's periodic vertical oscillation (half period) of about 31 ± 3 m.y. According to this theory, the Oort comet stands a strong chance of being upset once every 30 m.y. It is suggested that galactic triggering or forcing of terrestrial geological events may arise as a result of tidal distortion (close encounters) of the solar system with intermediate to large sized interstellar clouds of gas and dust which are sufficiently concentrated towards the galactic plane to produce the observed cyclicity (Rampino and Stothers, 1984). Among other consequences, a nearby interstellar cloud would gravitationally perturb the family of orbits of comets solar systems. The perturbation would increase the flux of comets (and commetary derived bodies) near the earth (Rampino and Volk, 1996). The process will lead to increase in probability of large body impact. It has been argued that some bolide impact on the earth's surface coincided with the time of polarity reversals of the geomagnetic field and may have severe consequences for the earth interior processes as well as environmental processes including biological mass extinction. Several researchers support the impact theory. But several others question it on the following grounds: (i) how surface impact energy is transmitted to the outer core? (ii) How the pattern of fluid motions in the outer core is disrupted as a consequence of this? and (iii) how could it induce geomagnetic polarity reversals? Debate is on, but not settled. However, it seems logical that, even if the earth's various interiors and external processes are self-sustaining in such an orderly manner, it is somewhat difficult to explain observed intensity of catastrophic cycle without involvement of external forcing/triggering. The intriguing concidences exist of impacts triggered mantle diapirism, accelerated sea floor spreading, catastrophic dynamics, cataclysmic volcanism, mountain building and subduction etc.

Recognition of the detectable role of orbital cycles in the earth's processes, in particular, in the climate changes has been one of the major revolutions in the earth science after the discovery of plate tectonic. It has been now widely recognised that orbital forcing is strongly impressed on several parameters of earth's system processes. The number of examples of ocean, atmosphere, cryosphere and stratigraphic phenomena attributed to orbital forcing is growing rapidly. Some of these examples include ice sheet and sea level changes, palaeomagnetics, amospheric CO_2 record, storm intensity, palaeosoles and Aeolian deposits, lake deposit and terrigenous clastic, shallow marine carbonates and calciturbidites (de Boer and Smith, 1994), volcanic record etc. The three main orbital cycles which have invariably appeared in these records are: precession cycle of about 23000 years (the longitude of perihelion with respect to the vernal equinoxes); the obliquity cycle with period of about 43000 years and eccentricity with main cycle of about 95–100 k.y. years. Besides these three main cycles, there are several other periodicities, e.g., some of these are beat frequencies, modulation cycles and/ or combination tones and higher order eccentricity cycles, which also appear in various climate records. Understanding the role of these periodicities in the climate spectra is important for prediction of the future palaeoclimate variability. Various studies indicate that relatively the climate variability on this time scale is stationary and are extremely useful for tuning the geological time scales.

The origin of eccentricity cycles in the climate record, however, is controversial. Although spectral analyses of various climate records display highest significant power at eccentricity cycles, theoretical calculations of eccentricity variations suggest that it should not have significant effect on the mean annual insolations. Several researchers have associated eccentricity cycles to non-linearity associated with ice sheet growth and decay, while others have argued that this might result from the climate system's non-linear response or modulation of obliquity and precession cycles. Alternatively, theoretical climate models suggest that it can also result from amplification by the climate system itself of weak astronomical forcing. If, as suggested in several studies, the climate system is non-linear and chaotic phenomena, at the times of 'criticality' even small variations might create instability leading to large amplitude in climate response at eccentricity frequency. However, the origin of eccentricity is still an open question.

It is important to note that orbital cycles have been changing during the geological time. It has been shown that the tidal friction is slowing down the earth's rotation, which is bound to have effect on both precession and obliquity. However, Berger and Loutre (1994) have shown that this process does not affect the eccentricity periods. It is, therefore, prudent that while interpreting the climate and /or tuning the geological record one should consider carefully these changes. Another important point may be noted that climate prediction is hampered by presence of chaos. This is due to the fact that external forcing of climate is a result of complex interactions among the numerous bodies of the solar system. Hence, the long-term extrapolation of the present system of motions back into the time may not be possible, because of chaotic/ behaviour effects.

With availability of precise and better resolved global climate record, researchers have discovered cyclicity on still shorter time scales (millenial scales ranging 1000–10000) years in the coupled earth system processes, in particular climate records. This indicates that climate system operates in at least a quasi-cyclically, if not in a strict periodic manner. The so-called Dansgaard-Oeschger (D–O) and Heinrich cycle of the order of 6000–7000 years have been invariably reported in the Greenland ice core records. The other main cycles of order of 2600–3000, 2200, 1450–1600, 900–1100, 600–700, 350–420, 180–200, 130–140, 80–90, 60–70 20–22, 9–12 years have also been suggested in various climate and other related phenomena. The possible role of solar and tidal effects/influences on climate and related earth's phenomena has been suggested. The stability as well as origin of these various cycles, however, is again a controversial issue but most of these periodicities are found globally. For instance, if we consider millenial scale cycle, the origin of 2200 years cycle found in the North Atlantic oscillations have been associated with ocean circulation. Several others, however, believe that this is of solar origin because a similar cycle has also been found in the proxy record of solar variability. Similarly, a dominant 1500 years cycle in various indices of climate change records appears to be of internal origin. Alternatively, some others believe that it might originate from tidal forcing or planetary conjunction and/or fifteenth harmonics of precessional cycle. Other prominent cycles of order of 420, 180–200, 80–90, 22, and 11 years (decadal, inter-decadal and century scales), which has appeared in the climate and other geophysical records, seems to relate nicely with solar cycles. But this is true only up to numerical correlation level. It appears that more efforts are required to strengthen this correlation physically for bringing it up to prediction level. A cycle of the order of 60–70 years reported in climate spectra, Indian monsoon, appears to originate from the ocean circulation while a similar periodicity of order of 61 years in geomagnetic field variations may be related to the earths axial rotation. Solar cycles and tidal cycles of different orders are also dominant in various short-term environmental records of varve, laminae, precipitation and drought, temperature, etc. These findings are quite significant for understanding the short-term environmental conditions.

Coming to the inter-annual and decadal variabilities, NAO and ENSO display strong source of climate and monsoon variability globally through tele-connection. Besides a strong source decadal variability, the NAO oscillations displays quasi-periodicity of 7.5–8, and 3.5 years. Similarly, the ENSO variability displays quasi-periodicities in a period range of 2.1–7 years. However, both the phenomena also display chaotic nature and periodicities are non-stationary. Hence, prediction is not readily possible, because of inherent complex non-linearity and modulation involved in this cyclic evolution pattern.

In addition to this, spectral analysis of calculated high-tide elevation series in the tees estuary reveals dominant (1year) annual and fortnightly (14.8 days) neap spring cycles detected in the tees estuary of North England. Further analysis of such record at hourly resolution reveals half-daily (semi-diurnal) tidal cycle with weak signal at diurnal inequality (Weedon, 2003).

Figure 10.1 gives the full spectrum of tentative cyclic evolution of earth–ocean–atmospheric processes ranging from hours to million years scale (Modified from Mitchell, 1977; Berger, 1980).

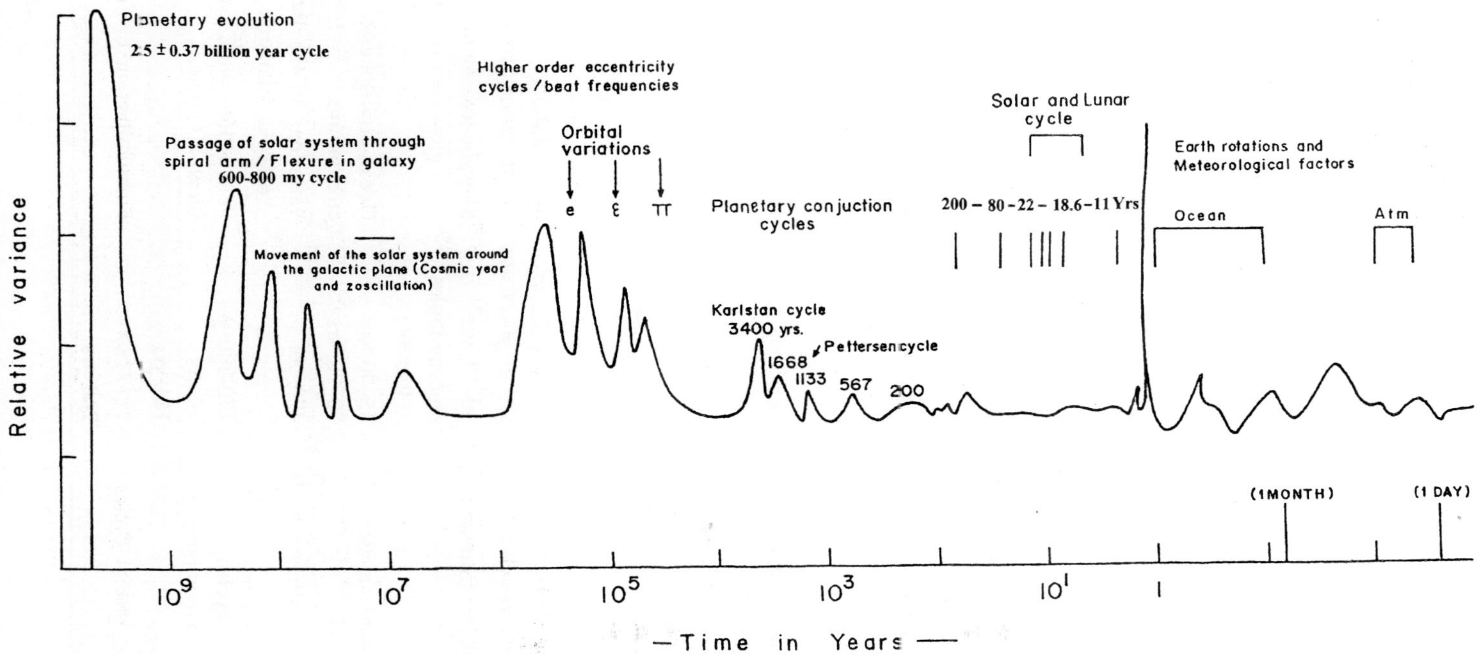

Fig. 10.1. A tentative hypothetical geo-spectrum of various earth's processes beginning from year to million years time scales. (Modified after Mitchell, 1976)

Table 10.2 gives various astronomical and tidal cycles prominently found in various earth-ocean-atmospheric phenomena on different time scales.

Table 10.2 Recognized Astronomical and Tidal Cycles in Earth's Processes on Different Time Scales

Type of cycle	*Duration*	*Cause*
Tidal	About 12 hrs	Attraction of moon and sun
Daily	24 hrs	Rotation of the Earth about its axis
Tidal	13.661 days	Rotation of the earth about the sun
Tidal	27.332 days	
Tidal	29.551 days	
Tidal	182.62 days	
Annual	1 yr	
Global atmospheric cycle and Temperatures cycle in European data series since 1970	2.2 yrs	Quasi-biennial variations (QBO)
Climate, drought and precipitation	18.61 yrs	Tidal
Precipitation in Europe since 1800	2.5 yrs	
Various climatic cycles ENSO cycles	2.1–7.5 yrs	Ocean and atmospheric interactions (Pacific phenomena)
Climate cycle (NAO) Inter-annual cycles	7.5 and 3.2 yrs 5–6 yrs, 8 yrs	North Atlantic Oscillations (NAO)
Decadal cycles (Sunspot periods)	10–12 yrs (and multiples thereof)	Sunspot maximum
Various climatic cycles	35–40 yrs 50–70 yrs	Ocean-atmospheric circulations
Century scale climate cycle	420 yrs	Variations in the intensity of solar radiation
Precipitation cycles	170-200 yrs	Fluctuations in C^{14} content
Advance of Greenland glaciers	78 yrs 181 yrs	Sunspot cycles
Climate and glacial cycles	71–77 yrs 208 yrs	Fluctuations in C^{14} content

(Contd ...)

Dansgaard-Oeschger (D-O) (Heinrich cycle)	189, 205 6000–7000 yrs 2200 yrs 1500 yrs 900 yrs 600–700 yrs	Solar sunspot cycles Ocean circulations Solar variability Ocean and atmospheric circulation
Millenial scales climatic cycles/Tidal cycles	1600 yrs, 1100 yrs, 567 yrs	Sea level and climate changes
Milankovitch cycles	21,000 yrs	Precession of the earth's axis
	41,000 yrs	Changes in the inclination of the earth's axis to the plane of the ecliptic
	92,000 yrs, 413,000 yrs	Eccentricity of the earth's orbit round the Sun
	15,000; 21,000; 31,000; 50,000; 54,000; 60,000; 64,000	Combination of various causes mentioned above
Sun's oscillation perpendicular to the galactic plane	26-37 m.y.	Dominant cycle in global bio-geo-ocean-atmospheric-chemical processes.
Flexure and tidal in the Galaxy	137-155 m.y.	Glaciations cycles.
Cosmic year cycle	285 m.y.	Long-term earth's processes.
Wilson cycle and Galactic cycle	400-450 m.y.	Sea level changes, glacial cycle
Tidally generated cycle in Galaxy	600-800 m.y.	Long-term magmatic cycle

References

Abbott, D.H. and Isley, A.E., 2002. Extraterrestrial influences on mantle activity. *Earth Planet Sci. Lett.,* Vol. 205, pp. 53–62.

Abbott, D.H. and Isley, A.E., 2003. Reply to comment on extraterrestrial influence on mantle plume activity by Andrews Glikson. *Earth Planet Sci. Lett.,* Vol. 215, pp. 429–432.

Ager, D.V., 1981. Major marine cycles in the Mesozoic, *J. Geol. Soc. Lon.*, Vol. 138, pp. 159–166.

Akaike, H., 1969. Fitting autoregressive models for prediction, *Ann. Inst. Stat. Math.*, 21, pp. 243–247.

Akaike, H., 1970. Statistical predictor identification, *Ann. Inst. Stat. Math.*, 22, pp. 203–217.

Akaike, H., 1974. A new look at the statistical model identification, IEEE Trans. *Autom. Control*, 19, pp. 716–723.

Algeo, T.J. and Wilkinson, B.H., 1988. Periodicity of mesoscale phanerozoic sedimentary cycles and the role of Milankovitch orbital modulation, *J. Geology*, 1996, pp. 313–322.

Alldredge, L.R., 1976. Effects of solar activity on annual means of geomagnetic components. *Journal of Geophysics Research*, Vol. 81, pp. 2990–2996.

Alldredge, L.R., 1977. Geomagnetic variations with periods from 13 to 30 years. *Journal of Geomag. Geoelectr.*, Vol. 29, pp. 123–135.

Allen, M.R. and Smith, L.A., 1996. Monte Carlo SSA: Detecting irregular oscillations in the presence of coloured noise. *Jour. Clim.*, Vol. 9, pp. 373–404.

Alley, R.B. and MacAyeal, D.R., 1994. Ice-rafted debris associated with binge/purge oscillations of the Laurentide Ice sheet. *Paleoceanography*, Vol. 9, pp. 503–511.

Alley, R.B., Anandakrishnan, S. and Jung, P., 2000. Stochastic resonance in the North Atlantic, *Paleoceanography*, Vol. 16, No. 2, pp. 190–198.

Alley, R.B. et al., 2003. Abrupt climate change. *Science*, Vol. 299, pp. 2005–2010.

Alvarez, L.W., 1987. Mass extinctions caused by large bolide impacts. *Phys. Today,* July, pp. 24–33.

Alvarez, L.W., Alvarez, W., Asaro, F. and Michel, H.V., 1980. Extra-terrestrial cause for the Cretaceous tertiary extinctions. *Science*, Vol. 208, pp. 1095–1108.

Alvarez, W. and Muller, R.A., 1984. Evidence from crater ages for periodic impact on the earth. *Nature*, Vol. 308, pp. 718–720.

Alvarez, W., Asaro, F., Michel, H.V. and Alvarez, L.W., 1982. Iridium anomaly aproximately synchronous with terminal Eocene extinctions. *Science*, Vol. 216, pp. 886–888.

Ammann, C.M. and Naveau Phillippe, 2003. Statistical analysis of tropical explosive volcanism occurrences over the last six centuries. *Geophysical Research Letters*, Vol. 30, No. 5, 1210, doi:10.1029/2002GL016388.

Anderson, D.L., 1998. The scales of mantle convection. *Tectonophysics*, Vol. 284, pp. 1–17.

Anderson, N., 1974. On the calculation of filter coefficients for maximum spectral analysis. *Geophysics*, Vol. 39, pp. 69–72.

Appenzeller, C., Sticker, T.F. and Anklin, M., 1998. North Atlantic Oscillation dynamics recorded in Greenland ice core. *Science*, Vol. 282, pp. 446–449.

Archer, A.W. and Johnson, T. W., 1997. Modelling of cyclic tidal rhythmites (Carboniferous of Indiana and Kansas, Precambrian of Utah, USA) as basis for reconstruction of intertidal positioning and paleotidal regimes. *Sedimentology*, Vol. 44, pp. 991–1010.

Arthur, M.A., Kump, L.R., Dean, W.F. and Schlanger, S.O., 1985. Variation in the global climate cycle during the cretaceous related to climate volcanism and changes in atmospheric CO_2. *In*: E.T. Sundquist and W.S. Broecker (eds.), The carbon cycle and atmospheric CO_2: Natural variations Archean to present. *Geophysical monograph*, 32, pp. 561–572.

Arthur, M.A., Kump, L.R., Dean, W.F. and Larson, R.L. 1991. *Super plume, super greenhouse*? Eos Trans Amer. Geophy. Union. Vol. 72. pp. 301.

Bahcall, J.N., 1986. The Galactic environment of the solar system. *In*: R. Smoluchowski, J. Bahcall and M. Mathews (eds.), *The Galaxy and the Solar System*. Univ. Arizona, Tucson, pp. 3–12.

Bahcall, J.N., Hut, P. and Tremaine, S., 1985. Maximum mass of objects that constitute unseen disk material. *Astrophys. J.*, Vol. 290, pp. 15–20.

Bai, T. and Sturrock, P. A., 1991. The 154 day and related periodicities of solar activity as subharmonics of a fundamental period. *Nature*. Vol. 350. pp. 141–142.

Bailey, M.E., 1983. The structure and evolution of the solar system comet cloud. *Mon. Not. R. Astron. Soc.*, Vol. 204, pp. 603–633.

Bailey, M.E., 1986. The mean energy transfer rate to comets in the Oort cloud and implications for cometary origin. *Mon. Not. R. Astron. Soc.*, Vol. 218, pp. 1–30.

Bailey, M.E., Wilkinson, D.A. and Wolfendale, A.W., 1987. Can episodic comet showers explain the 30-m.y. cyclicity in the terrestrial record ? *Mon. Not. R. Astron. Soc.*, Vol. 227, pp. 863–885.

Baksi, A.K., 1990. Search for periodicity in global events in the geologic record: Quo vadimus? *Geology*, Vol. 18, pp. 983–986.

Barlow, N.G., 1990. Application of the inner solar system cratering record to the earth. *Geolo. Soc. America. Spec. Paper*, Vol. 247, pp. 181–187.

Barnet, T. P., 1981a. Statistical relation between ocean/atmosphere fluctuations in the tropical Pacific. *Jour. Phy. Oceanography*, Vol. 11, pp. 1083–1088.

Barnet, T.P., 1981b. Interaction of the monsoon and Pacific trade wind system at interannual time scales I. The equatorial zone. *Month Weather Rev.*, Vol. 111, pp. 756–773.

Barnola, J.M. et al., 1987 Vostok ice core provides 160,000 years record of atmospheric CO_2. *Nature*, Vol. 329, pp. 408–414.

Barrel, J., 1917. Rhythms and movement of geologic time. *Geol. Soc. Amer. Bull.* Vol. 28, pp. 745–904.

Barron, E.J., 1985. A warm, equable Cretaceous: the nature of the problem. *Earth. Sci. Rev.* Vol. 300, pp. 249–262.

Barton, Charles E., 1983. Analysis of pale magnetic time-series techniques and applications. *Geophysical Surveys*, Vol. 5 pp. 335–368.

Bash, F.N., 1986. Present, past and future elocity of nearby stars: The path of the Sun in 10^8 years. *In*: R. Smoluchowski, J. Bahcall and M. Matthews (eds.), *The Galaxy and the Solar System*. Univ. Arizona, Tucson, pp. 35–46.

Bates, R.L. and Jackson, J.A. (eds.), 1980. *Glossary of Geology* (second edition) p. 139.

Bath, M., 1974. *Spectral analysis in geophysics*. Elsevier Scientic Pub. Comp. Amsterdam.

Battisti and Sarachik, 1995. Understanding and predicting ENSO. *Review of Geophy Suppl.*, (9 U.S. National Report to IUGG 1991–1994), pp. 1367–1376.

Beauchamp, K.G., 1974. *Walsh functions and their applications*, Academic Press.

Beaufort, Luc., de Garidel-Thoron, T., Mix, A.C. and Pisias, N.G., 2001. ENSO-like forcing on oceanic primary production during the late Pleistocene, *Science*, Vol. 293, September, 28, pp. 2440–2444.

Becker, L., Poreda, R.J., Basu, A.R., Pope, K.O., Harrison, T.M., Nicolson, C. and Laski, R., 2004. Bedout: A possible end Permian Impact crater offshore of Northwestern Australia. *Science*, Vol. 304, pp. 1469–1476.

Beerling, D., 2002. CO_2 and the end Triassic mass extinctions. *Nature*, Vol. 425, pp. 386–387.

Berner, R.A., Lasaga, A.C. and Garrels, R.M., 1983. The carbonate-silicate geochemical cycle and its effect on atmospheric carbon dioxide over the past 100 million years, *Am. J. Sci.*, 283, pp. 641–683.

Benton, M. J., 1995. Diversification and extinction in the history of life. *Science*, Vol. 268, pp. 52–58.

Benzi, R., Parisi, G., Sutera, A. and Vulpiani, A., 1982. Stochastic resonance in climatic change. *Tellus,* Vol. 34, pp. 10–16.

Berger, A.L., Melice, J.L. and Hinnov, L., 1991. A strategy for frequency spectra of Quaternary climate records, *Clim. Dyn.*, 5, pp. 227–240.

Berger, A. Li, X.S. and Loutre, M.F. 1999. Modeliing Northern Hemisphere ice volume over the last 3 m.y. *Quat. Sci. Rev.* Vol. 18, pp. 1–11.

Berger, A., 1976. Obliquity and precession for the last 5,000,000 years. *Astron. Astrophys.*, Vol. 51, pp. 127–135.

Berger, A., 1977. Support for the astronomical theory of climatic change. *Nature*, Vol. 269, pp. 44–45.

Berger, A., 1978a. Long-term variations of the daily insolation and quaternary climatic changes. *J. Atmos. Sci.*, Vol. 35, pp. 2362–2367.

Berger, A., 1978b. Numerical values of the elements of the earth's orbit from 5,000,000 years BP to 1,000,000 AP, Contribution No. 35, Universite Catholique de Louvain Inst d' astronomie et de Catholique, Glemaitre, Louvain-la-Neeuve, B-1348, Belgique.

Berger, A., 1980. The Milankovitch astronomical theory of paleoclimates: A modern review. *Vistas in Astronomy*, Vol. 24, pp. 103–122.

Berger, A., 1984. Accuracy and frequency stability of the earth's orbital elements during the quaternary. *In*: A. Berger, J. Imbric, J. Hays, G. Kukla and B. Saltzman (eds.), *Milankovitch and Climate*, Reidel, Dordrecht, pp. 3–39.

Berger, A., 1988. Milankovitch theory and climate. *Rev. Geophys.*, Vol. 26, pp. 624–657.

Berger, A., 1989a. Pleistocene climatic variability at astronomical frequencies. *Quat. Int.*, Vol. 2, pp. 1–14.

Berger, A., 1989b. The spectral characteristic of pre-quaternary climatic records, an example of the relationship between the astronomical theory and geo-sciences. In: Berger, A., Schneider, S. and Duplessy, J.C. (eds.). *Climate and Geo-sciences, a Challenge for Science and Society in the 21st Century*, Kluwer Academic Publishers, Dordrecht, pp. 47–76.

Berger, A. and Pestiaux, P., 1984. Accuracy and stability of the quaternary terrestrial insolation. *In*: *Milankovitch and climate*. A. Berger, J. Imbrie, J. Hays, G. Kukla and B. Saltzman (eds.) pp. 83–112.D, Reidel Publ. Comp. Dordrecht, Holland.

Berger, A., Loutre, M.F. and Dehant, V., 1989c. Influence of the changing lunar orbit on the astronomical frequencies of the pre-quaternary insolation patterns. *Paleoceanography*, Vol. 4, pp. 555–564.

Berger, A., Loutre, M.F. and Dehant, V., 1989d. Milankovitch frequencies for pre-quaternary. *Nature*, Vol. 342, p. 133.

Berger, A., Loutre, M.F., 1989e. Astronomical frequencies for pre-quaternary palaeoclimate studies. *Terra Nova*, Vol.1, pp. 474–479.

Berger, A. and Loutre, M.F., 1990. Origine des frequences des elements astronomiques intervenant dans Ie calcul de I'insolation. *Bulletin de la Classe des Sciences*. Academie Royale de Belgique, 6ᵉ serie, Vol. 1(1/3), pp. 45–106.

Berger, A., Loutere, M.F. and Lasker, J. 1992. Stability of the astronomical frequencies over the earth history for paleoclimate studies. *Science*, Vol. 255, pp. 560–565.

Berger, W.H., Schimmelmann, A. and Lange, C.B., 2004. Tidal cycles in the sediments of Santa Barbara Basin, *Geology*, Vol. 32, pp. 329–332.

Berggren, W.A., Kent, D.V., Flynn, J.J and Van Couvering J.A.,1985. Cenozoic Geochronology. *Bull.Geological society of Am.* Vol. 96, pp.1407–1418.

Berggren, W.A., Kent, D.V., Swisher, C.C., III and Aubry, M.P., 1995. A revised Cenozoic geochronology and chronostratigraphy. *In*: W.A. Berggren, D.V. Kent, M.P. Aubry, and J. Hardenbol, (eds.), *Time Scales and Global Stratigraphic Correlations: A Unified Temporal Framework for a Historical Geology*: Society for Economic Paleontologists and Mineralogists Special Publication, Vol. 54, pp. 129–212.

Bernett, T., 1990. A solar-ocean relation: Fact or fiction: *Geophys. Res. Lett.*, Vol. 16, pp. 803–816.

Berry, W.B.N and Barker, R.M., 1975. Growth increments in fossil and modern bivalues: *In*: G.D Rosenberg and S.K Runcorn (eds.), *Growth rhythms and the History of Earth's rotation*. John Wiley & Sons, New York, pp. 9–25.

Bhandari, N., Gupta, M., Pandey, J. and Shukla, P.N., 1994. Chemical profiles in K/T boundary section of Meghalaya, India: cometary, asteroidal or volcanic, Chemical Geology, Vol. 113, pp. 45–60.

Bigg, G.R., 1996. *The Ocean and Climate*. Cambridge University Press, Cambridge. pp. 1–266.

Bjerknes, J., 1969. Atmospheric teleconnections from the equatorial Pacific. *Month. Weath. Rev.*, Vol. 97, pp. 163–172.

Black et. al., 1999. Eight centuries of North Atlantic ocean atmosphere variability, *Science,* Vol. 286, pp. 1709–1713.

Blackman, R.B. and Tukey, J.W. 1958. The measurement of power spectra from the point of view of communication engineering, Dover, Mineola, New York.

Bloxham, J. and Gubbins, D., 1985. The secular variation of earth's magnetic field, *Nature*, Vol. 317, pp. 777–781.

Bloxham, J. and Gubbins, D., 1987. Thermal core-mantle interactions, *Nature*, Vol. 325, pp. 511–513.

Boiseau, M., Juillet-Leclere, A., Yiou, P., Salvat, B., Isdale, P. and Guillaume, M.,1998. Atmospheric and oceanic evidences of El Nino–Southern Oscillation events in the south central Pacific ocean from coral stable isotopic records over the last 137 years. *Paleoceanography*, Vol. 1, pp. 671–685.

Boiseau, M., Ghil, M. and Juillet-Locleres, A., (1999). Climatic trends and interdecadal variability from South-Central Pacific coral records, *Geophys. Res. Lett.*, Vol. 26, No. 18, pp. 2881–2884.

Bojariu, R. and Gimeno, L., 2003. Predictability and numerical modeling of the North Atalantic Oscillation. *Earth Sci. Rev.,* Vol. 63, pp. 145–168.

Bond, G. et al., 1992. Evidence for massive discharge of icebergs into the Northern Atalantic Ocean during the last glacial period. *Nature,* Vol. 360, pp. 245–249.

Bond, G. et al., 1997. A pervasive millennial-scale cycle in North Atlantic Holocene and glacial climates. *Science*, Vol. 278, pp. 1257–1266.

Bond, G. et al., 2001 Persistent solar influence on North Atlantic climate during the Holocene. *Science*, Vol. 294, pp. 2130–2136.

Bond, G., Broecker, W., Johnsen, S., Mc Manus, J., Labeyrie, L., Jouzel, J. and Bonani, G., 1993. Correlations between climate records from North Atlantic sediments and Greenland ice, *Nature*, 365, pp. 143-147.

Bond, G.C., Showers, W., Elliot, M., Evans, M., Lotti, R., Hajdas, I., Bonani, G. and Johnson, S., 1999. The North Atlantic's 1-2 kyr climate rhythm: relation to Heinrich events, Dansgaard/Oeschger Cycles and the Little Ice Age. *In*: P.U. Clark. R.S. Webb, L.D. Keigwin (eds.), *Mechanisms of Global Climate Change at Millennial Time-scales*. American Geophysical Union. Washington DC, pp. 35–58.

Bonifay et al., 1987. Study of the Holocene and Late Wurwian sediments at Lacdu Bouchet (Heute-Lore, France): First result. In: *Climate history periodicity and predictability* by M.R. Rampino et al., pp.88, Van Nostrand Reinhold Company, New York.

Box, G.E.P. and Jenkins, G.M., 1970. *Time series analysis, forecasting and control*, Holden-Day, Boca Raton, Fla.

Bradley, W.H., 1929. The varves and climate of green river epoch. *U.S. Geol. Surv. Prof. Paper.*, Vol. 158, pp. 87–110.

Braginsky, S.I., 1970. Torsional magnetohydrodynamic dynamo of the earth. *Geomag. Aeron.* 10, pp. 172–181.

Braginsky, S.I. and Fishman, V.M., 1977. 60-yr variations of the geomagnetic field and of the electrical conductivity of the mantle. *Geomag. Aeron.* 17, pp. 607–613.

Bray, J. R., 1968. Glaciation and solar activity since the fifth century BC and the solar cycle. *Science,* Vol. 220, pp. 672–674.

Bray, J.R., 1971. Solar-climate relationships in the Post-Pleistocene, *Science*, Vol. 171, pp. 1242–1243.

Broecker, W.S. et al., 1992. Origin of the northern Atlantic's Heinrich events. *Climate Dynamics*. Vol. 6, pp. 265–273.

Bucha, V., 1967. Intensity of the Earth's magnetic field during Archaeological times in Czechoslovakia. *Archaeometry* 10, pp.12–22.

Bucha, V., 1969. Changes in the Earth's magnetic moment and radiocarbon dating, *Nature,* Vol. 224, pp. 681–683.

Budyko, M.I. and Ronov, A.B., 1979. Chemical evolution of the atmosphere in the Phanerozoic (In Russian), *Geokhymiya,* Vol. 5 pp 643–653. (Geochemical Int. Engl. Tran., Vol. 15, pp. 1–9, 1979).

Buffet, B.A., 1992. Constraints and magmatic energy and mantle conductivity from rotation of the earth. *JGR*, vol. 97, pp.19581–19597.

Bullard, E.C., Freedman, C., Grellman, H. and Nixon, J., 1950. *The westward drift of the Earth's magnetic field*. Phil. Trans., Roy. Soc Lond. A., 243, pp. 67–92.

Burek, P. and Wanke, H., 1988. Impacts and glacio-eustasy, plate tectonic episodes, geomagnetic reversals: a concept to facilitate detection of impact events, *Phys. Earth. Planet. Int.*, Vol. 50, pp. 183–194.

Burg, J. P., 1967. Maximum entropy spectral analysis. Paper presented at the 37th Annual International Meeting of the Society of Exploration Geophysicists, Oklahoma City, Okla., (Published in *Modem Spectrum Analysis*, edited by D. G. Childers, pp. 42–48, IEEE Press, Piscataway, N. J., 1978.).

Burrough, W.J., 1992. *Weather cycles: real or imaginary*? Cambridge University Press, pp. 1–207.

Campbell, I.D. et al., 1998. Late Holocene ~1500 yr climate periodicities and their implications. *Geology*, Vol. 26, pp. 471–473.

Campsie, J., Johnson, G.L., Jones, J.E. and Rich, J.E., 1984. *Episodic volcanism and evolutionary crises*, EOS Trans., Am. Geophys. Un., Vol. 65, pp. 796–800.

Cane, M.A., 1985. *Predicting El Nino and the Southern Oscillation*, pp. 11–13.

Cane, M.A., 1986. El Nino. *Annul. Rev. Earth Planet Sci.*, Vol. 14, pp. 43–70.

Cane, M.A., 1992. Tropical Pacific ENSO models: ENSO as a mode of coupled system. *In*: K. E. Trenberth (ed.), *Climate system modelling*, Cambridge University Press Cambridge. pp. 583–614.

Capedes, Daniel. L., 1977. *McGraw-Hill Encyclopedia of the geological Sciences*. pp. 296.

Cartwright, D. E. et al., 1985. Changes in relative mean sea level. *EOS,* Vol. 66, November issue, p. 11.

Castagnoli, G.C., Bonino, G.and Provenzale, A., 1989. The 206 years cycle in tree ring radio carbon data in the thermoluminescence profile of a recent sea sediment. *Jour. Geophy. Res.,* Vol. 94, pp. 11971–11976.

Chakraborty, S. and Ramesh, R., 1998. Stable isotope variations in a coral (*Favia speciosa*) from the Gulf of Kutch during 1948–1989 A.D., *Environmental Implications*. Vol. 107, pp. 331–341.

Chamberlin, R. T., 1914. Diastrophism and formative processes VII. Periodicity of Paleozoic orogenic movement. *Journ. Geol.* Vol. 22, pp. 315–345.

Chance, A. and Kelly, P.M., 1979. An apparent periodicity in an index of volcanic activity. *Nature*, Vol. 280, pp. 671–672.

Chandrasekhar, S., 1961. *Hydrodynamic and hydromagnetic stability*. Oxford Univ Press.

Channell, J.E.T. et al., 1998. Orbital modulation of the earth's magnetic field intensity. *Nature,* Vol. 394, pp. 464–468.

Charles, C.D., Hunter, D.E. and Fairbanks, R.G., 1997. Interaction between the ENSO and the Indian monsoon in a coral record of tropical climate. *Science*, Vol. 277, pp. 925–927.

Chen, W.V. and Stegen, G.R., 1974. Experiment with maximum entropy power spectra of sinusoids. *Journ. Geophy. Res*, Vol. 79, pp. 3019–3022.

Chillingworth, D.R.J. and Holmes, P.J., 1980. Dynamical system and models for the reversal and the earth's magnetic field, *Math Journal*, Vol. 12, pp. 41–59.

Christine-Blick, N., 1988. Sea level history (Technical Comments). *Science* Vol. 241., pp. 596–597.

Christine-Blick, N., Mountain G.S. and Miller, K.G., 1988. Comment on a paper by Haq, B. U. et al., *Science,* Vol. 235, pp. 1156, 1987: Sea level history. *Science* Vol. 241, pp. 596.

Clark, P.U., Alley, R.B., Pollard, D., 1999. Northern Hemisphere ice sheet influences on global climate change. *Science* 286, 1104–1111.

Clark, P.U., Webb, R.S. and Keigwin, L.D (Eds) 1999. Mechanisms of global climate change at millennial time scales. *Geophysical Monograph*, Vol. 112, American Geophysical Union, Washington, pp. 1–394.

Caldeira, K. and Rampino, M.R. 1991. The mid Cretaceous superplume, Carbondioxide and global warming. *Geophy. Res. Lett.* Vol. 18, pp 987–990.

Clemens, S.C. and Tiedemann, R., 1997. Eccentricity forcing of Pliocene Early Pleistocene climate revealed in a marine oxygen-isotope record. *Nature* 385, 801–804.

Clement, A. C., Seager, R. and Cane, M. A., 1999. Orbital controls on the El Nino/Southern Oscillation and the tropical climate. *Paleoceanography*, Vol. 14, pp. 441–456.

Clube, S.V.M. and Napier, W.M., 1982a. Spiral arms, comets and terrestrial catastrophism. *Q. J. R. Astro. Soc.,* Vol. 23, pp. 45–66.

Clube, S.V.M. and Napier, W.M., 1982b. The role of episodic bombardment in geophysics, *Earth and Planetary Sci. Letters*, Vol. 57, pp. 251–262.

Clube, S.V.M. and Napier, W.M., 1986. Giant comets and the galaxy: Implications of the terrestrial record. *In*: R. Smoluchowski, J. Bahcall and M. Matthews (eds.), *The Galaxy and the Solar System.* Univ. Arizona, Tucson, pp. 260–285.

Cobb, K.M., Charles, C.D. and Hunter, D.E., 2001. A central tropical Pacific coral demonstrates Pacific, Indian and Atlantic decadal climate connections, *Geophys. Res. Lett.*, Vol. 28, No.11, pp. 2209–2212.

Coe, R.S., Gromme, C.S. and Mankinen, E.A., 1984. Geomagnetic and Paleointensities from excursion sequences in lavas on Oahu, Hawaii. *J. Geophys. Res.*, 89, 1059.

Cohen, T.J. and Lintz, P.R., 1974. Long term periodicities in the sunspot cycle. *Nature*, Vol. 250, pp. 398–400.

Cole, J., 1994. Reconstructing and predicting climate variability, *Geotimes*, November issue, pp. 12–18.

Cole, J.E. and Fairbanks, R.G., 1990. The southern oscillation recorded in the δ^{18} of corals from Tarawa Atoll, *Paleoceanography*, Vol. 5, No. 5, pp. 669–683.

Cole, J.E., Dunbar, R.B., McClanahan, T.R. and Muthiga, N.A., 2000. Tropical Pacific forcing of decadal SST variability in the western Indian ocean over the past two centuries. *Science*, Vol. 287, pp. 617–619.

Coleman, J.M. and Roberts, H. H., 1988. Sedimentary development of the Louisiana continental shelf related to sea level cycles: Part I– Sedimentary sequences. *Geomarine Letters*. Vol. 8, pp. 63–108.

Compton, J. S. and Mallinson, D. J., 1996. Geochemical consequences of increased late Cenozoic weathering rates and global CO_2 balance since 100 m.y. *Paleoceanography*, Vol. 1, pp. 431–446.

Condie, K.C. 1976. *Plate tectonics and crustal evolution*. Pergamon Press, New York, pp. 288.

Consolini, G. et al., 2000. Multifractility and punctuated equilibrium in the earth's magnetic field polarity reversals. *Geophy. Res.* Lett. Vol. 27, pp. 293–296.

Cook, A.E. and Robert P.H., 1970.*The Rikitake two disk dynamo system*. Proc Cambridge Philos. Soc., Vol 68, pp 547–569.

Cooper, M.R., 1990. Tectonic cycles in southern Africa, *Earth. Sci. Rev.*, Vol. 28, pp. 321–364.

Cortijo, Elsa et al., 1995. Sedimentary record of rapid climatic variability in the north Atlantic ocean during the last glacial cycle. *Paleoceanography*, Vol. 10, pp. 911–926.

Cortini, M. and Barton C. C., 1994. Chaos in geomagnetic reversal records: A comparison between earth's magnetic field data and model disk dynamo data. *Jour. Geophy. Res.* Vol. 99. pp. 18021–18033.

Courtillot, V. and Besse, J., 1987. Palaeomagnetism and core-mantle coupling, *Science*, Vol. 247, pp. 1140–1147.

Courtillot, V. and Cisowski, S., 1987. The Cretaceous-tertiary boundary events, external or internal causes? *EOS Trans. Am. Geophys. Un.*, Vol. 68, pp. 193–200.

Courtillot, V. and Besse, J., 1987. Magnetic field reversals: polar wander and core mantle coupling. *Science*, Vol. 237, pp. 1140–1148.

Courtillot, V. and Gaudemer, Y., 1996. Effects of mass extinctions on biodiversity. *Nature*, Vol. 381, pp. 146–148.

Courtillot, V.E. and Le Mouel J.L. 1976. On the long period variation of the earth's magnetic field from 2 months. *J. Geophys. Res.,* 81, pp. 2941–2950.

Courtillot, V.E., Le Mouel J.L. and Mayaun, P.N., 1977. Maximum entropy spectrum analysis of the Geomagnetic activity index aa over a 107 year interval. *J. Geophys. Res.*, 82, pp. 2641–2649.

Cox, A., 1968. Length of geomagnetic polarity reversals. *JGR*, Vol. 73, pp. 3247–3260.

Crain, I.K., Crain, P.L. and Plaunt, M.S., 1969. Long period Fourier spectrum of geomagnetic reversals, *Nature*, Vol. 223, pp. 282.

Creer, K.M., Thompson, R., Molyneux, L. and Mackereth, F.J.H, 1972. Geomagnetic secular variation recorded in the stable magnetic remanance of recent sediments. *Earth Planet. Sci. letters*, 14, pp. 115–127.

Creer, K.M., 1975. A tentative correlation between changes in the geomagnetic polarity bias and reversals frequency and earth's rotation through Phenerozoic time: G.D. Rosenberg and S.K. Runkarn ed. *Growth and rhythm history of earth's rotation*. Wiley, pp. 293–317.

Croll, J., 1864. On the physical cause of climate during geological epoch. *Phil. Mag.* Vol. 28, pp. 121–137.

Croll, J., 1875. *Climate and Time*. Appleton, New York.

Cronin, T.M., 1999. *Principles of Paleoclimatology*. Columbia. Univ. Press, New York, 560 pp.

Cronin, T.M and Raymo, M.E., 1997. Orbital forcing of deep-sea benthic species diversity, *Nature*, Vol. 385, pp. 624–627.

Crough S.T., 1978. Thermal origin of mid plate swells. *Geophy J. Rev. Astro Society*, Vol. 55, pp.451–469.

Crowell, J. C., 1978. Gondwana glaciation, Cyclothems, continental positioning and climate change. *Amer. Jour. Scie.* Vol. 278, pp 1345–1372.

Crowley and North, G.R., 1988. Abrupt climate change and extinction events in Earth history, *Science*, Vol. 240, pp. 996–1002.

Cullen, H.M. and Demenocal, P., 2000. North Atlantic influence in Tigris Euphrates stream flow. *Int. Jr. of Chm.* (in press).

Cullen, H.M., Rosanne, D.D' Arrrigo and Edward R, Cook, 2001. Multiproxy reconstructions of the north Atalantic. *Paleoceanography*. Vol. 16, pp. 27–39.

Currie, R.G., 1968. Geomagnetic spectrum of internal origin and lower mantle conductivity. *Journal of Geophysical Research,* Vol. 73, pp. 2779–2786.

Currie, R.G., 1973. Geomagnetic line Spectra-2 to 70 years. *Astrophy. Space Sci.*, 21, pp. 425–438.

Currie, R.G., 1976. The spectrum of sea level from 4 to 40 years, *Geophys. J.R. Astr. Soc.*, 46, pp. 513–520.

Currie, R.G., 1981. Evidence for 18.6 year M_N signal in temperature and drought conditions in North America since AD, 1800. *Jour. Geophy. Res* Vol. 86, pp. 11055–11064.

Currie, R.G., 1982. Evidence for 18.6 year Mn term in Japanese air pressure and geophysical implications, *Geophys. J.R. Astr. Soc.*, 69, pp. 321–327.

Currie, R.G. and Fairbridge, R.W., 1985. Periodic 186-year and cyclic 11-year induced drought and flood in the northeastern China and some global implications, *Quaternary Science Reviews*, Vol. 4, pp. 109–134.

Currie, R.G. and O'Brien, 1988. Periodic 18.6 years and cyclic 10–11 year signals in Northeast United States precipitation data. *Int Jour Climatology*, Vol. 8, pp. 255–281.

Domon, P.E., 1971. The relationship between late Cenozoic volcanism and tectonism signal and orogenic-epirogenic periodicity in Turekian K., (ed.) *Late Cenozoic Glacial Ages*, New Haven, Yale University Press, pp. 15–36.

Dansgaard, W. et al., 1993. Evidence for general instability of past climate from a 250 k.y. ice-core record. *Nature*, Vol. 364, pp. 218–220.

Das, P.K., 1988. Why monsoons fail? *Science Reporter*, May issue, pp. 256–274.

Davies, R.D., 1972. Observations of the outer spiral structure of the Milky way and its relation to the high velocity clouds. *Mon. R. Astron. Soc.*, Vol. 160, p. 266.

Davies, R.D., 1989. Catastrophes and evolutions: The 1988 BAAS mason meeting of the Royal Astronomical Society at Oxford. *In: Catastrophes and evolution: astronomical foundations,* pp. 1–11.

Davis, M., Hut, P. and Muller, R.A., 1984. Extinction of species by periodic comet showers, *Nature*, Vol. 308, pp. 715–717.

De Boer, P.L. and Wonders, A.A.H., 1984. Astronomically induced rhythmic bedding in Cretaceous pelagic sediments near Moria (Italy). *In*: Berger, A., Imbrie, J., Hays, J., Kukla, G. and Saltzman, B., (eds.), *Milankovitch and Climate*, Reidel Publ. Co., Dordrecht, pp. 177–190.

DeBlonde, G. and Peltier, W.R., 1991. A one-dimensional model of continental ice volume fluctuations through the Pleistocene: implications for the origin of the mid-Pleistocene climate transition. *J. Clim.* 4, pp. 318–344.

DeBoer, P.L. and Smith, D.G., 1994. Orbital forcing and cyclic sequences. *Spec. Pub. Int. Assoc. Sediment.* Vol. 19, pp. 1–14.

Deffeys, K.S., 1972. Plume convection with an upper-mantle temperature inversion. *Nature*, 240, pp. 539–544.

Dehant, V., Loutre, M.F. and Berger, A., 1990. Potential impact of the Northern hemisphere quartnery ice sheets on the frequencies of the astroclimatic orbital parameters. *Jour. Geophy. Res.*,Vol. 95, pp. 7573–7578.

Delworth, T.L. and Mehta, V.M., 1998. Simulated interannual to decadal vaiability in the tropical and sub-tropical North Atlantic, *Geophys. Res. Lett*, Vol. 25, No.15, pp. 2825–2828.

Dicke, R.H., 1978. Is there a chronometer hidden deep in the sun? *Nature*, Vol. 276, pp. 676–680.

Dicke, R.H., 1979. Solar luminosity and the sunspot cycle. *Nature*, Vol. 280, pp. 24–27.

Dickens, G.R. and Francis, J.M., 2003. Comment on a core for a comet impact trigger for the paleo/Eocene thermal maximum and carbon isotope excursion by Kent, et al., *Earth Planet Sci. Lett.* Vol. 211, pp.13–26.

Dima, M., Stefan, S., Dima, V. and Borsan, D. 2001. Interdecadal variability generated by interactions between Pacific and Atlantic oceans, *Geophys. Res. Lett.*, Vol. 28, No. 23, pp. 4459–4462.

Dimri, V.P. and Ravi Prakash, M., 2001. Scaling of power spectrum of extinction events in the fossil record, *Earth and Planetary Science Letters*, Vol. 186, pp. 363–370.

Doake, C.S.M., 1977. A possible effect of ice ages on the earth's magnetic field, *Nature*, Vol. 267, pp. 415–417.

Dodson, R., Fuller, M. and Kean, W.E., 1977. Palaeomagnetic records of secular variation from Lake Michigan sediment cores. *Earth and Planetary Science Letters*, Vol. 34, pp. 387.

Donn, W.L. and Shaw, D. M. 1977. Model of climate evolution based on continental drift and polar wandering. *Geol. Soc. Amer. Bull.* Vol. 88, pp. 390–396.

Dorman, F.H., 1968. Some Australian oxygen isotope temperatures and a theory for a 30 million-year world-temperature cycle, *Jour. Geology*, Vol. 76, pp. 297–313.

Dreher, J.W., Robert, D.H. and Lehar, J., 1986. Very large array observations of rapid non-periodic variations in OJ287, *Nature*, 320: 239–242.

Dunbar, R.B., 1994. Eastern Pacific sea surface temperature since 1600 AD: The $\delta^{18}O$ record of climate variability in Galapagos corals. *Palae oceanography*, Vol. 9, pp. 291–315.

Dunbar, R.B., 2000. Clues from corals, *Nature*, Vol. 407, pp. 956–957.

Duncan, R.H. and Richards, M.A., 1991. Hotspots mantle, plumes, Flood basalts and True polar wandering. *Review of geophysics*, Vol. 39, pp. 31–50.

Duvall, Jr., T.L. and Harvey, J.W., 1984. Rotational frequency splitting of solar oscillations. *Nature*, 310, pp. 19–22.

Eddy, J.A., Gilliland, R.L. and Hoyt, D.V., 1982. Changes in the solar constant and climatic effects. *Nature*, Vol. 300, pp. 689–693.

Eldholm, O., 1990. Paleogene North Atlantic magmatic-tectonic events: environmental implications, *Mem. Soc. Geo. Ital.*, Vol. 44, pp. 13–28.

Eldholm, O. and Thomas, E., 1993. Environmental impact of volcanic margin formation. *Earth Planet Sci. Lett.,* Vol. 117, pp. 147–167.

Elkibbi, M. and Rial, J. A., 2001. An outsider's review of of the astronomical theory of the climate: Is the eccentricity-driven insolation the main driver of the ice age? *Earth Scie Review,* Vol. 56, pp. 161–177.

Elsasser, W.M., 1946. Induction effect in terrestial magnetism, *I. Theory Phys. Rev.* Vol. 69. pp. 106–116.

Emery, W.J. and Thomson, R.E., 1997. *Data analysis method in physical oceanography*. Elsevier, Amsterdam, pp. 1–634.

Emiliani, C., Kraus, E.B. and Shoemaker, E.M., 1981. Sudden death at the end of Mesozoic. *Earth. Planet. Sci. Lett.*, 55, pp. 317–334.

Enfield, D.B., 1989(a). El Nino, Past and Present, *Reviews of Geophysics*, Vol. 27, 1, pp. 159–187.

Enfield, D.B., 1989(b). Is El Nino becoming more common? *Oceanography*, Vol. 1, pp. 23–27. Also, D.B. Enfield Luis Cid.s Toga Notes No.1, Fall 1990.

Eriksson, K.A. and Simpson, E.L., 2000. Quantifying the oldest tidal record: the 3.2 Ga Moodies Group, Barbneton Greenstone Belt, South Africa, *Geology*, Vol. 28, pp. 831–834.

Ernst, Richard E. and Buchan Keneth, L., 2003. Recognizing mantle plumes in the geological records. *Annu. Review of Earth and Planet Sci.*, Vol 116, pp. 469–523.

Erwin, D.H., 1993. *The great Paleozoic crises*: New York, Columbia Univ. Press, pp. 257.

Erwin, D.H., 1994. The Permo-triassic extinction. *Nature*, Vol. 367, pp. 231–236.

Erwin, D.H., Bowring, S.A. and Yugan, J., 2002. End Permian mass extinctions: A review. *Geol Soc. America. Spec Paper*, Vol. 356. pp. 363–383.

Evans, M.N., Kaplan, A. and Cane, M.A., 2000. Inter comparison of coral oxygen isotope data and historical sea surface temp (SST): Potential for coral-based SST field reconstructions, *Paleoceanography*, Vol. 15, pp. 551–563.

Fairbridge, R.W., 1984. Planetary periodicties and terrestrial climate stress. *In*: N.A. Maurer and W. Karlen (eds) *Climate change on a yearly to millennial basis*. pp. 509–520.

Fairbridge, R.W. and Hillaire-Marcel, C., 1977. An 8,000-yr paleoclimatic record of the "Double Hale" 45-yr solar cycle. *Nature*, Vol. 268, pp. 413–416.

Federov, A.V. and Philander, S.G., 2000. Is El Nino changing? *Science*, Vol. 288, pp. 1997–2001.

Felis, T., Patzold, J., Loya, Y., Fine, M., Natwar, A.H. and Wefer, G., 2000. A coral oxygen isotope record from the northern Red Sea documenting NAO, ENSO and North Pacific teleconnections on Middle East climate variability since the year 1750, *Paleoceanography*, Vol. 15, No. 6, pp. 679–694.

Fischer, A.G., 1981. Climatic oscillations in the biosphere in Nitecki, M. H. (ed.), *Biotic Crises in Ecological and Evolutionary Time*: New York, Academic Press, pp. 103–111.

Fischer, A.G., 1982. The carbon cycle—controls on atmospheric CO_2 and climate in the geological past, in *Climate Earth History*: Washington D.C., National Academy Press, pp. 55–67.

Fischer, A.G. and Arthur, M.A., 1977. Secular variations in the pelagic realm. *In*: Cook, H.E., and Enos, P. (ed.) *Deep Water Carbonate Environments*, SEPM Spec. Pub. Vol. 25, pp. 19–50.

Fischer, A.G. and Schwarzacher, W., 1984. Cretaceous bedding rhythms under orbital control? *In*: Berger, A., Imbrie, J., Hays, J., Kukla, G. and Saltzman, B. (eds.). *Milankovitch and Climate*, Reidel Publ. Co., Dordrecht, pp. 163–175.

Fischer, Arthur, 1988. What flips earth's field? *Popular Science*, January issue, pp.71–74 and 112.

Fisher, R.A., 1929. Test of significance in harmonic analysis. *Proc. Roy. Soc. Series A*, Vol. 125, pp. 54–59.

Fligge, M. and Solanki, S.K., 1999. The solar spectral irradiance since 1700. *Geophysical Research Letters*, 26, pp. 2465–2468.

Follmi, K. B., 1996. The Phosphorus cycle, Phosphogenesis and marine phosphate-rich deposits. *Earth. Sci. Rev.*, Vol. 40, pp. 55–124.

Forte, A. M. and Mitrovica, J. X., 1997. A resonance in the earth's obliquity and precession over the past 20 m.y. driven by mantle convection., *Nature,* Vol. 390, pp. 676–680.

Frakes, L.A., 1979.*Climate change, throughout geological time*. Amsterdam: Elsevier, 310 pp.

Friis-Christensen, E. and Lassen, K., 1991. Length of the solar cycle: An indicator of solar activity closely associated with climate. *Science*, Vol. 254, pp. 698–700.

Fronval, T., Jansen, E., Bloemendal, J. and Johnsen, S, 1995. Oceanic evidence for coherent fluctuations in Fennoscandian and Laurentide ice sheets on millennial timescales. *Nature*, 374: pp. 443–446.

Gaffin, S., 1987. Phase difference between sea level and magnetic reversal rate. *Nature*, Vol. 29, pp. 816–819.

Gaffin, S., 1989. Analysis of scaling in the geomagnetic polarity reversal record. *Phys. Earth. Planet. Int.,* Vol. 57, pp. 284–290.

Gallee, H., van Ypersele, J.P., Fichefet, T., Marsiat, I., Tricot, C., Berger, A., 1992. Simulation of the last glacial cycle by a coupled, sectorially averaged climate-ice sheet model: Part II. Response to insolation and CO_2 variation. *J. Geophys. Res.* 97, pp. 15713–15740.

Gallego, D., Garica, R., Hernandez, E., Gimeno, L. and Ribera, P. (2001). An ENSO signal in the North Atlantic subtropical area, *Geophys. Res. Lett.,* Vol. 28, No. 15, pp. 2939–2042.

Ganpathy, R., 1982. Evidence for a major meteorites impact on the earth 34 million years ago: Implication for Eocene extinctions. *Science*, Vol. 216, pp. 885–886.

Garric, G. and Hubber, M., 2003. Quasi-decadal variability in paleoclimatic records: Sunspot cycles or intrinsic oscillation? *Paleoceanography*, Vol. 18, pp. 13; 1–8.

Garidel-Thorn, Thibault de et al. (2001). Millenial-Scale dynamics of the East Asian Winter Monsoon during the last 2,00,000 years palaeoceanography Vol. 16, No. 5, pp. 491–502.

Genthon, C., 1987. Vostoke ice core: climate response to CO_2 and orbital forcing changes over the last climate cycle. *Nature*, Vol. 329, pp. 414–418.

Ghil, M., 1994. Cryothermodynamics: the chaotic dynamics of paleoclimate. *Physica D* 77, pp. 130–159.

Ghil, M. J.and LeTreut, H., 1981. A climate model with cryodynamics and geodynamics, *J. Geophys. Res*., 86, pp. 5262–5270.

Ghil, M. and Childress, S., 1987. Topics in Geophysical Fluid Dynamics: Atmospheric Dynamics, Dynamo Theory and Climate Dynamics, *Dynamo Theory and Climate Dynamics*. Springer, Berlin, pp. 485.

Ghil, M. and Vautard, R., 1991. Interdecadal oscillations and the warming trend in global temperature time series, *Nature*, Vol. 350, No. 6316, pp. 324–327.

Ghil, M. et al., 2002. Advance spectral methods for climate time series. *Review of Geophy*, Vol. 40, pp. 1–41.

Gilbert, G. K., 1895. Sedimentary measurement of geological time. *Jour. Geol.*, Vol. 3, pp. 351–376.

Gimeno, L., et al., 2003. Changes in the relationship NAO-Northern hemisphere temperature due to solar activity. Earth *Planet Sci. Lett.*, Vol. 206, pp. 15–20.

Giovanelli, R., 1984. *Secrets of the Sun*. Cambridge University Press.

Glasby, G.P and Kasahara, J., 2001. Influence of tidal effects on the periodicity of earthquake activity in diverse geological settings with particular emphasis on submarine hydrothermal systems. *Earth-Science Reviews*, Vol. 52, pp. 261–297.

Glass, B.P. and Heezen, B.C., 1967. Tektites and geomagnetic reversals, *Scientific American*, Vol. 217, pp. 32–38.

Gleissberg, W., 1958. The eighty-year sunspot cycle. *J. Br. Astr. Ass.*, Vol. 75, pp. 227–231.

Glikson, A., 2003. Comments on extraterrestrial influences on mantle plume activity by D.H. Abbott and A. E. Isley, *Earth Planet Sci. Lett.* Vol. 205, pp. 53–62.

Glikson, A.Y., 1996. Mega impact and mantle melting episodes: test of possible correlations. *AGSO Aust Jour. Geol. Geophy.* Vol. 16, pp. 355–376.

Glikson, A.Y., 1999. Oceanic mega impacts and crustal evolution. *Geology*, Vol. 27, pp. 387–390. Also see, Asteroid/comet mega impacts may have triggered major episodes of crustal evolution. *EOS*, pp. 49–54.

Grabau, A.W., 1940.The rhythm of ages. Peking, Vetch, pp. 561.

Gradstein et. al., 1988. Comments on paper by Haq et. al. (1987), Vol. 235, pp. 1156–1167, *Science,* 241, 1601.

Gradstein, F.M., Agterberg, F.P., Ogg., J.G., Hardenbol, J., van Veen P., Thierry, J. and Huang. Z., 1994. A Mesozoic time scale: *Jour. Geophys. Res.*, Vol. 99 B12, pp. 24051–24074.

Graham, N.E. and White, W.B. (1995). The El Nino cycle: A natural oscillator of the Pacific Ocean – atmospheric system, *Science*, Vol. 240, pp. 1293–1302.

Gribbin, J., 1978. *Astronomical influences, long-term effects in climate change.* Gribbin (edt.) Cambridge University Press, pp.133–138.

Gribbin, J., 1989. The end of ice ages? *New Scientist,* January, Vol. 17, pp. 166–174.

Grieve, R.A.F., 1982. The record of impact on Earth: Implications for a major Cretaceous/tertiary impact event. *In*: L.T. Silver and P.H. Schultz (eds.), *Geological Implications of Impacts of Large Asteroids and Comets on the Earth.*, Geol. Soc. Am. Spec. Pap., Vol. 190, pp. 25–38.

Grieve, R.A.F., 1989. Hyper velocity impact cratering: A catastrophic terrestrial geologic process. *In*: *Catastrophes and Evolution, Astronomical Foundations*, S. V. M. Clube. Cambridge Univ. Press. Cambridge. pp. 57–79.

Grootes, P.M. and Stuvier, M., 1997. Oxygen δ^{18}/δ^{16} variability in Greenland snow and ice with 10^{-3} to 10^{-5}-year time resolution. *J.Geophy. Res.*, 102, pp. 26, 455–470.

Gu, D. and Philander, S.G.H., 1997. Interdecadal climate fluctuations that depend on exchanges between the tropics and extra-tropics. *Science*, Vol. 275, pp. 805–807.

Gubbins, D., 1981. The influence of extrinsic pressure changes on the earth's dynamo, *Phys. Earth Planet.* Inter., Vol. 33, pp. 255–259.

Gubbins, D., 1987. Mechanism for geomagnetic polarity reversals, *Nature*, Vol. 326, pp. 167–169.

Gubbins, D. and Bloxham, J., 1987. Morphology of the geomagnetic field and implications for the geodynamo, *Nature*, Vol. 325, pp. 509–511.

Haigh, J.D., 1996. The impact of solar variability on climate. *Science*, Vol. 272, pp. 981–984.

Hale, G.E., 1924. The law of sunspot polarity. *Proc. Natn. Acad. Sci.*, USA, Vol. 10, pp. 53–55.

Hallam, A., 1977. Secular changes in marine inundation of USSR and North America through the Phanerozoic, *Nature*, Vol. 269, pp. 769–772.

Hallam, A., 1984. The causes of mass extinctions, *Nature*, Vol. 308, pp. 686–687.

Hallam, A., 1989. The case for sea level changes as a dominant causal factor in mass extinction for marine invertebrates *Phil. Trans. Roy. Soc. London*, Ser B, Vol. 325, pp. 437–455.

Hallam, A., 1994. The earliest triassic as an anoxic event, and its relationship to the end Paleozoic mass extinction, *In*: Embry, A.F., Beauchamp, B. and Glass, D.J. (ed.). Pangea: *Global Environments and Research*. Canadian Soc. of Petrol. Geologists. Memo., Vol. 17, pp. 797–804.

Hallam, A. and Wignall, P. B., 1999. Mass extinctions and sea level changes. *Earth Sci. Rev.*, Vol. 48, pp. 217–250.

Hamilton, W.L., 1973. Tidal cycles of volcanic eruptions fortnightly to 19 years period. *Journ. Geophy Res*, Vol. 78, pp. 107–119.

Hammond, S.R., Seyb, S.M. and Theyer, F., 1979.Geomagnetic polarity transitions in two oriented sediment core from the northwest Pacific, *Earth Planet. Sci. Lett.*, Vol. 44, pp. 167.

Hameed, S. et al., 1983. An analysis of periodicities in the 1470 to 1974 Beijing precipitation record. *Geophy, Res. Letts.* Vol. 10, pp. 436–439.

Hanson, K., Brier, G.W. and Maul, G.A., 1989. Evidence of significant non-random behaviour in the recurrence of strong El Nino between 1525 and 1988. *Geophys. Res. Lett.*, Vol. 16, pp. 1181–1184.

Haq, B.U., Hardenhol, L. and Vail, P.R., 1987. Chronology of fluctuating sea levels since the triassic: *Science*, Vol. 235, pp. 1156–1167.

Haq, B.U. and Vail, P.R., 1988. Hardenbol, J. and Van Wagoner, J.G. Reply to comments by Haq et. al., *Science*, 241, pp. 596–597.

Harland, W.B., 1981. Chronology of the earth's glacial and tectonic record, *Jour. Geol. Soc. Lond.*, Vol. 138, pp. 197–203.

Harland, W.B., Armstrong, R.L., Cox, A.V., Craig, L.E., Smith, A.G. and Smith, D.G. 1990. *A Geologic Time Scale*, Cambridge University Press, pp. 263.

Harmuth, H.F., 1970. *Transformation of information by orthogonal functions.* Springer Verlag, New York. pp. 306.

Hay, W.W., 1985. Potential errors in the estimates of carbonate rocks occumulation through geological time. *In*: E.T. Sundquist and W.S. Broecker (eds.). The carbon cycle and atmospheric CO_2: Natural variation from Archean to present. *A.G.U. Publ. Monograph*, Vol. 32, pp. 573–583.

Hays, J. and Pitman, W., 1973. Lithospheric plate motion, sea level changes and climatic and ecological consequences, *Nature*, 246, pp. 18–22.

Hays, J.D., Imbrie, J. and Shackleton, N.J., 1976. Variations in the earth's orbit: Pacemaker of the ice ages. *Science*, Vol. 194, pp. 1121–1132.

Hazen, R. and Trefil, J., 1991. Science's 20 greatest hits take their lump. *Science*, Vol. 231, pp. 1308–1309.

Heckel, P.H., 1986. Sea level curve for Pensylvanian eustatic marine transgressive-regressive, depositional cycles along mid-continent outcrop belt, North America, *Geology*, 14, pp. 330–334.

Hein, J.R. et al., 1992. Variation in the time scale composition of a central ferromanganese crust: Palaeoceanographic implications, *Palaeoceanography,* 7, pp. 63–77.

Heinrich, H., 1988. Origin and consequences of cyclic ice rafting in the north east Atalantic ocean during the past 130,000 years. *Quart. Res.*, Vol. 29, pp. 143–152.

Heisler, J. and Termaine, S., 1986. The influence of the galactic tidal field on the Oort comet cloud *Icarus*, Vol. 65, pp. 13–26.

Herbert, T.D. and Fischer, A.G., 1986. Milankovitch climatic origin of mid-Cretaceous black shale rhythms in central Italy. *Nature*, Vol. 321, pp. 739–743.

Herschel, J.F.W., 1932. On the astronomical causes which may influence geological phenomena. Trans *Geol. Soc.* 2nd series Vol. 3, pp. 393–990.

Hess, H.H., 1962. History of Ocean basins. *In*: Engel A.E.J., James H.L. and Leonard, B.F., (eds) *Petrological studies*: a volume to honour A.F. Buddington, pp. 599–620. New York, Gel. Soc. Amer.

Heusser, L. E. and Sirocko, F., 1997. Millennial pulsing of environmental change in southern California from the past 24 k.y a record of ENSO event? *Geology*, Vol. 25, pp. 243–246.

Hilderbrand, A.R. and Boynton W.V., 1987. (*Abstract*) Eighth lunar and planetary science conference. pp. 427–428.

Hilgen, F.J., 1989. Periodicities of $CaCO_3$ cycles in the Pliocene of Sicily: discrepancies with the quasi-periods of the earth's orbital cycles? *Terra Nova*, Vol. 1, pp. 409–415.

Hilgen, F.J., 1991a. Astronomical calibration of Gauss to Matuyama sapropels in the Mediterranean and implication for the geomagnetic polarity time scale. *Earth Planet. Sci. Lett.*, Vol. 104, pp. 226–244.

Hilgen, F.J., 1991b. Extension of the astronomically calibrated (polarity) time scale to the Miocene/Pliocene boundary. *Earth Planet. Sci. Lett.*, Vol. 107, pp. 349–368.

Hilgen, F.J. et al., 1999. *Present status of astronomical (polarity) time scale for the Mediterranean late Neogene*. Phil. Trans. Roy. Soc. Lond. Vol. 357, pp. 1931–1947.

Hills, J., 1985. The passage of a 'Nemesis-like' object through the planetary system. *Astron. J.*, Vol. 90, pp. 1876–1882.

Hills, J., 1986. Deflection of comets and other long-period solar companions into the planetary system by passing stars. *In*: R. Smoluchowski, J. Bahcall and M. Matthews (eds.), *The Galaxy and the solar system*. Univ. Arizona, Tucson, pp. 397–408.

Hinov, L.A. and Park, J., 1998. Detection of astronomical cycles in the stratigraphic record by frequency modulation (FM) analysis. *Jour. Sed. Res.*, Vol. 68, pp. 524–539.

Hodell, D.A., Elmstrom, K M. and Kennett, J. P. 1986. Latest Miocene changes of global ice volumes, sea level changes and the Messinian salinity crisis, *Nature*, 320, pp. 411–414.

Hodell, A. et al., 2001. Solar forcing of drought frequency in the Maya lowlands. *Science*, Vol. 292, pp. 1367–1370.

Hoffert, M.I., Caldeira, K., Covey, C., Duffy, P.B. and Santer, B., 1999. Solar variability and anthropogenic climate change, *Nature*, 401, pp. 764.

Hoffman, A., 1988. Patterns of family extinction depend on definition and geological timescale, *Nature*, Vol. 315, pp. 659–662.

Hoffman, A., 1989. Mass extinctions : the view of sceptic., *Jour. Geol. Soc.*, London. Vol. 146, pp. 21–35.

Hoffman, Kenneth A., 1988. Ancient magnetic reversals: Clues to the geodynamo. *Scie. American*, pp. 50–60.

Holmes, A., 1929. *Radioactivity and earth movements*. Trans. Geol. Soc. Glasgow. Vol 18, pp. 559–606.

Holschneider, M., 1995. *Wavelets: An Analysis Tool*, Oxford University Press, Oxford.

Hong, Y.T. et al., 2001. A 6000-year record of changes in drought and precipitation in northeastern China based on a ^{13}C time series from peat cellulose. *Earth Planet Sci. Lett.*, Vol. 185, pp. 111–119.

Hsu, K.J., 1989. Catastrophic extinctions and the inevitability of the improbable. *Jour. Geol. Soc. London*, Vol. 146, p. 749–754.

Huang, J., Higuchi, K. and Shabbar, A. (1998). The relationship between the North Atlantic Oscillation and El Nino-Southern Oscillation, *Geophys. Res. Lett.*, Vol. 25, No.14, pp. 2707–2710.

Hubbard, B.B., 1996. *The World According to Wavelets: The Story of a Mathematical Technique in the Making*. A.K. Peters, Wellesley, Massachussetts, pp. 1–264.

Huffer 1999, *Nature*, Vol. 401, p. 764.

Hurrell, J.W. (1995). Decadal trends in the North Atlantic oscillation: regional temperatures and precipitation, *Science*, Vol. 269, pp. 676–679.

Hurrel, J.W. and Van Loon., H., 1997. Decadal variations in climate associated with the North Atlantic oscillations. *Climate Change*, 36, pp. 301–326.

Hut, P., 1984. How stable is an astronomical clock that can trigger mass extinctions on earth? *Nature*, Vol. 311, pp. 638–641.

Hut, P., Alvarez, W., Elder, W.P., Hansen, T., Kauffman, E.G., Keller, G., Shoemaker, E.M. and Weissman, P.R., 1987. Comet showers as a cause of mass extinctions. *Nature*, Vol. 329, pp. 118–126.

Ifeachor, E.C. and Jervis, B. W., 1993. *Digital signal processing: A practical approach,* Addison Wesley Harlow, pp. 1–760.

Imbrie, J., 1982. Astronomical theory of the Pleistocene ice ages: A brief historical review. *Icarus*, Vol. 50, pp. 408–422.

Imbrie, J. and Imbrie, K.P., 1979. *Ice Ages: Solving Mystery*, Enslow Publ., Short Hills, NJ.

Imbrie, J.and Imbrie, J. Z., 1980. Modeling the climate response to orbital variations, *Science*, Vol. 207, pp. 943–953.

Imbrie, J., Hays, J., Martinson, D.G., McIntyre, A., Mix, A.C., Morley, J.J. et al., 1984. The orbital theory of Pleistocene climate: Support from a revised chronology of the marine ^{18}O record. *In*: *Milankovitch and Climate,* Berger, A., Imbrie, J., Hays, J., Kukla, G. and Saltzman, B. (eds.), Reidel Publ. Co., Dordrecht, pp. 269–305.

Imbrie, J., McIntyre, A. and Mix, A., 1989. Oceanic response to orbital forcing in the late quaternary: Observational and experimental strategies. *In*: *Climate and Geo-Sciences, A Challenge for Science and Society in the 21st Century*, Berger, A., Schneider, S. and Duplessy, J.Cl. (eds.), Kluwer Academic Publishers, Dordrecht, pp. 121–164.

Imbrie, J., Berger, A., Shackleton, N.J., 1993. Role of orbital forcing: a two-million-year perspective. *In*: Eddy, J.A., Oeschger, H. (eds.), *Global changes in the perspective of the past.* Wiley, New York, pp. 263–279.

Ingle, S. and Coffin, M. F., 2004. Impact origin for the greater Ontong Java plateau? *Earth planet. Sci. Lett.,* Vol. 218, pp. 123–134.

Innanen, K.A., Patrick, A.T. and Duley, W.W., 1978. The interaction of the spiral density wave and sun's galactic orbit. *Astrophys. Space. Sci.*, Vol. 57, pp. 511–515.

Iorio, M., Tarling, D. H. and D'Argenio, B., 1998. Periodicities of magnetic intensities in magnetic anomaly profiles:The Cretaceous of the central Pacific. *Geophy Jour Intl.*, Vol. 133, pp. 233–244.

Irving, E. and Pullaiah, G., 1976. Reversals of the geomagnetic field, magnetostratigraphy and the relative magnitude of palaeosecular variation in the Phanerozoic. *Earth Sci. Rev.*, Vol. 12, pp. 35–64.

Isley, A.E. and Abbot, D.H., 2002.Implications of the temporal distribution of high Mg magmas for mantle plume volcanism through time. *The Jour Geol*, Vol. 110, pp. 141–158.

Ito, K., 1980. Chaos in the Rikitake two-disk dynamo system.*Earth and Planet Sci Lettt.* Vol. 51, pp. 451–456.

Ivanov, B.A. and Melosh, H.J. 2003. Impact do not initiate volcanic eruptions: eruptions close to the crater. *Geology*, Vol. 31, pp. 869–872.

Ivanov, S.S., 1994. Self-similarity of the sequence of magnetic field inversion. *Geomagnetism and Aeronomy*, Vol. 33, pp. 716–720.

Jablonski, D., 1986a. Background of mass extinctions: The alteration of macroevolutionary regimes. *Science*, Vol. 231, pp. 129–133.

Jablonski, D., 1986b. Causes and consequences of mass extinctions: A comparative approach. *In*: D.K. Elliot (ed.), *Dynamics of Extinction,* Chichester. Wiley, pp. 183–229.

Jacobs, J.A., 1994. *Reversals of the Earth's Magnetic Field*: Second edition Cambridge Univ. Press. First edn, 1984, Bristol, Adam Hilger, Ltd, pp. 320.

Jeffreys, H., 1924. *The earth*. 278 pp, Cambridge, Cambridge University Press.

Jenkins, G.M. and Watts, D.G., 1968. Spectral analysis and its Applications (Holden-Day, San Francisco).

Jin, F.F., Neelin, J.D. and Ghil, M., 1994. El Niño on the Devil's staircase: Annual sub harmonic steps to chaos. *Science*, Vol. 264, pp.70–72.

Jin, Rong-Sheng and Thomas, D.M., 1977. Spectral line similarity in the geomagnetic dipole field variations and length of Day fluctuations, *Jour. Geophys. Res.* 82, pp. 828–834.

Joly, 1923. *The surface history of the earth*: Oxford, Clarendon Press, pp. 192.

Jones, A.P., Price, G.D., Price, N.J., Decarli, P.S. and Clegg, R.A. 2002. Impact induced melting and development of large igneous provinces. *Earth Planet Sci. Lett.* Vol. 202, pp. 551–561.

Jones, 1977. Thermal interaction of the core and mantle and long-term behaviour of the geomagnetic field. *Jour. Geophy Res*. Vol.36, pp. 229–255.

Jouzel, J., 1999. Calibrating the isotopic paleothermometer, *Science,* 286, pp. 910–911.

Jouzel, J. et al., 1987. Vostok ice core: a continuous isotopic temperature record over the last climate cycle (160,000 years). *Nature*, Vol. 329, pp. 403–407.

Kaiho, K., 1994. Global changes of Paleogene aerobic/benthic foraminifera and deep sea circulation. *Paleogeography, Paleoclimatology, Paleoecology*, Vol. 83, pp. 65–85.

Kaiho, K., 1998. Global forcing of deep-sea benthic foraminiferal test size during the past 120 m.y., *Geology*, Vol. 26, No. 6, pp. 491–494.

Kaiho, K. and Saito, S., 1994. Oceanic crust production and climate during the last 100 m.y. Terra Nova vol. 6, pp. 376–384.

Kailas, S.V. and Narasimha, R., 2000. Quasi-cycles in monsoon rainfall by wavelet analysis, *Current Science*, 78(5), March, Vol. 10, pp. 592–595.

Kallen, E., Crafoord, C., Ghil, M., 1980. Free oscillations in a climate model with ice-sheet dynamics. *J. Atmos. Dyn.* 36, pp. 2292–2303.

Kapalan, A., Cane, M.A., Kushnir, Y., Clement, A.C., Blumenthal, M.B. and Rajgopalan, B., 1998. Analysis of global sea surface temperature 1856–1991. Jour. *Geophy. Res.* Vol. 98, pp. 20, 246–259.

Kaporovich, I.G., Makarova, Z.V., Petrova, G.N and Rybak, R.S. (1966). The transitional state of a geomagnetic field in the Paleocene on the tertiary of Turkmenia and Azerbaidzhan. *Izv. Earth. Phys*, 1, 59

Kasahara, J. 2002. Tides, Earthquakes and Volcanoes. *Science*, Vol. 297, pp. 348–349.

Kashiwaya, K. et al., 2001. Orbit related long term climate cycles revealed in a 12 m.y continental record from lake Baikal. *Nature*, Vol. 410, pp. 71–74.

Kawai, N., Otofuji, Y. and Kobayashi, K. (1976).Paleomagnetic study of deep –sea sediments using thinsections. *J.Geomag.Geoelect.*, 28, pp. 395.

Kelly, P.M., 1977. Volcanic dust veils and North Atlantic climate change. *Nature,* Vol. 268, pp. 616–617.

Kelly, P.M. and Wigley, T.M.L., 1990. The influence of solar forcing trends on global mean temperature since 1861. *Nature*, Vol. 347, pp. 460–462.

Kemper, E., 1987. Das Klima der Kreide- Zeit. *Geol. J.*, Vol. A96, pp. 5–185.

Kent, D.V. and Opdyke, N.D., 1977. Paleomagnetic field intensity recorded in a Brunhes epoch deep-sea sediment core, *Nature*, Vol. 266, pp. 156–159.

Kent, D. V., Cramer, B. S., Lanci, L., Wang, D., Wright, J.D. and Vander Voo, R., 2003a. A case for a comet trigger for the Palecene/Eocene thermal maximum and carbon isotope excursion. *Earth Planet. Sci. Lett.* Vol. 211, pp 13–26.

Kent, D.V., Cramer, B.S., Lanci, L., Wang, D., Wright, J.D. and Vander Voo, R., 2003b. Reply to comment on paper "A case for a comet trigger for the Palecene/Eocene thermal maximum and carbon isotope excursion". *Earth Planet. Sci. Lett.* Vol. 211, pp. 13–26. by Dickens, G. R. and Francis, J. M., 2003. *Earth Planet Sci. Lett.* Vol. 211, pp. 13–26.

Kerr, R.A., 1996. Ice rhythms: core reveals a plethora of climate cycles. *Science*, Vol. 274, pp. 499–500.

Kerr, R.A., 2000. A north Atlantic climate pacemaker for the centuries, *Science*, Vol. 288, pp. 1984–1986.

Kilston, S. and Knopoff, L., 1983. Lunar-Solar periodicities of large earthquakes in Southern California, *Nature*, Vol. 304, pp. 21–25.

Kitz, Norman., 1996. Radiation, mutation and extinction, *American Jour. Sci.*, Vol. 296, pp. 954–964.

Klitgaard-Kristensen, D., Sejrup, H. P. and Haflidason, H., 2001. The last 18 k.y. fluctuations in Norwegian sea surface conditions and implications for the magnitude of climate change:Evidence from the North sea. *Paleoceanography*, Vol. 16, pp. 455–467.

Knutson, T.R and Manabe, S., 1994. The impact of increased CO_2 and ENSO like phenomena in a coupled ocean-atmosphere. *Geophysical Research Letters*, Vol. 2, pp. 2295–2298.

Kodera, K., 2003. Solar influence on the spatial structure of the NAO during the winter 1900–1999. *Geophys. Res. Lett.*, Vol. 30, No. 4, pp. (24)1–4.

Koeberl, C. and MacLeod, Kenneth G., 2002. (eds.). Catastrophic events and mass extinctions: Impact and beyond. Spec Paper, *Geolo. Soc. America*, Boulder, Colorado, U.S.A.

Kono, M., 1987. Rikitake two- disk dynamo and paleomagnetism. *Geophy. Res Lett.* Vol 14, pp. 21–24.

Kumar, P. and Foufoula-Georgiou, E., 1994. Wavelet analysis in geophysics: An introduction. *In*: *Wavelets in Geophysics*, edited by E. Foufoula-Georgiou, and P. Kumar, pp.1–43, Academic, San Diego, California.

Kumar, P. and Foufoula-Georgiou, E., 1997. Wavelet analysis for Geophysical applications, *Reviews of Geophysics*, AGU, Vol. 35, No. 4, pp. 385–412.

Kutzbach, J.E., 1981. Monsoon climate of the early Holocene: climate experiment with the earth's orbital parameters for 9000 years ago. *Science* Vol. 214, pp. 59–61.

Labitzke, K. and van Loon, H., 1988. Association between the 11–year solar cycle, the QBO and the atmosphere. Part I: The troposphere and stratosphere of the Northern Hemisphere in winter. *J. Atmos. Terres. Phy.*, Vol. 50, pp. 197–206.

Labrecque, J.L., Kent, D.V. and Cande, S.C., 1977. Revised magnetic polarity time scale for late cretaceous and Cenozoic time: *Geology*, Vol. 5, pp. 330–335.

Lacoss, R.T., 1971. Data adaptive spectral analysis methods. *Geophysics* Vol. 36, pp. 661–675.

Lamb, H.H., 1972. *Climate—Present, Past and Future*. Methuen and Co., London.

Lambert, D., 1988. *The Cambridge Guide to the Earth*. Cambridge University Press.

Lamy, F., Hebbeln, D., Rohl, U. and Wefer, G., 2001. Holocene rainfall variability in southern Chile: a marine record of latitudinal shift of the southern Westerlies. *Earth. Planet. Sci. Lett.*, Vol. 185, pp. 369–382.

Larson, R.L., 1991. Geological consequences of super plumes. *Geology*, Vol. 19, pp. 963–966.

Larson, R.L., 1991. Latest pulse of earth: Evidence for mid-Cretaceous superplume. *Geology*, Vol. 19, pp. 547–550.

Larson, R.L. and Olson, P., 1991. Mantle control magnetic reversal frequency. *Earth Planet Sci. Lett.*, Vol. 107, pp. 437–447.

Laskar, J., 1989. A numerical experiment on the chaotic behaviour of the solar system. *Nature*, Vol. 338, pp. 237–238.

Laskar, J., 1990. The chaotic motion of the solar system. A numerical estimate of the chaotic zones. *Icarus*, Vol. 88, pp. 266–291.

Laskar, J., 1999. The limits of earth orbital calculations for geologic time scale use. *Phil. Tras. Roy. Soc. Lond.*, Vol. 357, pp. 1735–1759.

Lassen, K. and E.Friis-Christensen, 1995. Variability of the solar cycle length during the past five centuries and the apparent association with the terrestrial climate. *Journal of Atmospheric and Terrestrial Physics*, Vol. 57, No. 8, pp. 835–845.

Lean and Rind, 2001. Sun-Climate connections: Earth's response to a variable sun, *Science*, Vol. 292, pp. 234–236.

LeTreut, H. and Ghil, M.J., 1983. Orbital forcing, climatic interactions and glaciation cycles, *J. Geoph. Res.*, 8, pp. 5167–5190.

Levender, Amy. et al., 1996. Productivity cycles of 200–300 years in the Atlantic Peninsula region: Understanding linkages among the sun, atmosphere, ocean, sea ice and biota. *Geol. Soc. America. Bull*: Vol. 108, pp. 1626–1644.

Lewin, R., 1988. Pattern and process in extinction. *Science* Vol. 241, pp. 26.

Linsley, B.K. and Thunell, R.C., 1990. The record of deglaciation in the Sulu Sea: Evidence for the Younger Dryas event in the tropical western Pacific, *Palaeooceanography 7*, Vol. 5(6), pp. 1025–1039.

Linsley, B.K., Wellington, G. M. and Schrage, D.P., 2000. Decadal sea surface temperature variability in the sub-tropical South Pacific from 1726 to 1997 A.D., *Science*, Vol. 290, pp. 1145–1148.

Lisitzin, E., 1974. *Sea level changes*, Elsevier, Amsterdam.

Loper, D.E., 1984. Structure of the core and lower mantle, *Adv. Geophys.*, Vol. 26, pp. 1–34.

Loper, D.E., 1991. Mantle plumes. *Tectonophysics*, Vol. 187, pp. 73–384.

Loper, D.E. and Stacey, F.D., 1983.The dynamical and thermal structure of deep mantle plumes. *Phys. Earth Planet Inter*, Vol. 33, pp. 304–317.

Loper, D.E. and Eltayeb, I.A.,1985. On the stability of the D' layer, *Geophy. Astro. Physics*. Fluid Dynamics, Vol. 36, pp. 229–255.

Loper, D.E. and McCartney, K., 1986. Mantle plumes and the periodicity of magnetic reversal. *Geophysical Research Letters*, Vol. 13, pp. 525–528.

Loper, D.E., McCartney, K. and Buzyana, G., 1988. A model of correlated episodicity in magnetic field reversals, climate and mass extinctions, *The Journal of Geology*, Vol. 96, pp. 1–15.

Lorenz, E.N., 1993.The essence of chaos. London UCZ Press Limited.

Loutre, M.F. and Berger, A., 1989. Pre-quaternary amplitudes in the expansion of the obliquity and climatic precession. *Scientific Report* 1989/4, Institut d'Astronomie et de Geophysique G. Lemaitre, Universite Catholique de Louvain, Louvain-la-Neuve.

Loutre, M.F., Berger, A., Bretagnon, P., Blanc, P.L., 1992. Astronomical frequencies for climate research at the decadal to century time scale. *Climate Dynamics*, Vol. 7, pp. 181–194.

Lowrie, W. and Alvarez, W., 1981. One hundred million year polarity history, *Geology*, Vol. 9, pp. 392–397.

Luc-Beaufort, 1994. Climate importance of the modulations of the 100 k.y. cycle inferred due to 16 m.y. long Miocene records, *Palaeoceanography*, 9, pp. 821–834.

Luc Beaufort, et. al., 1997. Insolation cycles as a major control of equatorial Indian Ocean primary production, *Science*, Vol. 278, pp. 1451–1454.

Lull, 1929.The pulse of life. *In*: *Organic evolution*. New York, Macmillan, pp. 693–698.

Lumsden, D.N., 1985. Secular variations in the dolomite abundance in deep marine sediments. *Geology*, Vol. 13, pp. 766–769.

Lutz, T.M., 1985. The magnetic record is not periodic, *Nature*, Vol. 317, pp. 404–407.

Lutz, T.M. and Watson, G.S., 1988. Effect of the long-term variation on the frequency spectrum of the geomagnetic reversal record, *Nature*, Vol. 334, pp. 240–242.

Lyle, M., Leinen, M., Owen, R.M. and Rea, D.K., 1987. Late tertiary history of hydrothermal deposition at the east Pacific rise, 19°S: correlation to volcano-tectonic events. *Geophy. Res. Lett.* Vol. 14, pp. 995–998.

Maasch K.A. and Saltzman, B., 1990. A low-order dynamical model of the global climatic variability over the full Pleistocene. *J. Geophys. Res.* 95, 1955–1963.

Macdougall, J.D., 1988. Seawater Strontium isotopes, acid rain and the Cretaceous-Tertiary boundary, *Science*, Vol. 239, pp. 485–487.

Machado, F., 1967. Geological evidence for a pulsating gravitation. *Nature*, Vol. 214, pp. 1317–1318.

Mackereth, F.J.H., 1971. On the variation in direction of the horizontal component of remnant magnetization in Lake Sediments, *Earth Planet Sci. Letters*, Vol. 12, pp. 332–338.

Maksimov, I.V., Saruhanjan, E.I. and Smirnov, N.P., 1970. Oceans and Cosmos (in Russian) (Gidrometeoizdat, Leningrad).

Mallat, S., 1998. *A wavelet tour of signal processing*. Academic Press, London, pp. 1–577.

Mann, M.E. and Lees, J.M., 1996. Robust estimation of background noise and signal detection in climatic time series. *Climate Change*, Vol. 33, pp. 409–445.

Mantanari, A., Bagatin, A.C. and Farinella, P., 1998. Earth cratering record and impact energy flux in the last 150 m.y. *Planet. Spac. Sci.*, Vol. 46, pp. 271–281.

Marshal, Charles R., 1998. Mass extinction probed. *Science*, Vol. 392, pp. 17–19.

Marzocchi, W. and Mulargia, F., 1990. Statistical analysis of the geomagnetic reversal sequences, *Phys. Earth Planet. Inter.* Vol. 61, pp. 149–164.

Marzocchi, W. and Mulargia, F. 1990. Periodicity in geomagnetic reversals, IUGG, Vienna 11–24, August, 1991, *Program Abst.*, p. 20.

Marzocchi, W. et al., 1995. Ritikate's geodynamo model analysed in term of classical time series statistics. *Phy. Earth. Planet. Inter.* Vol. 88, pp. 83–88.

Matese, J.J., Whitman, D.G., Innanen, K.A. and Voltanen, M.J., 1995. Periodic modulation of the Oort cloud comet flux by the adiabatically changing galactic tide, *Icarus*, Vol. 116, pp. 255–268.

Mathews, R.K., 1988. Comments on paper by Haq et al. (1987). *Science* 235, pp. 1156–1167, *Science* 241, pp. 597–599.

Mathew, R.K. and Poore, R. J., 1980. Tertiary D180 record and glacio-eustatic sea level fluctuations, *Geology*, 8, pp. 501–504.

Maunder, E.W., 1922. The prolonged sunspot minimum 1645–1715. *J. Br. Astr. Ass.*, Vol. 32, pp. 140–145.

Mayewski, P.A., Meeker, L.D., Twickler, M.S., Whitlow, S., Yang, Q., Lyons, W.B. and Prentice, M., 1997. Major features and forcing of high latitude northern hemispheric circulation using a 110,000-year long glaciochemical series. *J. Geophy. Res.*, 102, pp. 26,345–26,366.

Mazaud, A., Laj, C., de Seze, L. and Verosub, K.L., 1983. 15-m.y. periodicity in the frequency of geomagnetic reversals since 100 m.y., *Nature*, Vol. 304, pp. 328–330.

Mazaud, A., Laj, C., de Seze and Verosub, K.L., 1984. Reply to McFadden, *Nature*, Vol. 311, p. 396.

Mazaud, A. and Laj, C., 1991. The 15 m.y. geomagnetic reversals periodicity: a quantitative test. *Earth Planet Sci. Lett.* Vol. 107. pp. 689–696.

Mc Phaden, M.J. and Yu, X., 1999. Equatorial waves and the 1997–98 El Nino. *Geophys. Res. Lett.*, Vol. 26, pp. 2961–2964.

McCrea, W.H., 1975. Ice ages and the Galaxy. *Nature,* Vol. 255, pp. 607–609.

McCrea, W.H., 1981. Long time scale fluctuations in the evolution of the Earth. *Proc. Roy. Soc. London* A, Vol. 375, pp. 1–35.

McElhinny, M.W., 1971. Geomagnetic reversals during phanerozoic, *Science*, Vol. 182, pp. 157.

McFadden, B.L., 1984, 15-m.y. periodicity in the frequency of geomagnetic reversals since 100 m.y., *Nature*, Vol. 311, p. 396.

McFadden, P.L. and Merrill, R.T., 1984. Lower mantle convection and geomagnetism, *J. Geophys. Res.*, Vol. 89, pp. 3354–3362.

McFadden, P.L., 1987. A periodicity of magnetic reversals: Comments, *Nature*, Vol. 330, pp. 27.

McFadden, P.L. and Merrill, R.T., 1995. History of earth's magnetic field and possible connection to core mantle boundary processes. *JGR*, Vol. 100, pp. 307–316.

McManus, J.F., Bond, G.C., Broecker, W.S., Johnsen, S., Labeyrie, L. and Higgins, S.M., 1994. High resolution climate records from the North Atlantic during the last interglacial. *Nature*, Vol. 371, pp. 326–329.

McManus, J.F., Bond, G.C., Broecker, W.S., Fleisher, M.Q. and Higgins, S.M., 1998. Radiometrically determined fluxes in the sub-polar North Atlantic during the last 140, 000 years. *Earth Planet. Sci. Lett.*, Vol. 135, pp. 29–43.

Meehl, G.A. et al., 1993. Tropical Pacific interannual variability and CO_2 climate change, *Journal of Climate*, Vol. 6, pp. 42–63.

Meier, M. F., 1990. Reduced rise in sea level. *Nature*, Vol. 343, pp. 115–116.

Melnyk, D.H. and Smith, D.G., 1989. Outcrop to subsurface cycle correlation in the Milankovitch frequency band: Middle Cretaceous, Central Italy. *Terra Nova*, Vol. 1, pp. 432–436.

Melosh, H.J., 1989. Impact cratering : A geological process. Oxford Univ. Press, New York, pp. 245.

Members of scientific planning working group, 2001. *Earth, Ocean and Life: Scientific investigation of the earth system using multiple drilling programs and new technologies*, Integrated ocean drilling program, Initial Science Plan, pp. 2003–2013.

Merrill, R.T. and McFadden, P.L., 1988. Secular variation and the origin of geomagnetic field reversals, *J. Geophys. Res.*, Vol. 93, pp. 589.

Merrill, R.T. and McFadden, P.L., 1990. Paleomagnetism and the nature of the geodynamo, *Science*, Vol. 248, pp. 345–350.

Milankovitch, M., 1941. *Canon of insolation in the ice age problem.* (English translation by Israel program for scientific translation, Jerusalem 1969) R. Serbian Acad. Spec. Publ. 132.

Minobe, S., (1997). A 50–70 year climatic oscillation over the North Pacific and North America, *Geophys. Res. Lett.*, Vol. 24, No. 6, pp. 683–686.

Mitchell, J.M. Jr., 1976. Climatic variability and its causal mechanism. Quat. Res. Vol. 6, pp. 681–693.

Mitrovica, J. X., Forte, A. M. and Pan, R., 1997. Glaciation induced variations in the earth precession frequency, obliquity and insolation over the last 2.6 m.y. *Geophy Jour. Int.* Vol. 128, pp. 270–284.

Mitton, S.M. (ed.), 1977. *Cambridge Encyclopedia of Astronomy*. Cambridge University Press.

Mix, A.C., Le, J., Shackleton, N.J., 1995a. Benthic foraminiferal stable isotope stratigraphy of site 846:0–1.8 Ma. *In*: Pisias, N.G., Mayer, L., Janecek, T., Palmer-Julson, A., van Andel, T.H. (Eds.), *Proc. ODP, Sci. Res.*, Vol. 138, pp. 839–856.

Mix, A.C., Pisias, N.G., Rugh, W., Wilson, J., Morley, A. and Hagelberg, T. 1995b. Benthic foraminiferal stable isotope record from site 849, 0–5 m.y.: local and global climate changes: *In*: Pisias N.G., L. Mayer, T. Tanecek, A. Palmer-Julson, T.H. Van Andel (eds). Proc. O.D.P. Scientific results 138, College Station, T_X (ocean drilling program) 371–412.

Molnar, P. and Cane, M. A., 2002. El Nino's tropical climate and teleconnections as a blueprint for pre-ice age climate. *Paleoceanography*, Vol. 17, pp. 11; 1–11.

Morgan, J.P., Reston, T.J. and Ranero, C.R., 2004. Contemporaneous mass extinction, flood basalt and impact signal: are mantle plume induced lithosphere gas explosion the causal link? *Earth Planet. Sci. Lett.*, Vol. 217. pp 263–284.

Morgan, W. J., 1971. Convection plumes in the lower mantle. *Nature*, Vol. 230, pp. 42–43.

Morlet, J., Arens, G., Fourgeau, E. and Girard, D., 1983. Wave propagation and sampling theory, 2. Sampling theory and complex waves, *Geophysics*, Vol. 47, pp. 203–221.

Morris, D.E. and Muller, R.A., 1986. Tidal gravitational forces: The infall of 'new' comets and comet showers. *Icarus*, Vol. 65, pp. 1–12.

Mulkus, M.V.R., 1966. Precession of the earth as a cause of geomagnetism *Science,* Vol. 160, pp. 259–264.

Muller, R.A. and McDonald, G.J., 1995 Glacial cycles and orbital inclination. *Nature*, Vol. 377, pp. 107–108.

Muller, R.A. and McDonald, G.J., 1997a. Glacier cycles and Astronomical forcing, *Science*, Vol. 277, pp. 215–218.

Muller, R.A. and McDonald, G.J., 1997b. Simultaneous presence of orbital inclination and eccentricity in proxy climate records from ocean drilling program site 806. *Geology*, Vol. 25, pp. 3–6.

Muller, R.A. and Morris, D.E., 1986. Geomagnetic reversals from impacts on the earth, *Geophys. Res. Lett.*, Vol. 13, pp. 1177–1180.

Murakami, M. and Miyazaki, S., 2001. Periodicity of the strain accumulation detected by permanent GPS array: Possible relationship to seasonality of major earthquakes' occurrences. *Geophysical Research Letters*, 28(15), pp. 2983–2986.

Murakoshi, N., Nakayama, N. and Masuda, F., 1995. *Diurnal inequality pattern of tide in the upper Pleistocene.*

Naidu, P.D. and Malmgren, B.A. A 2,200 years periodicity in the Asian monsoon system. *Geophys. Res. Lett.* Vol. 22, pp. 2361–2364.

Nair, R.R., Ittekkot, V., Manganini, S.J., Ramaswamy, V., Haake, B., Degens, E.T., Desai, B.N. and Honjo, S., 1989. Increased particle flux to the deep ocean related to monsoons, *Nature*, Vol. 338, pp. 749–751.

Nance, R.D., Worsley, T.R. and Moody, J.B., 1988. The super continental cycle. *Sci. American*, July issue, p. 4451.

Napier, W.M. and Clude, S.V.M., 1979. A thoery of terrestrial catastrophism. *Nature*, Vol. 282, pp. 455–459.

Napier, W.M., 1989. Terrestrial Catastrophism and galactic cycles: *In*: Catastrophes and Evolution, Astronomical Foundation (ed.) S.V.M. Clube, Cambridge Univ. Press, Cambridge. pp. 133–167.

Napier, W.M., 1998. NEO_S and impacts: The galactic connection. *Celestial Mechanics and Dynamical Astronomy*. Vol. 69, pp. 59–75.

Napoleone, G. and Ripepe, M., 1989. Cyclic geomagnetic changes in Mid-Cretaceous rhythmites, Italy. *Terra Nova*, Vol. 1, pp. 437–442.

Neelin, J.D., 1990. A hybrid coupled general circulation model for El Nino studies, *J. Atmos. Sci.*, Vol. 47, pp. 674–693.

Neelin, J.D. and Latif, M., 1998. El Nino dynamics, *American Institute of Physics Today*, December issue, pp. 32–36.

Neff, U. et al., 2001. Strong coherence between solar variability and the monsoon in the Oman between 9 and 6 k.y. ago. *Nature*, Vol. 411, pp. 290–293.

Negi, J.G. and Tiwari, R.K., 1983. Matching long term periodicities of geomagnetic reversals and galactic motions of the solar system, *Geophys. Res. Lett.*, Vol. 10, pp. 713–716.

Negi, J.G. and Tiwari, R.K., 1984. Periodicities of paleomagnetic and paleoclimatic variations: a Walsh spectral approach. *Earth Planet Sci. Lett.* Vol. 70, pp. 137–147.

Negi, J.G. and Tiwari, R.K., 1988. Climate and sea level variation cycles during the last 8,000 years and their relationship with the global human history. *Proceedings of First Indian Conference on Marine Archaeology of Indian Ocean Countries*, Ed: S.R. Rao, pp. 79–82.

Negi, J.G., Tiwari, R.K. and Rao, K.N.N., 1990a, 'Clean' spectral analysis of long-term sea-level changes, *Tera Nova*, Vol. 2, pp. 138–141.

Negi, J.G. and Tiwari, R.K., 1990b. Periodicities in natural remanent magnetization and paleoclimatic record decoded by Walsh spectrum. *Pure and Appli. Geophysics.*, Vol. 132, pp. 767–779.

Negi, J.G., Tiwari, R.K. and Rao, K.N.N., 1993. Comparison of Walsh and Fourier spectroscopy of geomagnetic reversals and non-sinusoidal paleoclimate time series. IEEE. *Trans. Geosci. Remot. Sensing.* Vol. 31, pp. 127–137.

Negi, J.G., Tiwari, R.K. and Rao, K.N.N., 1996. Clean periodicity in secular variations of dolomite abundance in deep marine sediments. *Marine. Geol.*, Vol. 133 pp. 113–121.

Nicolis, C., 1982. Stochastic aspects of climatic transitions response to a periodic forcing, *Tellus*, Vol. 34, pp. 1–9.

Noren, A. J. et al., 2002. Millenium scale storminess variability in the northern eastern United States during the Holocene epoch, *Nature*, Vol. 419, pp. 821–824.

Nowroozi, A.A., 1967. Table for Fisher's test of significance in harmonic analysis. *Geophy Jour. Roy Astro. Soc.* Vol. 12, pp. 512–520.

O'Keefe, J.D. and Ahrens, T.J., 1987. The interaction of the Crctaceous-tertiary extinction bolide with the atmosphere, ocean and solid earth. *In*: T. Silver and P.H. Schultz (eds.), Geological implications of the impact of large asteroids and comets on the earth. *Geol. Soc. Am. Spec. Pap.*, Vol. 190, pp. 103–120.

Officer, C.B. and Drake, C.L., 1983. The Cretaceous-tertiary transition, *Science*, Vol. 219, pp. 1383–1390.

Officer, C.B. and Drake, C.L., 1985. Terminal Cretaceous environmental events, *Science*, Vol. 227, pp. 1161–1167.

Officer, C.B., Hallam, A., Drake, C.L. and Devine, J.D., 1987. Late cretaceous and paroxysmal Cretaceous/Tertiary extinctions, *Nature*, Vol. 326, pp. 143–149.

Okal, E. and Anderson, D.L., 1975. On the planetary theory of sunspots. *Nature*, Vol. 253, pp. 511–513.

Oliver, R., Ballester, J.L.and Baudin, F., 1998. Emergence of magnetic flux on the sun as the cause of 158 day periodicity in sunspot areas. *Nature*, Vol. 394, pp. 552–553.

Olsen, P. and Nam, I.S., 1986. Formation of seafloor swells by mantle plumes. *JGR*, Vol. 91, pp. 7181–7191.

Olsen, P.E., 1986. A 40-million year lake record of early Mesozoic orbital climatic forcing. *Science*, Vol. 234, pp. 842–848.

Olsen, P. E. and Kent, D. V., 1999. *Long period Milankovitch cycles from the late Triassic and early Jurrassic of eastern North America and their implications for calibration of the early Mesozoic time scale and the long-term behaviour of the planet.* Phil. Trans. Roy. Soc Lond. A. Vol. 357, pp. 1761–1788.

Oort, J.H., 1950. The structure of the cloud of comets surrounding the solar system and a hypothesis concerning their origin. *Bull. Astron. Inst. Neth.*, Vol. 11, pp. 91–110.

Opik, E.J., 1965. Climate change in cosmic perspective, *Icarus*, Vol. 4, pp. 289.

Opik, E.J., 1976. *Interplanetary Encounters*. Elsevier, Amsterdam.

Otnes, R.K. and Enochson, L., 1978. Applied time series analysis V 1, *Basic techniques*, Wiley, Chichester, pp. 1–449.

Overpeck, J.T. 1995. Paleoclimatology and climate system dynamics Rev. *Geophy. Supplement*, pp. 863–871. U.S. Nat. Rep. IUGG 1991–1994.

Owen, R.M. and Rea, D. K. 1985. Sea floor hydrothermal activity links climate to tectonics: the Eocene CO_2 greenhouse. *Science*, V27 7 pp. 167–169.

Pal, P.C. and Creer, K.M., 1986. Geomagnetic reversal spurts and episodes of extraterrestrial catastrophism. *Nature*, Vol. 320, pp. 148–150.

Paleo-Tokyo Bay: Reconstruction from tidal deposits and growth-lines of fossils vivalves, *Spec Publ Int Assoc. Sed.*, Vol. 24, pp. 289–300.

Palike, H., Shackleton, N.J. and Rohl, U. 2001.Astronomical forcing in late Eocene marine sediments. *Earth Planet Sci. Lett*, Vol. 193, pp. 589–602.

Pandey, O.P. and Negi, J.G., 1987. Global volcanism, biological mass extinctions and the galactic vertical motion of the solar system, *Geophys. J.R. Astr. Soc.*, Vol. 89, pp. 857–867.

Pardo-Iguzquiza, E., Chica Olmo, M. and Rodriguez–Tovar, F. J., 1994, CYSTRATI: a computer program for spectral analysis of stratigraphic succession, *Compt. Geosci*, Vol. 20 pp. 511–584.

Park, J. and Herbert, T.D., 1987. Hunting for the paleoclimatic periodicities in a geologic time series with an uncertain time scale. *Jour. Geophysical research*, Vol. 92, pp. 14027–14040.

Park, J., Lindberg, C.R. and Vernon, F.L.I., 1987. Multitaper spectral analysis of high-frequency seismograms, *J. Geophys. Res.*, 92, 12,675–12,684.

Park, J., et. al., 1993. Late Cretaceous precessional cycles in double time: A warm earth Milankovitch response, *Science*, Vol. 261, pp. 1431–1434.

Parker, E.N., 1999. Sunny side of global warming. *Nature*, Vol. 399, pp. 416–417.

Paterne, M. et al., 1990. Fluctuations of companion explosive volcanic activity (South Italy) during past 190,000 years, as determined by marine tephrochronology. *Earth Planet Scie Lett.*, Vol. 98, pp. 166–174.

Patterson, T.R. and Fowler, A.D., 1996. Evidence of self organization in planktic foraminiferal evolution: implications for interconnectedness of paleoecosystems. *Geology*, Vol. 24, pp. 215–218.

Paul, A. and Schulz, M., 2002. Holocene climate variability on centennial-to-millenial time scales: 2. Internal feedbacks and external forcings as possible causes. *In*: G. Wefer, W.H. Berger, K.E. Behre, E. Jansen (Eds.), *Climate development and history of the North Atlantic Realm*, Springer-Verlag, Berlin, pp. 55–73.

Percival, D.B. and Walden, A.T., 1993. *Spectral analysis for physical application multitaper and conventional univariate techniques*. Cambridge University Press, Cambridge, pp 1–583.

Pestiaux, P.,Duplssy, J.C. and Berger, A., 1988. Paleoclimatic variability at frequencies ranging from 10^{-4} cycle/year to 10^{-3} cycle/year. Rampino M. R., Sanders. J., Newman, W.S. and Konigsson, L.K. (eds.) Evidence for non-linear behaviour of the climate system. *In*: *Climate, History, Periodicity and Predictability*, pp. 285–299.

Peterson, L.C., Overpeck, J.C., Kipp, N.G. and Imbrie, J., 1991. A high resolution lake quarternary upwelling record from the anoxic cariaco basin, Venezuela., *Paleoceanography*, Vol. 6, pp. 99–119.

Philander, G. (1989). El Nino and La Nina, *American Scientist*, Vol. 77, September–October, pp. 451–459.

Philander, S.G. and Fedorov, A., 2003. Is El Nino sporadic or cyclic? *Ann. Review of Earth and Planet Scie.* Vol. 31, pp. 579–594.

Pike, J. and Kemp, A.E.S., 1997. Early Holocene decadal scale ocean variability recorded in Gulf of California laminated sediments. *Paleoceonography*, Vol. 12, pp. 227–238.

Pirrazzoli, P.A., 1989. Present and near future global sea level changes. *Paleogeography, Paleoclimatology, Paleoecology (Global and planetary changes section)*. Vol. 75, pp. 241–258.

Pisias, N.G., 1978. Palaeoceanography of the Santa Barbara Basin during the last 8000 years. *Quaternary Research*, Vol. 10, pp. 366–384.

Plotnick, R. and McKinney, M., 1993. *Ecosystem Organization and Extinction Dynamics*. Paios, Vol. 8, pp. 202–212.

Poirier, J.P., 1992. *Introduction to the physics of the Earth's interior*. Cambridge Univ. Press. New York.

Pollard, D., 1982. A simple ice-sheet model yields realistic 100 k.y. glacial cycles. *Nature* 296, pp. 334–338.

Poore, H.J., Dowesett, R.Z., Verardo S. and Quinn, T. M., 2003. Millenial to century scale variability in the Gulf of Mexico Holocene climate records. *Paleoceanography*, Vol. 18, pp. 26; 1–13.

Poore, R.Z., Quinn T.M., and Vorardo, S., 2004. Century scale movement of the Atlantic inter tropical convergence zone linked to solar variability. *Geophy. Res. Lett.* Vol. 31, L12214 doi:10.1029/2004 GL 019940.

Porter, S.C. and Zhisheng, A., 1995. Correlation between climate events in the North Atlantic and China during the last glaciations. *Nature,* Vol. 375, pp. 305–308.

Price, N.J., 2001. *Major Impacts and Plate Tectonics: A Model for Phanerozoic Evolution of Earth's Lithosphere.*, Routledge, London and New York, p. 354.

Priestley, M.B., 1981. *Spectral Analysis and Time Series*, Academic Press, London, pp. 1–890.

Prokoph, A. and Barthelmes, F., 1996. Detection of non-stationarities in geological time series: Wavelet transform of chaotic and cyclic sequences. *Comp. and Geosci.*, Vol. 22, pp. 1097–1108.

Prokoph, A., Flower, A.D. and Patterson, R.T., 2000. Evidence for periodicity and non-linearity in a high-resolution fossil record of long term evolution, *Geology*, Vol. 28, No.10, pp. 867–870.

Prokoph, A., Earnst, R.E. and Buchan, K.L., 2004. Time series analysis of large igneous provinces: 3500 m.y. to present. *The Journ. Geol.* Vol. 112, pp. 1–22.

Rahmstorf, S., 2003. Timing of abrupt climate change: A precise clock. *Geophys. Res. Lett.,* Vol. 30, pp. 17; 1–4.

Rampino, M.R., 1987. Impact cratering and flood basalt volcanism, *Nature*, Vol. 327, p. 468.

Rampino, M.R. and Stothers, R.B., 1984a. Terrestrial mass extinctions, cometary impacts and the sun's motion perpendicular to the galactic plane. *Nature*, Vol. 308, pp. 709–712.

Rampino, M.R. and Stothers, R.B., 1984b. Geological rhythms and cometary impacts, *Science*, Vol. 226, pp. 1427–1431.

Rampino, M.R. and Stothers, R.B., 1988. Flood basalt volcanism during the past 250 million years, *Science*, Vol. 241, pp. 663–668.

Rampino, M.R. and Caldeira, K., 1993. Major episodes of geologic changes: Correlation, time structure and possible causes. *Earth. Planet. Sci. Lett.,* Vol. 114, pp. 215–227.

Rampino, M.R. and Volk, T. 1996. Multiple impact event in the Paleozoic: Collision with a string of comets or asteroids? *Geophy. Res Lett.* Vol. 23, pp. 49–52.

Rampino, M.R. and Caldeira, K., 1997a. Episodes of terrestrial geologic activity during the past 260 m.y.: A quantitative approach. *In*: J. L. Remo (ed.) *Annals of the New York Academy of Sciences, Near Earth Objects*: The United Nations International conference Vol. 822, pp. 144–159.

Rampino, M.R. and Caldeira, K., 1997b. J. L. Remo (ed.). A unified theory of near earth objects: In: *Annals of the New York Academy of Sciences*: The United Nations International Conference, Vol. 822, pp. 403–430.

Rampino, M.R. and Adler, A.C., 1998. Evidence for abrupt latest Permian mass extinction of foraminifera: result of tests for the Signore- Lipps effect. *Geology*. Vol. 26, pp. 415–418.

Rasmusson, E.M., Wang, X. and Ropelewski, C.F. 1990. The biennial component of ENSO variability, *J. Mar. Syst.*, Vol. 1, pp. 71–96.

Raup, D.M., 1985a. Magnetic reversals and mass extinctions, *Nature*, Vol. 314, pp. 341–343.

Raup, D.M., 1985b. Rise and fall of periodicity. *Nature*, Vol. 317, pp. 384–385.

Raup, D.M., 1986. Biological extinction in the earth history. *Science*, Vol. 231, pp. 1528–1533.

Raup, D.M. and Sepkoski, J.J., 1984. Periodicity of extinctions in the geologic past, *Proc. Natl. Acad. Sci.* USA, Vol. 81, pp. 801–805.

Raup, D.M. and Sepkoski, 1986. Periodic extinction of families and genera: *Science*, Vol. 31, pp. 833–836.

Raup, D.M. and Sepkoski, Jr., J.J., 1988. Testing for periodicity of extinction. *Science*, Vol. 241, pp. 94–96.

Raymo, M.E., 1997. The timing of major climate terminations, Paleoceanography 12 (4), pp.577–585.

Raymo, M.E., 1998. Millenial scale climate instability during the early Pleistocene epoch. *Nature*, Vol. 392, pp. 699–702.

Rea, D.K., Zachos, J.C., Owen, R.M. and Gingerich, P.D. 1990. Global hange at the Paleocene-Eocene boundary: Climatic and evolutionary consequences of tectonic events, *Paleogeogr, Paleoclim, Paleoecol.*, Vol. 79, pp. 117–128.

Retallack, G.J., 1995. Permian–Triassic life crisis on land. *Science*, Vol. 267, pp. 77–80.

Retallack, G.J., 2002. Triassic –Jurrassic atmospheric CO_2 spike. *Nature*, Vol. 415, pp. 387–389.

Reynolds, R.W. and Smith, T.M., 1994. Improved global sea surface temperature analyses, *Journal of Climate*, Vol. 7, pp. 929–948.

Rial, J.A., 1999. Pace making the ice ages by frequency modulation of earth's orbital eccentricity. *Science*, Vol. 285, pp. 564–568.

Rial, J.A. and Anaclerio, C.A., 2000. Understanding the non-linear responses of climate system to orbital forcing. *Quat. Sci. Rev.* Vol. 19, pp. 437–440.

Richards, et al., 1989. Flood basalts and hot spot tracks: Plume heads and tails. *Science*, Vol 246, pp. 103–107.

Ridgwell, A. J., Watson, A. J. and Raymo, M. E., 1999. Is the spectral signature of 100 k.y. glacial cycle consistent with a Milankovitch origin? *Paleoceanography*, Vol. 14, pp. 437–440.

Rikitake, T., 1958. Oscillations of a system of disk dynamos. Proce. Cambridge Philo's Ser, Vol. 54, pp. 89–105.

Rind, D., 1999. Complexity and climate. *Science*, Vol. 284, pp. 105–107.

Rittentour, T.M., Brigham-Crette, J. and Mann, M.E., 2000. El Nino-like climate teleconnections in New England during the late Pleistocene, *Science*, Vol. 288, pp. 1039–1042.

Roberts, P.H., 1988. Future of geodynamo theory. *Geophys. Astrophy. Fluid dynamics*, Vol. 42, pp. 3–31.

Robert, D.H., Lehar, J. and Dreher, J.W., 1987. Time series analysis with CLEAN. I. Derivation of spectra. *Astron. J.*, 93(4): 968–989.

Rogers, J.G., 1984. The association between the North Atlantic Oscillations and the Southern Oscillation in the Northern Hemisphere, *Mon. Weath. Rev.*, Vol. 112, pp. 1385–1404.

Roig, Fidel A. et al., 2001. Climate variability 50,000 years ago in mid-latitude Chile as reconstructed from tree rings. *Nature*, Vol. 410, pp. 567–570.

Rosen, R.G. et al., 1976. Earth tides, volcanos and climatic change. *Nature*, Vol. 261, pp. 680–682.

Royer, J.F., Deque, M. and Pestiaux, P. 1983. Orbital forcing of the inception of the Lourentide ice sheets, *Nature*, Vol. 304, pp. 43–46.

Rubicam, D.P., 1995. Has climate changed the earth's tilt ? *Paleoceanography*, Vol. 10, pp. 365–372.

Ruddiman, W.F., 1977. Late quarternary deposition of ice rafted sand in the subpolar North Atalantic (latitude 40° N to 65°N) *Geol. Soc. Am. Bull.*, Vol. 88, pp. 1813–1827.

Ruddiman, W.F. and McIntyre 1981. Oceanic mechanisms for amplification of the 23,000 years ice volume cycle. *Science*, Vol. 212, pp. 617–627.

Ruzmaikin, A. and Feynman, J., 2002: Solar Influence on a Major Mode of Atmospheric Variability. *J. Geophys. Res.*, 107, 4209–4219, doi:10.1029/2001JD001239.

Saito, T. and Kobayashi, K., 1989. Long period secular variations of the earth's magnetic field revealed by Pacific deep-sea sediment cores. *Jour of Geomag. and Geoelect.*, Vol. 14, pp. 147–159.

Saito, T. et al., 1998. Quarternary geomagnetic field intensity: constant periodicity or variable period. *Geophy Res Lett.*, Vol. 25, pp. 2221–2224.

Saravanan, R., 1990. A multiwave model of the quasi-biennial oscillation, *J. Atmos. Sci.*, Vol. 54, pp. 2465–2474.

Sarkar, A., Ramesh, R., Somayajulu, B.L.K., Agnihotri, R., Jull, A.J.T. and Burr, G.S., 2000. High resolution Holocene monsoon record from the eastern Arabian sea. *Earth Planet, Sci. Lett.*, Vol. 177 pp. 209–218.

Scalo, J. M., 1985. Fragmentation and hierarchical structure in the interstellar medium. *In*: D.C. Black and M. S. Mathews (eds.) *Protostars and Planet*, ILL University Arizona, Tucson pp. 201–296.

Schaaf, M. and Thurow, J., 1997. Tracing short cycles in long records: the study of inter-annual to inter-centennial climate change from long sediment records, examples from the Santa Barbara Basin. *J. Geol. Soc. London,* Vol. 154, pp. 613–622.

Schimel, D. and Sulzman, E., 1995. Variability in the earth climate system: Decadal and longer timescales, *Reviews of Geophysics*, Supplement, July, pp. 873–882.

Schlesinger, M. E. and Ramankutti, N., 1994. An oscillation in the global climate system of period 65–70 years. *Nature*, Vol. 367, pp. 723–726.

Schuchist, C., 1932.The periodicity of oceanic spreading mountain making and pale oceanography. *In*: *Physics of the earth, V. Oceanography*, Natl. Res. Council., Washington, Bull, Vol. 85, pp. 1317–1318.

Schulz, M., 2002. On the 1470-years pacing of Dansgaard-Oeschger warm events. *Paleoceanography*, Vol. 17, pp. 1014–1029.

Schulz, M. and Stattegger, K., 1997. Spectrum: spectral analysis of unevenly spaced paleoclimate time series. *Comput. Geosci.*, Vol. 23 pp. 929–945.

Schwaarzschild, B., 1987. Do asteroid impacts trigger geomagnetic reversals? *Physics Today*, February issue, pp. 17–20.

Schwabe, S.H., 1843. *Astronomische Nachrichten*, Vol. 20, No. 495, pp. 234–235.

Schwartz, M. and Shaw, L., 1975. *Signal Processing: Discrete Spectral Analysis, Detection and Estimation*, McGraw-Hill, Kogakusha Ltd.

Schwarzacher, W., 1993. Cyclostratigraphy and the Milankovitch Theory, Elsevier, Amsterdam, pp. 1–225.

Schwartz, R.D. and James, P.B., 1984. *Periodic mass extinction and sun's oscillation about the galactic plane. Nature*, Vol. 308, pp. 712–713.

Scovillle, N.Z. and Sanders, D.B., 1986. Observational constraints on the interaction of giant molecular clouds with the solar system. *In*: R. Smoluchowski, J. Bachal and M. Mathews (eds.), *The Galaxy and the Solar System*. Univ Arizona,Tucson, pp. 69–82.

Scuderi, L.A., 1993. A 2000 year Tree Ring record of annual temperatures in Sierra Nevada Mountains, *Science*, Vol. 259, pp. 1433–1436.

Sepkoski, 1988. Extinctions and Life; Los Alamos Science fellow coloquium, pp. 37–49.

Sepkoski, J.J. Jr., 1989. Periodicity in the extinction and the problem of catastrophism in the history of life. *Jour. Geol. Soc. London*, Vol. 146, pp. 7–19.

Seyfert, C.K. and Sirkin, L.A., 1979. *Earth history and plate tectonics*, Harper and Row, New York.

Shackleton, N.J., 2000. The 100,000-year ice-age cycle identified and found to lag temperature, carbon dioxide and orbital eccentricity. *Science* 289, 1897–1902.

Shackleton, N.J. and Pisias, N.G., 1985. Atmospheric carbon dioxide orbital forcing and climate: *In*: E.T. Sundquist and W. S. Broecker. The Carbon Cycle and Atmospheric CO_2. Natural Variations from Archean to Present. *AGU Monograph,* Vol. 32, p. 303.

Shackleton, N.J. and Imbrie, J. 1990. *The δ18 spectrum of oceanic deep water over a five decade band. Climate change*, Vol. 16, pp. 217–230.

Shackleton, N.J. and Opdyke, N.D., 1993. Oxygen isotope and paleomagnetic stratigraphy of equatorial Pacific core V28-238: Oxygen isotope temperatures and ice volumes on a 105 and 106 year scale. *Quat. Res.* 3, pp. 39–55.

Shackeleton, N.J. and Crowhurst, S. 1997. Sediment fluxes based on an orbitaly tuned time scale 5 m.y. to 14 m.y. site 926., *Proc. Ocean. Drill. Prog.* Vol. 154, pp. 69–82.

Shaviv, N. J. and Veizer J., 2003. Celestial driver of Phanerozoic climate? *Geol. Soc. America. Today*. July issue, pp. 4–10.

Sheehan, P. et al., 1991. Sudden extinction of the dinosaurs: Latest Cretaceous, upper great plains, *U.S.A. Science*, Vol. 254, pp. 835–839.

Sheridan, Robert E., 1983. Phenomena of pulsation tectonics related to the breakup of the eastern north American continental margin. *Tectonophysics*, Vol. 94, pp. 169–185.

Sheridan, Robert E., 1987. Pulsation tectonics as the control of long-term stratigraphic cycles. *Paleoceanography*. Vol. 2, pp. 97–118.

Shindell, D.T., Rind, D., Balachandran, N., Lean, J. and Lonergan, P., 1999. Solar cycle variability, ozone and climate. *Science*, Vol. 284, pp. 305–308.

Shoemaker, E.M., 1983. Asteroid and comet bombardment of the earth. *Annu. Rev. Earth Planet. Sci.*, Vol. 11, pp. 461–494.

Shoemaker, E.M., 1984. Large body impacts through geologic time. *In*: H.D. Holland and A.F. Trendall (eds), *Patterns of change in earth evolution*. Springer, Berlin, pp. 15–40.

Shoemaker, E.M., Williams, J.G., Helin, E.E. and Wolfe, R.F., 1979. Earth crossing asteroids: Orbital classes, collision rates with the earth and origin. *In*: T. Gehrels (eds.), *Asteroids*. Univ. Arizona, Tucson, pp. 253–282.

Shoemaker, E.M. and Wolfe, R.F., 1984a. Evolution of the Uranus-Neptune planetesimal swarm. *Lunar Planet. Soc.*, Vol. 15, pp. 780–781 (abstract).

Shoemaker, E.M. and Wolfe, R.F., 1984b. Mass extinctions, crater ages and comet showers. *In*: R. Smoluchowski, J. Bachal and M. Mathews (eds.), *The Galaxy and the Solar System*. Univ. Arizona, Tucson, pp. 338–386.

Signore, P. W., III and Lipps, J.H., 1982. Sampling bias, gradual extinction pattern and catastrophes in the fossil record. *In*: Silver, L.T. and Shultz, P.H. (eds.), Geological implication of impacts of large asteroids and comets on the earth. *Geol. Soc. Amer. Spec. Paper*, Vol. 190, pp. 291–296.

Simpson, J.F., 1966. Evolutionary pulsation and geomagnetic polarity, *Bull. Geol. Soc. Amer.*, Vol. 77, pp. 197–204.

Simshoni, M., 1971. On the Fisher's test of significance in harmonic analysis. *Geophy. Jour. Roy. Astro. Soc.*, Vol. 23, pp. 373–377.

Singh, N. and Sontake, N.A., 1996. The instrumental period rainfall series of the Indian region: A documentation. Contribution from Indian Institute of Tropical Meteorology, Pune, *Research Report* No. PR-067 ISSN pp. 252–1075.

Sklovsky, I. S., 1968. *Supernovae*. Wiley, New York, p. 444.

Slepian, S., 1978. Prolate spheroidal wave functions, Fourier analysis and uncertainty, V, The discrete case, Bell. *Syst. Tech.* J., 57, pp. 1371–1430.

Sloan et al., 1996. Gradual dinosaur extinction and simultaneous ungulate radiation in the Hell Greek Formation, *Science*, Vol. 234, pp. 1173–1175.

Sloan, L.C. and Huber, M., 2001. Eocene Oceanic response to orbital forcing on precessional time scale. *Paleoceanography*, Vol. 16, pp. 101–111.

Smith, A.G. 1973. Phanerozoic world maps. Spec. paper, *paleontology*, Vol. 12, pp. 1–42.

Smith, A.G. and Briden, J. C., 1977. *Mesozoic and Cenozoic paleocontinental maps*, Cambridge Univ. Press, *Cambridge*.

Smith, D.G., 1981. *Cambridge Encyclopedia of Earth Sciences*. Cambridge University Press.

Smoluchowski, R. and Torbett, M.V., 1984. The boundary of the solar system. *Nature*, Vol. 311, pp. 38–39.

Sonett, C. P. and Williams, G.E., 1985. Solar periodicities expressed in varves from glacial Skilak lake, Southern Alaska. *Jour. Geophy. Res.*, Vol. 90, pp. 12, 019–26.

Sonett, C.P., Finney, A.S. and Williams, C.R., 1985. Solar periodicities expressed in varves from glacial Skilak lake, Southern Alaska. *Jour. Geophy. Res.,* Vol. 90, 12, pp. 19–26.

Sonett, C.P., Finney, A.S. and Williams, C.R., 1988. The lunar orbit in the late Precambrian and the Elatina Sandstone laminae. *Nature*, Vol. 335, pp. 806–808.

Sonett, C.P. and Finney, A.S., 1990. The spectrum of radiocarbon. *Philos. Trans. Roy. Soc. Lond.*, Vol. 330 A: pp. 413–426.

Stacey, F.D. and Loper, D.E.,1983. The thermal boundary layer interpretation of "D" and its role as a plume science. *Phys. Earth. Plan. Inter.*, Vol. 33, pp. 44–55.

Stanley, S.M., 1984. Temperature and biotic crises in the marine realm. *Geology*, Vol. 12, pp. 205–208.

Stanley, S.M., 1988. Paleozoic mass extinctions: shared patterns suggest global cooling as a common cause. *Ame. Jour. Sci.*, Vol. 288, pp. 334–352.

Steiner, J., 1967. The sequence of geological events and the dynamics of the Milky Way Galaxy, *Geol. Soc. Australia Jour.*, Vol. 14, pp. 99–132.

Steiner, J. and Grillmair, E., 1973. Possible galactic causes of periodic and episodic glaciation. *Geol. Soc. Amer. Bull.* Vol. 84, pp. 1003–1018.

Stigler, S.M., 1987. Aperiodicity of magnetic reversals? *Nature*, Vol. 330, pp. 26.

Stigler, S.M. and Wagner, 1987. A substantial bias in non-parametric test for periodicity in geophysical data. *Science*, Vol. 238, pp. 940–945.

Stigler, S.M. and Wagner, 1988. Response to the comment on testing for periodicity of extinction. *Science,* Vol. 241, pp. 96–99.

Stocker, T.F., 1994. The variable ocean, *Nature*, Vol. 367, 20 January, pp. 221–222.

Stocker, T.F., 2002. North-South connections, *Science*, Vol. 297, pp. 1814–1815.

Stothers, R.B., 1979. Solar activity cycles during classical antigue. *Astronomy and Astro Phys.*, Vol.77, pp.121–127.

Stothers, R.B., 1985. Terrestrial record of the solar system's oscillation about the Galactic plane. Ch:3, pp. 55.

Stothers, R.B., 1986. Periodicity of the earth's magnetic reversals. *Nature*, Vol. 322, pp. 444–446.

Stothers, R.B.,1987. Beat relationships between Orbital Periodicities in Insolation theory, *Journal of the Atmospheric Sciences*, Vol. 44, No. 14, pp. 1875–1876.

Stothers, R.B. and Rampino, M.R., 1990. Periodicity in flood basalt, mass extinction and impacts : a statistical view and a model. *Geol. Soc. Amer. Spec., paper*, Vol. 247, pp. 181–187.

Stuiver, M., 1978. Radiocarbon Time scale tested against magnetic and other dating methods, *Nature*, Vol. 273, pp. 271–274.

Stuiver, M. and Quay, P.D., 1980. Changes in atmospheric Carbon-14 attributable to a variable sun. *Science*, Vol. 207, pp. 405–408.

Stuiver, M. and Braziunas, T.F., 1989. Atmospheric ^{14}C and century-scale solar oscillations. *Nature*, Vol. 338, pp. 405–408.

Stuiver, M. and Braziunas, T.F., 1993. Modeling atmospheric ^{14}C influences and ^{14}C ages of marine samples to 10,000 B.C., *Radiocarbon*, Vol. 35, pp. 137–189.

Stuiver, M. and Reimer, P.J., 1993. Calib Rev., 3, *Radiocarbon*, Vol. 35, pp. 215–230.

Stuiver, M., Braziunas, T.F., Grootes, P.M. and Zielinski, G.A., 1997. Is there evidence for solar forcing of climate in the GISP2 oxygen isotope record? *Quarternary Research*, Vol. 48, pp. 259–266.

Sutton, J., 1963. Long term cycle in the evolution of the continent. *Nature*, Vol. 198, pp. 731– 735.

Szbo, A., Lepping, R.P. and King, J. H., 1995. Magnetic field observation of the 1.3 year solar wind oscillation. *Geophy. Res. Lett.* Vol. 22, pp. 1845–48.

Tammy, M., Rittenour, Julie Brigham-Grette and Mann, Michael, E., 2000. El Nino likes climate teleconnections in New England during the late Pleistocene, *Science*, Vol. 288, pp. 1039–1042.

Tamrazyan, G.P., 1967. The global history and geological regularities of the earth's development as a reflection of its cosmic origin (as a sequence of interaction in the course of galactic movement of the solar system: Ostrva Vysoka, Skola, Banska, Sbornik Vol. 13, pp. 5–24.

Tenkate, W.G. and Sprenger, A., 1989. On the periodicity in a Calcilutite-marl succession (SE, Spain). *Cretaceous Research*, Vol. 10, pp. 1–37.

Thackeray, A.D., 1971. Survey of principle characteristics of the magellanic clouds. In: A.B. Muller (ed.), *Magellanic Clouds*, Astrophys. Spac. Sci. Library, Vol. 23, Reidel, Dordrecht, p. 3.

Thaddeus, P., 1986. Molelcular clouds and periodic events in the geologic past. *In*: R. Smoluchowski, J. Bahcall and M. Mathews (eds.), *The Galaxy and the Solar System*. Univ. Arizona, Tucson, pp. 61–68.

Thaddeus, P. and Chanan, G.A., 1985. Impacts, molecular clouds, and the motion of the sun perpendicular to the galactic plane. *Nature*, Vol. 314, pp. 73–75.

Thompson, R. and Turner, G.M., 1979. British Geomagnetic Master curve 10000-0 yr B.P. for dating European sediments. *Geophys. Res. Letters*, Vol. 6, pp. 249–252.

Thomson, D.J., 1982. Spectrum estimation and harmonic analysis. *Proc. IEEE*, Vol. 70, pp. 1055–1096.

Tiedemann, R. and Franz, S.O., 1997. Deep water circulation, chemistry and terrugenious sediment supply in the equtorial Atlantic during the Pliocene 3.3–2.6 m.y. and 5–4.5 m.y., *Pro. Ocean. Drill. Prog. Sci. Res.* Vol. 154, pp. 299–318.

Tiwari, R.K.,1985. *Spectral analysis of geophysical time series*, Unpublished Ph.D. thesis, B.H.U, Varanasi, pp. 1–180.

Tiwari, R.K., 1987a. Higher order eccentricity cycles of the late and mid Miocene climate variations. *Nature*, Vol. 327, pp. 219–221.

Tiwari, R.K., 1987b. A Walsh spectral comparison of oxygen and carbon isotopes variations of Pleistocene borehole (EUREICA 67–135) from Gulf of Mexico and their orbital significance. *Marin. Geol.*, Vol. 78, pp. 167–174.

Tiwari, R.K., 2003. Mega geocycles: Echoes of astronomical events. *Jour. Geol. Soc. India*, Vol. 62, pp. 181–190.

Tiwari, R.K., Negi, G.J and Rao, K.N.N., 1994. Strange Attractor in Nonlinear Fluctuations of Length of the Day (LOD) Time Series. *In*: *Geophysical monograph* 83, IUGG, 75th AGU, Nonlinear dynamics and predictability of geophysical phenomena, Willianm I. Newman, Andrei Gabrielov, Donald L. Turcotte, Eds, Vol. 18, pp. 61–67.

Tiwari, R.K., Negi, J.G. and Rao, K.N.N., 1997. Detection of long term modulation orbital cycles in the sea level oscillations using clean algorithm of spectral analysis. *Geofizika*, Vol. 14, pp. 133–147.

Tiwari, R.K. and Rao, K.N.N., 1998. Correlated variations and periodicity of global CO_2, biological mass extinctions and extra-terrestrial bolide impacts over the past 250 million years and possible geodynamics implications, *Geofizika*, Vol. 15, pp. 103–117.

Tiwari, R.K. and Rao, K.N.N., 1999a. Periodicity of phosphorus burial rate. *Nature*, Vol. 400, pp. 31–32.

Tiwari, R.K. and Rao, K.N.N., 1999b. Solar and tidal reverberations of deglaciation records from the tropical western Pacific: a clean spectral approach. *Geofizika*, Vol. 16–17, p. 3341.

Tiwari, R.K. and Rao, K.N.N., 2003. A statistically significant long-term characteristic time scale of test size variation of calcareous trochospiral benthic foraminifera (CTBF) during the past 120 m.y. *Geophy. Res. Lett.*, Vol. 30, pp. 21–1 to 21–4.

Tiwari, R.K. and Rao, K.N.N., 2004. Signature of ENSO signals in the coral growth rate record of Arabian sea and Indian monsoons. *Pur. Appl. Geophy.* Vol. 161, pp. 413–427.

TOGA Notes, No.1, Fall 1990. *Statistical Analysis of El Nino/Southern Oscillation over the Last 500 Years.*

Toon, O.B., Pollack, J.B., Ackerman, T.P., Turco, R.P., Mckay, C.P. and Lui M.S. 1982. Evolution of an impact generated dust cloud and its effect on the atmosphere. *In*: T. Silver and P. H. Schultz (Eds), Geological implications of the impact of large Asteroids and comets on the earth. *Geol. Soc Amer. Spec. Paper*. Vol. 190: pp.187–200.

Torbett, M.V., 1986a. Injection of Oort cloud comet to the inner solar system by galactic tidal fields. *Mon. Not. R. Astron. Soc.*, Vol. 223, pp. 885–895.

Torbett, M.V., 1986b. Dynamical influence of galactic tides and molecular clouds on the Oort cloud of comets. *In*: R. Smoluchowski, J. Bahcall and M. Mathews (eds.). *The Galaxy and the Solar System*. Univ. Arizona, Tucson, pp. 147–172.

Torbett, M.V., 1989. Solar sysrem and galactic influences on the stability of the earth. *Paleogeography, Paleoclimatology, Paleoecology (Global and Planetary Changes Section)*. Vol. 75, pp. 3–33.

Torbett, M.V. and Smoluchowski, R., 1980. Sweeping Jovian resonances and the evolution of the asteroids, *Icarus*, Vol. 44, pp. 722–729.

Torbett, M.V. and Smoluchowski, R., 1984. Orbital stability of an unseen companion linked to periodic extinction events, *Nature*, Vol. 311, pp. 641–642.

Torrence, C. and Compo, G.P., 1998. A practical guide to wavelet analysis. *B. Am. Meteor. Soc.*, Vol. 79, pp. 61–78.

Tremaine, S., 1986. Is there evidence for solar companion? *In*: R. Smoluchowski, J. Bahcall and M. Mathews (eds.), The *Galaxy and the Solar System*. Univ. Arizona Tucson, pp. 409–416.

Trenberth, K.E. and Ottobliesner, 2003. Paleoclimate toward Integrated Reconstruction of Past Climates, *Science*, Vol. 300: 589–590.

Tziperman, E., Stone, L., Cane, M.A and Jarosh, H., 1994. El Nino chaos: overlapping of resonances between the seasonal cycle and the Pacific ocean–atmosphere oscillator. *Science*, Vol. 264, pp. 72–74.

Ulrych, T.J., 1972. Maximum entropy spectrum of long period geomagnetic reversals, *Nature*, Vol. 235, pp. 218–219.

Ulrych, T.J. and Bishop, T.N., 1975. Maximum entropy spectral analysis and autoregressive decomposition, *Rev. Geophys. Space. Phys.*, Vol. 13, pp. 183–200.

Umbarave, J.N.F (1947). The pulse of the earth. Martinus Nijhoff. The Hague, pp. 200.

Uppenbrink, J. (1999). The north Atlantic oscillation, *Science*, Vol. 283, pp. 948–949.

Urban, F.E., Cole, J.E. and Overpeck, J.T., 2000. Influence of mean climate change on climate variability from a 155 year tropical Pacific coral record. *Nature*, Vol. 407, pp. 989–993.

Vail, P.R., Mitchun, Jr., R.M. and Thompsons III, 1977. Seismic stratigraphy and global changes of sea level part 4, Global cycles of relative changes of sea level: *Am. Ass. Petrol. Geol. Mem.*, Vol. 6, pp. 83–97.

Valdes, P.J. and Glover, R.W., 1999. Modelling the climate response to orbital forcing. *Philos. Trans. R. Soc.*, London 357, pp. 1873–1890.

Vallis, G.K., 1986. El Nino: A chaotic dynamical system? *Science*, April 1986, pp. 243–246.

Van Geel, B., Raspopov, O.M., Renssen, H.,Vander Plicht, J., Dergachev, V. A. and Meijer, H.A.J., 1999. The role of solar forcing upon climate change. *Quarternary Science Review*, Vol. 18, pp. 331–338.

Van Tassell, J., 1987. Upper Devonian Catskill Delta margin cyclic sedimentation: Brallien Scherr and Foreknobs Formations of Virginia and West Virginia, *Geol. Soc. Am. Bull.*, Vol. 99, pp. 414–426.

VanZijl, J.V.S., Graham, K.W.T. and Hales A.L.,1962. The paleomagnetism of the stormberg lava II. The behaviour of magnetic field during a reversal. *Geophy. Jour. Roy. Soc.*, Vol. 7, pp. 169–182.

Vautard, R. and Ghil, M., 1989. Singular spectrum analysis in non-linear dynamics, with applications to paleoclimatic time series, *Phys.* D, 35, 395–424.

Vestine, E.H. and Kahle, A.B., 1968. The westward Drift and Geomagnetic secular change. *Geophysics J. Roy. Astr. Soc.*, Vol. 15, pp. 29–37.

Vine, F.J. and Mathews, D.H., 1963. Magnetic anomalies over oceanic ridges, *Nature*, Vol. 199, pp. 947–949.

Vio, R., Cristiani, S., Lossi, O. and Provenzale, A., 1992. Time series analysis in astronomy: An application to quasar mass variability studies. *Astron. J.*, 391: 518–530.

Vogt, P.R.,1972. Evidence for global syncronism in mantle plume convection and possible significance for geodesy. *Nature*, Vol. 240, pp. 338–342.

Vogt, P.R., 1975. Change in reversal frequency at times of tectonic change: evidence for coupling between core and upper mantle process. *Earth. Planet. Sci. Lett*, Vol. 25, pp. 313–321.

Walker, J.C.G. and Zahnle, K.J., 1986. Lunar nodal tide and distance to the moon during the Precambrian. *Nature*, Vol. 320, pp. 600–602.

Wang, H., Hughes, R.E., Steele, J.D., Lepley, S.M. and Tian, J., 2003. Correlation of climate cycles of Middle Mississippi Valley loess and Greenland ice. *Geology*, Vol. 31, No, 2, pp. 179–182.

Wanner, et. al., 2001. North Atlantic Oscillation-concepts and studies, Surveys in *Geophysics*, Vol. 22, pp. 321–382.

Webster, P.J. and Palmer, T.N., 1997. The past and the future of El Nino, *Nature*, Vol. 390, pp. 562–564.

Weedon, G.P., 2003. *Time Series Analysis and Cyclostratigraphy*. Camdridge University Press.

Weedon, G.P., Jenkyns, H.C.,Coe, A.L. and Hesselbo, S.P., 1999. Astronomical calibration of the Jurassic time scale from cyclostratigraphy in British mudrock formation, *Phil. Trans. Roy. Soc Lond.*, Vol. 357, pp. 1787–1813.

Weedon, G.P., Schakeleton, N.J. and Pearson, P.N., 1997. The Oligocene time scale and cyclostratigraphy on the Ceara rise western equatorial Atalantic. Proc. *Ocean. Drill. Prog. Sci. Res.*, Vol. 154, pp. 101–114.

Weinberg, M.D., Shapiro, S.L. and Wasserman, I., 1987. The dynamical fate of wide binaries in the solar neighbourhood. *Astrophy. Jour.*, Vol. 312, pp. 367–389.

Weissman, P.R., 1982. Terrestrial impact rates for long and short period comets. *Geol. Soc America, Spec. Paper,* Vol. 190, pp. 15–24.

Weissman, P.R., 1985. Terrestrial impacters of geological boundary events: Comets or asteroids, *Nature*, Vol. 314, pp. 517–518.

Weissman, P.R., 1986. The Oort cloud and the galaxy: Dynamical interactions. *In*: R. Smoluchowski, J. Bahcall and M. Matthews (eds.), *The Galaxy and the Solar System*. Univ. Arizona, Tucson, pp. 204–237.

Wetherill, G.W. and Shoemaker, 1982. Collision of astronomically observable bodies with the earth. *Geol. Soc America, Spec. Paper,* Vol. 190, pp. 1–13.

White, R. and Mckenziz D.,1989. Magmatism at rift zones:The generation of volcanic i.e. continental and margin flood basalts. *Jour. Geophy. Research.* Vol.94, pp.7685–7729.

Whitmire, D.P. and Jackson, A.A., 1984. Are periodic mass extinction driven by a distant solar Companion. *Nature*, Vol. 308. pp. 36–38.

Whitmire, D.P. and Matese, J.J., 1985. Periodic comet showers and planet X. *Nature*, Vol. 313, pp. 36–38.

Whyte, M.A.,1977. Turning points in Phenerozoic history. Nature 267, pp. 579–582.

Wicander, R. and Monroe, James S., 2000. *Historical Geology: Evolution of the Earth and Life through Time*, Second edition, West Publishing Company. New York, U.S.A.

Wigley, T.M.L., 1976. Spectral analysis and astronomical theory of climate change, *Nature*, Vol. 264, pp. 629–631.

Williams, G.E., 1975. Possible relation between periodic galaciation and the flexure of the Galaxy: *Earth and Planet. Sci. Letter,* Vol. 26, pp. 361–369.

Williams, G.E., 1981a. *Megageocycles*. Hutchinson Ross Pub. Comp. Stroudsburg, Pennsylvania.

Williams, G.E., 1981b. Sunspot period in the late Precambrian glacial climate and solar planetary relations. *Nature* 291, pp. 624–628.

Williams, G.E., 1988. Cyclicity in the late Precambrian Elatina formation, South Australia: Solar or tidal signature? *Climate Change,* Vol. 13, pp. 117–128.

Williams, G.E., 1989. Late Precambrian tidal rhythmites in South Australia and the history of the Earth's rotation, *Journal of Geological Society*, London, Vol. 146, pp. 97–111.

Williams, G.E. and Sonette, C.P., 1985. Solar signature in sedimentary cycles from the Late Precambrian Elatina formation, Australia, *Nature*, Vol. 318, pp. 523–527.

Williams, D.F., Peck, J., Karabanov, E.G., Prokopenko, A., Kravchinsky, V. et al., 1997. Lake Baikal record of continental climate response to orbital forcing during the past 5 m.y. *Science* Vol. 278, pp. 1114–1117.

Wilson, J.T., 1966. Did Atlantic close and re-open? Nature Vol. 211, pp. 676–681.

Wilson, R.C. and Hudson, H.S., 1991. The sun's luminosity over a complete cycle. *Nature*, Vol. 351, pp. 42–44.

Winograd, Isaac, J., Coplen, T.B., Landwehr, J.M., Riggs, A.C., Ludwig, K.R., Szabo, B.J., Kolesar, P.T., Revesz, K.M., 1992. Continuous 500, 000 year climate record from vein calcite in Devils Hole, Nevada. *Science* 258, pp. 255–260.

Wolf, R., 1856. Die sonnenflecken. *Astron. Mitt*, Zurich, Vol. 61, p. 1856.

Wollin, et al., 1971. Climate change and geomagnetism, *Nature*, Vol. 232, pp. 549–550.

Wollin, G., Ryan, W.B.F., Ericson, D.B., 1978. Climatic changes, magnetic intensity variations and fluctuations of the eccentricity of the earth's orbit during the past 2000,000 years and a mechanism which may be responsible for the relationship. *Earth Planet. Sci. Lett.*, Vol. 41, pp. 395–397.

Wood, K.D., 1972. Sunspots and Planets. *Nature*, Vol. 240, pp. 91–93.

Woodworth, P.L., 1984. A worldwide search for the 11-year solar cycle in mean sea-level records, *Journal of Atmospheric Sciences*, pp. 743–753.

Worsley, T. R. et al., 1985. Proterozoic to recent tectonic tuning of bio-geo-chemical cycles: In Sundquist E. T. and Broecker W.S. (eds) The carbon cycle and atmospheric CO_2: Natural variations Archean to present. *Geophysical monograph* 32, pp. 561– 572.

Wunsch, C., 2000. On sharp spectral lines in the climate record and the millenial peak. *Paleoceanography*, Vol. 15, pp. 417–424.

Yabushita, S., 1991. A statistical test for periodicity hypothesis in the crater formation rate. *Mon. Not. Roy. Astron. Soc.*, Vol. 250, pp. 481–485.

Yamazaki, T. and Oda, H., 2002. Orbital influence on Earth's Magnetic field: 100, 000-year periodicity in Inclination, *Science*, Vol.295, pp. 2435–2438.

Yasunari, T., 1989. A possible link of the QBOs between the stratosphere, troposphere and sea surface temperature in tropics. *Jour. Met. Soc. of Japan*, Vol. 67, pp. 483–493.

Yiou, P., Genthon, C., Ghil, M., Jouzel, J., LeTreut, H., Bamola, J.M., Lorius, C., Korotkevitch, Y.N., 1991. High-frequency Paleovariability in climate and CO_2 levels from Vostok ice records. *J. Geophys. Res.* 96 (B12), 20365–20378.

Yiou, P., Jouzel, J., Johnsen, S. and Rognvaldsson, O. E., 1995. Rapid oscillations in Vostok and GRIP ice cores, *Geophys. Res., Lett.*, Vol. 22, pp. 2179–2182.

Yu, Z. and Ito, E., 1999. Possible forcing of century scale drought frequency in the northern Great plains. *Geology*, Vol. 27, pp. 263–266.

Yuen, D.A. and Scubert, A.,1976. Mantle plumes a boundary layer approach for Newtonian and non Newtonian temperature dependent rheologies. *GJR*, Vol.81, pp. 2499–2501.

Yukutake, T., 1973. Fluctuations in the earth's rate of rotation related to changes in the Geomagnetic dipole field, *J. Geomag. Geolectr*, Vol. 25, pp. 195–212.

Yule, G.U., 1927. On a method of investigating periodicities in disturbed series, Philos. Trans. R. Soc. London, Ser. A, Vol. 226, pp. 267–298.

Zachos, J.C., Arthur, M.A. and Dean, W.E., 1989. Geochemical evidence for suppression of pelagic marine productivity at the Cretaceous/Tertiary boundary. *Nature*, Vol. 337, pp. 61–64.

Zachos, J.G., Shackeleton, N.J., Revenough, J.S., Palike, H. and Flower, B. P., 2001. Climate response to orbital forcing across the Oligocene- Miocene boundary. *Science,* Vol. 292, pp. 274–278.

Index

D

O

P

R

S